Sea Stars, Sea Urchins, and Allies

Sea Stars, Sea

Urchins, and Allies

Echinoderms of Florida and the Caribbean

Gordon Hendler
John E. Miller
David L. Pawson
Porter M. Kier

Smithsonian Institution Press

Washington and London

© 1995 by the Smithsonian Institution
All rights reserved

Copy Editor: Eileen D'Araujo
Supervisory Editor: Deborah L. Sanders
Designer: Linda McKnight
Production Supervisor: Kenneth Sabol

Library of Congress Cataloging-in-Publication Data
    Sea stars, sea urchins, and allies : echinoderms of Florida and the Caribbean / Gordon Hendler . . . [et al.].
      p. cm.
    Includes index.
    ISBN 1-56098-450-3
    1. Echinodermata—Florida—Identification. 2. Echinodermata—Caribbean Area—Identification.
    I. Hendler, Gordon.
    QL383.1.U6S43 1995
    593.9′09759—dc20   94-31424

British Library Cataloguing-in-Publication Data is available

Manufactured in Hong Kong
02 01 00 99 98 97 96 95 5 4 3 2 1

♾ The paper used in this publication meets the minimum requirements of the American National Standard for Permanence of Paper for Printed Library Materials Z39.48-1984.

For permission to reproduce illustrations appearing in this book, please correspond directly with the owners of the works, as listed in the individual captions. The Smithsonian Institution Press does not retain reproduction rights for these illustrations individually, or maintain a file of addresses for photo sources.

Identification of illustrations used to separate major sections of this book is as follows: title page, *Diadema antillarum*; p. xii, *Nemaster rubiginosus*; p. 6, mangrove habitat; p. 28, *Ophiopsila riisei*; p. 43, *Nemaster discoideus*; p. 59, *Astropecten articulatus*; p. 89, *Amphipholis gracillima* (left) and *Amphiodia atra* (right); p. 197, *Echinometra lucunter lucunter*; p. 251, *Ocnus suspectus*.

CONTENTS

vii PREFACE

ix ACKNOWLEDGMENTS

1 INTRODUCTION

7 SHALLOW-WATER MARINE COMMUNITIES
   of the Florida Keys and the Bahama Islands

17 GENERAL FEATURES OF THE PHYLUM
    ECHINODERMATA

21 OBSERVING, COLLECTING, AND
    PRESERVING ECHINODERMS

29 PHOTOGRAPHING ECHINODERMS

34 COMPOSITION AND SIGNIFICANCE OF
    SCIENTIFIC NAMES

36 SYSTEMATIC LIST OF THE ECHINODERMS
    of the Florida Keys and the Bahama Islands
    Found at Depths Less Than 30 m (98 ft)

CLASS CRINOIDEA   43
Feather Stars

48 FAMILY COMASTERIDAE
54 FAMILY COLOBOMETRIDAE
56 FAMILY ANTEDONIDAE

CLASS ASTEROIDEA   59
Sea Stars

66 FAMILY LUIDIIDAE
71 FAMILY ASTROPECTINIDAE
74 FAMILY ASTERINIDAE

CONTENTS

75	FAMILY OPHIDIASTERIDAE		222	FAMILY ECHINOMETRIDAE
81	FAMILY ASTEROPSEIDAE		227	FAMILY ECHINONEIDAE
82	FAMILY OREASTERIDAE		228	FAMILY CLYPEASTERIDAE
84	FAMILY ECHINASTERIDAE		232	FAMILY MELLITIDAE
			238	FAMILY SCHIZASTERIDAE
			241	FAMILY BRISSIDAE

CLASS OPHIUROIDEA 89
Brittle Stars, Basket Stars

96	KEY TO THE FAMILIES OF BRITTLE STARS
98	FAMILY OPHIOMYXIDAE
100	FAMILY GORGONOCEPHALIDAE
105	FAMILY OPHIURIDAE
111	FAMILY OPHIOCOMIDAE
123	FAMILY OPHIONEREIDIDAE
129	FAMILY OPHIODERMATIDAE
142	FAMILY HEMIEURYALIDAE
143	FAMILY OPHIACTIDAE
151	FAMILY AMPHIURIDAE
180	FAMILY OPHIOTRICHIDAE

CLASS HOLOTHUROIDEA 251
Sea Cucumbers

259	FAMILY CUCUMARIIDAE
266	FAMILY SCLERODACTYLIDAE
272	FAMILY PHYLLOPHORIDAE
279	FAMILY STICHOPODIDAE
282	FAMILY HOLOTHURIIDAE
302	FAMILY CAUDINIDAE
303	FAMILY SYNAPTIDAE
313	FAMILY CHIRIDOTIDAE

329	APPENDIX: *Data on Looe Key National Marine Sanctuary Station and the Distribution of Echinoderm Species by Station*
335	GLOSSARY
343	LITERATURE CITED
383	INDEX TO SCIENTIFIC NAMES

CLASS ECHINOIDEA 197
Sea Urchins, Sand Dollars, Heart Urchins

206	FAMILY CIDARIDAE
208	FAMILY DIADEMATIDAE
214	FAMILY ARBACIIDAE
216	FAMILY TOXOPNEUSTIDAE

PREFACE

Echinoderms (sea urchins, sea cucumbers, sea stars, brittle stars, and feather stars) are remarkable and beautiful members of the shallow-water Caribbean biota. It is no wonder that they invariably fascinate scientists, students, and even casual observers of marine life. But despite their allure and importance, a handbook for their identification and life history has been lacking.

The only general taxonomic treatment of Caribbean echinoderms (H. L. Clark 1933) is out of print, out of date, and very poorly illustrated. Newer field guides to marine invertebrates, replete with color photographs, have proliferated (e.g., Zeiller 1974; Voss 1976; Colin 1978; Kaplan 1982; Sefton and Webster 1986; Humann 1992). However, they consider a minority of the echinoderm fauna and provide little biological and taxonomic information.

Therefore, we designed this book to be a comprehensive resource for the identification and appreciation of shallow-water echinoderms of the Florida Keys and the Bahama Islands and to be a useful reference for a majority of the species of the Caribbean and the Gulf of Mexico. Our aim is to cast light on what is known and what is not yet known and, thereby, to foster the study of these intriguing animals. A review of the full echinoderm fauna of the tropical western Atlantic could not have been presented in a compact handbook of this size, but our study area, the Florida Keys and the Bahama Islands, includes most shallow-water echinoderm species common in the greater Caribbean region. Of course, this does not guarantee that we have covered every echinoderm species in the Florida Keys and Bahamas area. In the future, others certainly will be encountered, and new species will be discovered and described.

ACKNOWLEDGMENTS

This book was begun during a survey of the echinoderms of Looe Key National Marine Sanctuary (LKNMS), Florida, a project funded in part by the National Marine Sanctuary Program of the National Oceanic and Atmospheric Administration (NOAA). John Miller originally conceived the idea for the project, directed the Florida Keys fieldwork, and took color photographs that are the core of this publication. He was responsible for the accounts of asteroids and holothuroids, Gordon Hendler for ophiuroids, David Pawson and Porter Kier for echinoids, and Miller, Pawson, and Hendler for crinoids. As the study progressed, the roles of the four collaborators altered as they changed jobs, institutions, and careers. The order of authorship, as it now stands, reflects overall responsibility for developing the completed manuscript.

Financial support for research was granted through Harbor Branch Oceanographic Institution, Inc. (HBOI), Fort Pierce, Florida; NOAA, Washington, D.C. (Cooperative Agreement No. NA-84-AA-H-CZ017); the National Museum of Natural History (NMNH) of the Smithsonian Institution, Washington, D.C. (catalog numbers of specimens from the collections of this museum are prefixed by the letters USNM); the Smithsonian Marine Station at Link Port (SMSLP), Fort Pierce, Florida; the Natural History Museum of Los Angeles County (LACM), Los Angeles, California; and the Caribbean Coral Reef Ecosystems Program of the Smithsonian Institution, Washington, D.C. A portion of the publication costs was provided by NOAA and HBOI.

We are grateful to many individuals for their assistance with this handbook. Barbara Littman and Arnold Powell (NMNH) and Paula Mikkelsen (HBOI) participated in fieldwork at LKNMS. Paula Mikkelsen also helped in numerous ways in compiling the manuscript, as did Barbara Littman and Robert Peck (LACM). Ailsa Clark (British Museum, Natural History) and Maureen Downey (NMNH) shared their then-unpublished manuscript used in completing the sea star chapter. Karen J. Friedmann provided translations of Danish text. Carroll Curtis, Nancy Foster, Ralph Lopez, and William Thomas of NOAA offered encouragement and assistance related to research at Looe Key and with the Cooperative Agreement. Looe Key Sanctuary Manager, Billy Causey, provided logistical support during our survey in the Keys. The late Charles Cutress and Bertha M. Cutress (University of Puerto Rico) were unfailingly helpful during field trips to Puerto Rico. Robert Grumet contributed the use of his boat and home in Miami for fieldwork in Biscayne Bay. The crews of the research vessels *Sea Diver, Seward Johnson,* and *Edwin Link* facilitated our research cruises to the Florida Keys, the Bahama Islands, and the Lesser Antilles. The Department of the Interior approved our research in the Dry Tortugas and provided quarters at Fort Jefferson National Monument.

Loans of museum specimens were arranged through Cynthia Ahearn (NMNH), David Camp and Sandra Farrington-LaGant (Marine Research Laboratory, Florida Department of Natural Resources), Margit Jensen (Zoological Museum [UZM], Copenhagen), Erik Martin (Applied Biology, Jensen Beach, Florida), Robert Woollacott (Museum of Comparative Zoology, Harvard University [MCZ]), and the late Gilbert Voss (Rosenstiel School of Marine and Atmospheric Science, University of Miami). We also are grateful to scientists who identified organisms associated with echinoderms: Frederick Bayer (NMNH), Stephen Cairns (NMNH), Fenner Chace (NMNH), Masahiro Dojiri (Hyperion Treatment Plant, Playa del Rey, California), Darryl Felder (University of Southwestern Louisiana), Rosalie Maddocks (University of Houston), Marian Pettibone (NMNH), Edward Ruppert (Clemson University), David Russell (Washington College), Paul Scott (Santa Barbara Museum of Natural History), and Anders Warén (Swedish Museum of Natural History). George Steyskal (NMNH) shared his nomenclatorial expertise. Assistance was also rendered by librarians of NMNH, LACM, and the Allan Hancock Foundation of the University of Southern California.

We sincerely appreciate the suggestions offered by colleagues who read various parts of the manuscript: Daniel B. Blake (University of Illinois at Urbana-Champaign), Bertha M. Cutress (University of Puerto Rico), John H. Dearborn (University of Maine, Orono), William Lyons (Florida Department of Natural Resources), Charles G. Messing (NOVA University, Florida), and Rich Mooi (California Academy of Sciences).

Line drawings were prepared by Charles G. Messing (Figures 5, 11, 12, 107, 108, 109, 137, 153, 154, 165), Paula Mikkelsen

(Figures 138, 191), Rich Mooi (Figures 30, 31, 32), and Mary Ann Nelson-Poole (Vero Beach, Florida) (Figure 141). Illustrations were labeled by Patricia Condit (Burke, Virginia), and she also prepared the map of the Caribbean area (Figure 1). Tom Smoyer (HBOI) handled considerable photographic copying work and provided the photos of sea urchin tests for Figures 134 and 135. Dick Meier (LACM) provided photographs for other urchins (Figure 136). The Smithsonian Institution's Photographic Division, NMNH Branch, provided photographic copying services. Pamela Blades-Eckelbarger (HBOI) and Julie Piraino (SMSLP) assisted in SEM photography. All photographs of whole animals in the text were made by John Miller (HBOI) unless otherwise stated in the photo credits. We thank Bertha M. Cutress, Edward Ruppert, Richard S. Fox (Lander College), David L. Meyer (University of Cincinnati), Chip Clark (NMNH), William D. Lee (SMSLP), Paul H. Humann (Hollywood, Florida), and Walter Stearns (Ocean Arts, Inc., Miami, Florida) for allowing us use of their photographs.

This book would not have been completed without the talents and expertise of the Smithsonian Press staff. In particular, we thank our editors, Peter Cannell, Eileen D'Araujo, and Deborah Sanders, and the designer, Linda McKnight, for their patience and dedication to this project.

Voucher material collected during our fieldwork at LKNMS and several other localities throughout the Caribbean has been deposited at NMNH, LACM, and Harbor Branch Oceanographic Museum. This publication is Contribution No. 1013 of HBOI and No. 351 of SMSLP.

INTRODUCTION

This is a handbook of the echinoderms found in the Florida Keys and the Bahama Islands, and it will be useful throughout the Caribbean region and the Gulf of Mexico (Figure 1). It treats in detail echinoderms occurring at scuba depths (30 m [98 ft] or less) in the Florida Keys and the Bahamas. Complete discussion and illustrations are provided for 144 species, many of which are widespread and abundant, and other species are described briefly. Consequently, the book covers about 80% of the echinoderm species inhabiting shallow tropical and subtropical waters of the western Atlantic north of Brazil. The procedure for using this information to identify species is described in this chapter.

For each species account, we have distilled information from scientific publications and summarized previously unpublished observations compiled by our research team over the past 10 years from museum collections and field studies throughout the Caribbean. The initial fieldwork connected with this handbook was conducted within and near Looe Key National Marine Sanctuary, an 18.2-km^2 (5.3 square nautical miles) area approximately 11 km (6 nautical miles) south of Ramrod Key, Florida. A map of the sanctuary (Figure 191), illustrating the position of sampling sites ("stations"), may be found in the Appendix along with lists of station data and of the echinoderm species collected or observed at each station.

Within each chapter, the species of a class of echinoderms are organized phylogenetically by family, then alphabetically according to genus and species. The species accounts include information on habitat, distribution (geographic and depth range), biology (i.e., life history), and remarks distinguishing among similar species. For most species, color photographs depict living individuals in their

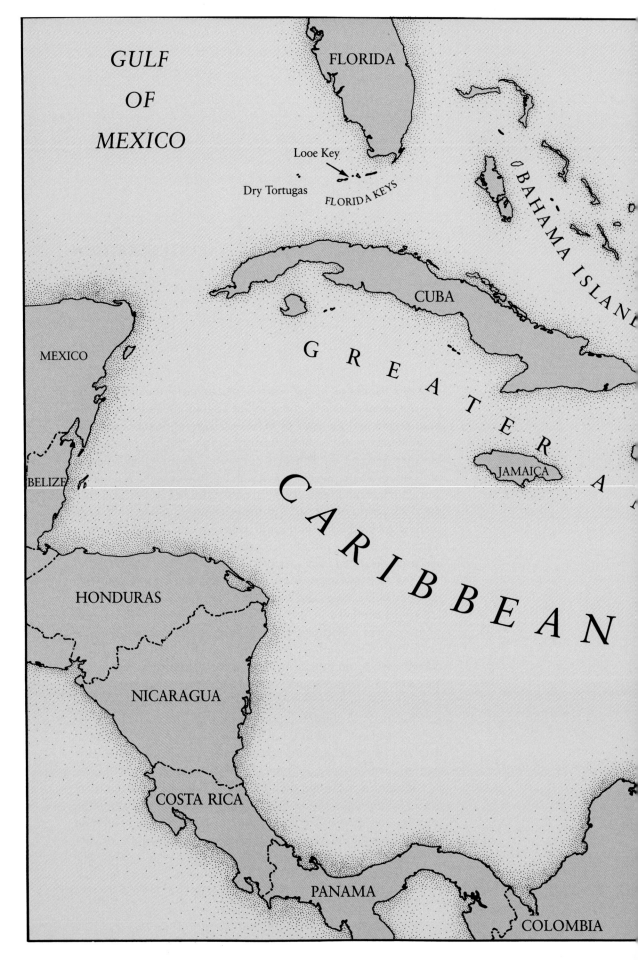

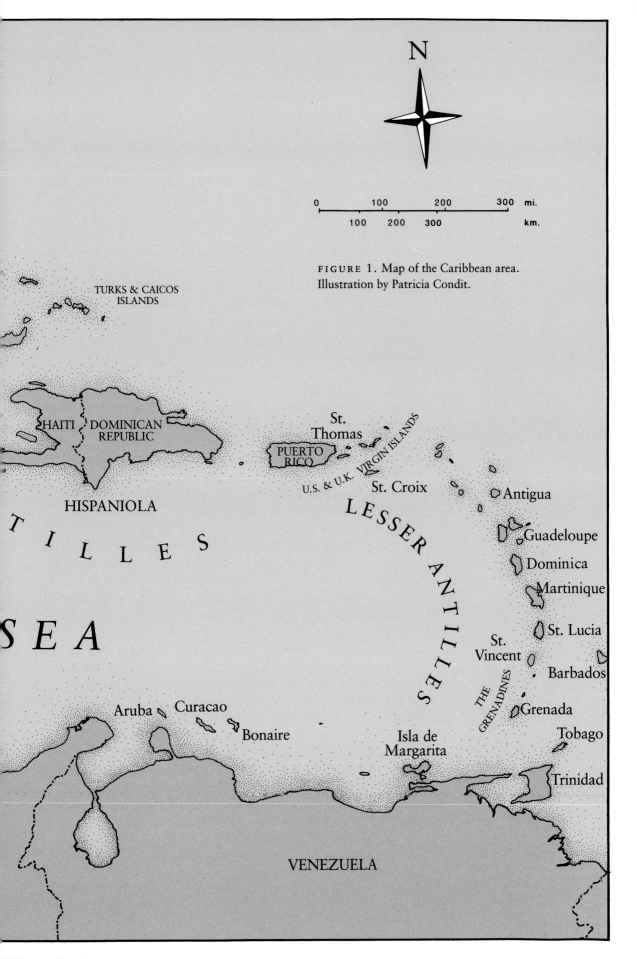

FIGURE 1. Map of the Caribbean area. Illustration by Patricia Condit.

natural habitats or against a black background in an aquarium. When a living animal was not available, or where color photography was unnecessary, black-and-white photography or line drawings are provided. At the end of each species account, a list of selected references offers entry into the taxonomic literature and suggests sources of additional illustrations; one of the citations specifies the earliest description of the species. A citation followed by a scientific name in brackets indicates that the reference employs an outdated name for the species.

Throughout the book, we have used as little scientific jargon as possible. However, for a topic of this nature, some special terminology is necessary. To aid the reader, a glossary of scientific terms is provided. In addition, labeled drawings showing key anatomical features appear near the beginning of the chapters on each class. Measurements are usually expressed in both metric units and their U.S. equivalents.

USING THIS BOOK TO IDENTIFY ECHINODERMS

The first step in identifying an echinoderm is to determine to which class it belongs. Figure 2 can be used to decide whether a specimen is a sea urchin, a sea cucumber, a sea star, a brittle star, or a feather star. After the class has been established, turn to the appropriate chapter.

Then, refer to the illustrations to find the species most closely resembling the unknown specimen. Verify the initial identification by referring to the written description, carefully noting similarities and differences between the description and the specimen. This will yield an accurate identification of most of the sea urchin, sea star, and feather star species of the Florida Keys and the Bahamas. Because the skeletons (tests) of sea urchins can persist long after the animal has died and are collected frequently, photographs of naked tests are provided (Figures 134–136) for their identification.

In the case of brittle stars, even species belonging to different families can be superficially similar. Therefore, the family should be established by using the identification key provided, before consulting the photographs and descriptions. It is often possible to recognize closely related brittle star species by their color patterns, but the reliability of identification will improve if characteristics in the written description (especially arm spines, mouthparts, and disk plates) are examined with a low-power ("dissecting") microscope. The identification of amphiurid brittle stars, which closely resemble one another, is simplified with a special key and close-up photographs of each species (Figures 100–106).

The sea cucumbers present special problems in identification because photographs of live individuals seldom resemble preserved specimens. It simply is not possible to identify some holothuroids

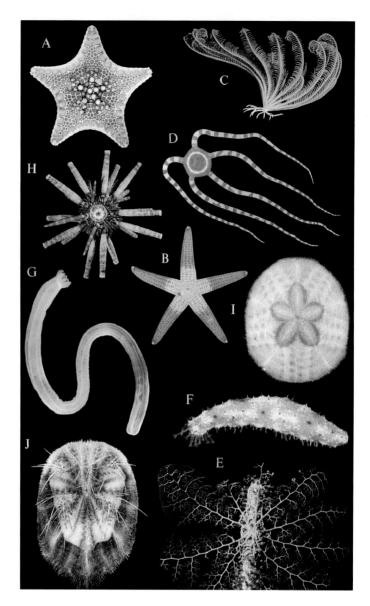

FIGURE 2. Representatives of five classes of echinoderms. Sea stars: *A, Goniaster tesselatus*; *B, Hacelia superba*; Feather stars: *C, Crinometra brevipinna*; Brittle stars: *D, Ophioderma rubicundum*; *E, Astrophyton muricatum*, a basket star; Sea cucumbers: *F, Holothuria floridana*; *G, Leptosynapta tenuis*; Sea urchins: *H, Eucidaris tribuloides*; *I, Clypeaster subdepressus*, a sand dollar; *J, Plagiobrissus grandis*, a heart urchin. Species shown in A, B, and C live in deep water or outside the range covered in this book.

without examining the skeletal ossicles of their body wall. Granted, the necessity to view ossicles with a compound microscope will preclude many readers from examining them. However, we present a simple procedure for preparing ossicles. We also provide a synoptic series of ossicle photographs made with scanning electron microscopy (Figures 139, 178–190). Accurate identifications of sea cucumbers can be made by using these photographs in conjunction with a simple, temporary ossicle preparation and a student microscope.

If a specimen appears to fit more than one description and the illustrations are not conclusive, consider the habitat in which the species was encountered. Species within a genus can occur in distinctly different habitats; thus, information on where the echinoderm was found can be a clue to its identity. When in doubt, it may be necessary to consult additional literature or an echinoderm specialist.

SHALLOW-WATER MARINE COMMUNITIES
of the Florida Keys and the Bahama Islands

The region under scrutiny encompasses the Florida-Bahama Province: the archipelagos of the Florida Keys and the Bahama Islands, covering about 518,000 km^2 (200,000 mi^2) from the Dry Tortugas to the eastern tip of the Bahama Islands. The Florida reef tract and the associated bays along the shallow continental shelf of Florida consist of over 100 km (62 mi) of discontinuous bank reefs enclosing a lagoon 5–10 km (3–6 mi) wide with over 6,000 patch reefs. The Bahama Islands consist of 700 islands and uncounted reefs.

This extensive territory is uniquely influenced by Atlantic seas, the northward-flowing Gulf Stream, and by twice-daily tidal flux from Florida Bay through the Florida Keys. However, the overriding patterns and processes that shape this province are found throughout the Caribbean region and are paralleled in similar communities around the world.

In warm-water regions such as the Florida-Bahama Province, three intimately interacting habitats, seagrass beds, mangroves, and coral reefs, are dominant biological units. Stands of mangrove trees and beds of seagrass on sandy bottom are often intimately associated with reefs.

The seagrasses (Figure 3B) are a specialized group of flowering plants that can live permanently submerged in seawater. They populate substrates that are too unstable, and tolerate sediment loads too abrasive, for the colonization and survival of most corals. Seagrass growth and reproduction are largely vegetative, presumably because of the uncertainties of underwater pollination and seedling survival. These grasslike plants provide substratum, shelter, and nutrients for a profusion of minute plants and myriad invertebrates. The seagrasses and associated organisms are consumed by fish, sea turtles,

FIGURE 3. Coral reef and associated marine habitats. *A*, Mangrove: red mangrove trees with underwater prop roots covered by plants and animals (photo by C. Clark); *B*, Seagrass Bed: turtle grass and other marine plants on sand with some clumps of algae and sponges; *C*, Back Reef: patches of sand and coral rubble and scattered soft corals and sponges; *D*, Reef Flat: predominantly coral rubble, some of which supports soft corals such as sea fans; *E*, Fore Reef: a spur composed of living and dead stony coral, covered with soft corals, sponges, and algae; *F*, Deep Sand Bed: carbonate sand and scattered rocky outcrops.

F

and manatees. In Florida, over 100 species of algae may grow on the blades of turtle grass, and a single square meter of a grass bed can support 25,000 tiny snails of a single species. Dense seagrass meadows act as a baffle to entrap plant debris and the detritus produced by resident animals. Leaves of some seagrasses grow 5–10 mm (0.2–0.4 in) per day, and the abundant output of seagrass beds is harvested by resident fauna and transported to nearby reefs, mangroves, and even offshore deep-sea communities.

Mangrove trees are uniquely capable of thriving in areas where their leaves bake in the sun and their roots are submerged in seawater and poorly oxygenated mud (Figure 3A). Along broad stretches of tropical shores, vast stands of mangroves populate marine coastlines and estuarine river drainages. The different species of mangrove trees are often distributed, according to their physiological tolerances, across a series of ecological zones. They may lie adjacent to reefs, but occupy shoals with extremes of salinity and temperature that are too severe for extensive coral growth.

Wherever they gain a foothold, mangroves have the potential of proliferating into canopied forests that shelter and stabilize the coast and furnish rookery sites for birds and nursery grounds for shrimps, crabs, and fish such as mullet, gray snapper, red drum, and tarpon. Mangroves are home to some mammals and reptiles, many insects and spiders, and a diverse spectrum of marine invertebrates. The confusing warble of birds, buzz of insects, and staccato chatter of pistol shrimp barely hint at the abundance and productivity of life-forms in mangrove stands.

Less than half a hectare (1 acre) of red mangroves shed an estimated 2,700 kg (6,000 lb) of leaves each year. The mangrove roots form a living sieve that intercepts, accumulates, and stabilizes sediment and decomposing vegetation. In time, this accretion of material may produce a seaward extension of the shoreline. A complex web of interactions connects the abundant detritivore and herbivore populations of mangroves, but the system is more than self-supporting. In southern Florida, massive quantities of soluble and particulate organic materials from the mangroves are the first link of a far-reaching food chain supporting the rich fisheries of Florida Bay. Furthermore, nutrient-rich seagrass meadows and mangroves provide nursery habitats for the young of numerous animals (including commercially important shrimps and fish) that spend their adult lives elsewhere.

Coral reefs are the pièce de résistance of the shallow-water marine environment (Figure 3C–F). The beauty and bounty of their resident biota are a feast for the eye and a challenge for the intellect. A reef's structure and organization derive from the characteristics of reef-building corals, which are highly specialized, colonial organisms. Coral reefs are characterized by an astounding biomass and productivity despite the surprisingly low nutrient content of their surrounding waters.

The resistant reef architecture can bear the full brunt of waves, and its ramparts shelter seagrass and mangrove communities. Despite their impressive durability, however, reefs are restricted to a narrow range of ecological conditions and are replaced by algae beds, seagrass meadows, or mangroves in less favorable circumstances. Reef-building corals require a stable substratum, sunlight, and water that is well circulated, has temperatures between 18 and 36°C (64–97°F), has near-oceanic salinity of 27–40 ppt (parts per thousand), and is relatively free of suspended sediment. These conditions are satisfied on the eastern margins of islands in the Florida Keys and the Bahamas, where clean, well-mixed oceanic waters prevail.

Each coral species has a typical growth form (e.g., massive, branching, encrusting, leafy) and a delicately sculptured skeleton of unique design. The skeleton is a white, calcareous material, perhaps more familiar as paperweights, aquarium ornaments, and landfill than as part of a once-living animal. Each of the repeated patterns of fine grooves in the skeleton corresponds to one of the legion of polyps composing the living coral. Each polyp, like a sea anemone, bears an upraised crown of tiny stinging tentacles, and each secretes calcium carbonate skeleton at its base. As the colony grows, new polyps bud from older ones, and the sister polyps remain interconnected. Thus, they fashion a living mosaic that cloaks the stony skeleton with a soft veneer. Through time, a cumulative secretion of mineral augments the mass of skeleton, over which the living tissue becomes a relatively thin covering.

An army of tiny symbionts is responsible for the success of many reef corals. Microscopic blue-green algae live within the skeleton, and microscopic protists (single-celled organisms called "zooxanthellae") inhabit the polyp tissues. The zooxanthellae occur in densities around 5 million/cm^2 (32 million/in^2), numerous enough that their photosynthetic pigments render living corals a golden-brown hue. The vital interaction beween the polyps and their symbionts is complex; the zooxanthellae provide much of the energy for coral growth. However, the major primary producers of the reef are small free-living algae ("scuzz," "turf," and calcareous algae) at the base of the food web. Part of the secret of reef success is that the system retains much of the organic material it produces.

The accretive growth of reef corals creates a primary framework, and the cumulative deposition of associated "shelled" animals and calcareous plants furnishes the reef's substructure and massive volumes of unconsolidated rubble and sand. Reef formations, in concert with the mangroves and seagrasses that extend and stabilize soft-bottom habitats, determine the topography of the marine and terrestrial terrain. The distributions of animals and plants, in and about the reef framework, are constrained by environmental stresses and by interactions among organisms. For example, the grazing of certain fish and sea urchins limits certain algal species to deep-water refuges and clears plant-free halos around reefs and mangroves. Space on the reef is at a premium. Encrusting organisms such as corals and sponges engage in a life-or-death competition, gaining substrate, light, and food by overtopping, drilling, smothering, stinging, or poisoning neighboring colonies. Mobile territorial organisms, such as damselfish, literally defend their turf—tiny plots of filamentous algae. Sand-dwelling animals may rework and resuspend fine sediments, smothering or bulldozing their neighbors. This is a tactic employed by mound-building shrimps, and it appears to be used by burrowing sea urchins on the sand plains of Florida.

Thus, competitive networks, physical disturbance, and predation regulate the diversity within reefs, seagrass beds, and mangroves. Moreover, because of their proximity, these three habitats are shaped by physical and biological interactions with each other. Key organisms can migrate between habitats, some in the course of a lifetime and others every day. For example, at dusk schools of colorful fish (such as grunts) stream through passes in the reef to feed in seagrass beds. Those not falling prey to larger fish return to the reef at dawn to wait out the day. Nutrients from the seagrass beds, imported in the form of fish excreta, enrich the reef with essential chemicals. Coral heads with resident fish schools grow faster than those without fish.

A BRIEF HISTORY OF THE FLORIDA-BAHAMA PROVINCE

The history of the Florida Keys and the other landmasses of the Caribbean region is bound up in the evolution of reef, mangroves, and seagrass communities. Through time these communities have changed drastically in composition and configuration. They will continue to alter in the centuries ahead, their mutable boundaries responding to environmental perturbations. Unfortunately, the geological history of the Bahama Islands remains obscure. However, a dramatic illustration of the dynamics of the reef is well documented in the geologic record of southern Florida and illuminates the present-day situation.

The Bahama Islands, drastically altered by tectonic, eustatic, and climatic changes, are the eroded and drowned remains of an ancient, shallow lagoonal and backreef environment. In the Pliocene and Pleistocene periods (5 million years ago to 10,000 years ago), the banks were atoll-shaped structures, rimmed with living reef. Today, they consist of flat-topped limestone platforms covered by carbonate sands, with only their seaward shores supporting luxuriant coral growth. During the Pleistocene (Ice Ages) period, there were several glacial and interglacial cycles, and 15,000 years ago so much water was bound up as ice in high-latitude glaciers that global sea levels fell 100 m (328 ft) or more. Caribbean continental shelves were exposed to the air, with devastating effect on the shallow-water communities.

During the last glacial period, the exposed marine sediments and reefs deposited during the preceding interglacial period were cemented and fossilized, producing two distinctive geological formations in southern Florida. The Key Largo Limestone extends from Miami Beach to Key West and forms the backbone of the Upper and Middle Florida Keys. It originated from reef corals and other calcareous components. Navigational canals blasted through the limestone have revealed corals (*Montastrea annularis* Ellis & Solander), one with branches 2.4 m (8 ft) long. The Miami Oolite of the Lower Keys, which overlies the Key Largo Limestone, was derived from sands that precipitated directly from seawater and accumulated as sandbars in an ancient lagoon.

During periods of low sea level, water covering the Florida continental shelf was shallow enough to permit reef growth. Extensive stands of elkhorn coral (*Acropora palmata* Lamarck) growing on the shelf margin were bathed by warm Gulf Stream waters. They were exterminated during rising sea levels and today form relict reefs or prominences that support deep-water corals.

In the course of the last major Ice Age, the ocean eroded a nearly vertical sea cliff around the continental shelf in areas such as the Bahamas. Today that steep rock face occurs at a depth of 50–60 m (164–197 ft) below present sea level. Offshore islands with extremely high water clarity can support reef corals at those depths. Generally, though, the submerged cliff is occupied by tenacious sponges and encrusting and endolithic algae that can cope with extraordinarily low light and nutrient levels and that take advantage of relaxed grazing and predation pressures. Gradual slopes and deeper topographies accumulate a layer of sediment, creating another set of conditions and supporting an entirely different fauna.

Before sea level rose to its current stand, South Florida and Florida Bay were high and dry, and present-day offshore reefs, such as Looe Key, were at the edge of the Florida mainland. Then, about 6,000–10,000 years ago, warming and glacial melting elevated sea levels. In what is now Hawk Channel (Figure 191G), rising waters created a swamp that intruded between the mainland and the evolving keys. By 3,500 years ago the ocean inundated Everglades-like terrestrial communities of Florida and Biscayne bays, paving the way for the mudbanks and mangrove keys of the present day. Bay waters, seasonally chilled or diluted with fresh water, streaming through reef passes, stressed the luxuriant Pleistocene coral reefs, presumably leading to the discontinuous reef terrain that we see today.

The remains of Pleistocene reefs, now barely above water, form the Florida Keys—an archipelago that separates Florida and Biscayne bays from the sea. The Keys and sections of mainland coast where urbanization, demolition, and pollution have not prevailed are populated by stands of mangroves and bordered by seagrass beds. Hawk Channel separates the Keys from the offshore living reefs. It is a mile or two wide and mostly less than 12 m (39 ft) deep, but filled with mud that in places is 6 m (20 ft) thick. Living reefs, the "rocks," "reefs," and "shoals" off the Florida coast, are built on Pleistocene formations. Reef corals colonized the inundated hard bottom areas, and coral growth followed the rising sea, reinforcing and modifying the preexisting terrain. The Marquesas and Dry Tortugas reefs at the southwestern tip of the Keys are ring-shaped reefs, possibly overgrowths on Pleistocene reef structures. On the present reefs, mangrove trees are sparse, and seagrass beds are more extensive than coral cover at shallow depths.

The inescapable lesson of current ecological monitoring is that significant alterations of shallow-water reef communities can occur rapidly, as well as in the span of geological time. Whether the world continues to warm and the coastline continues to retreat and how the growth of corals tracks sea-level fluctuation remain to be seen. As Looe Key reef has developed over the past 6,000 years, it has "backstepped" as sea level has risen, growing landward over a storm-tossed bed of rubble. Looe Key was a dry sand island when its namesake, H.M.S. *Looe,* ran aground in 1744; within 100 years it was reduced to shoals, and at present it is submerged. The diminished growth of *Acropora palmata,* as evidenced by coral growth rings, and the deterioration of the island in the last 200 years suggest that the reefs of Looe Key could someday lose their race with the sea (compare: Porter and Meier 1992; Jackson 1992).

MAJOR REEF ZONES AND HABITATS

Today, a profile across the Florida Keys (based on the configuration of the Looe Key National Marine Sanctuary bank reef) intersects a series of habitats that are typical of the Caribbean region. Along the shore of the Keys, arching prop roots of the red mangrove tree (*Rhizophora mangle* Linnaeus) support lush encrustations of sponges, hydroids, anemones, oysters, and algae; populations of brittle stars and sometimes sea stars and sea urchins cling to the roots and the surrounding banks. Landward, the black mangrove (*Avicennia nitida* Jacquin), with upturned "snorkel roots" (pneumatophores), is common.

The echinoderms that populate mangrove channels and nearby

seagrass beds are a select subgroup of the more diverse assemblage found on the reef. In surrounding waters, the muddy sediments of mangroves and the sandy seagrass beds are home to burrowing echinoderms: sea urchins (heart urchins and sand dollars), long-armed brittle stars, and suspension-feeding sea cucumbers that are mostly absent from hard reef substrata. Farther from shore, accumulations of lagoon muds, which derive from decayed calcareous algae, are populated by burrowing heart urchins accompanied by mound-building ghost shrimps (*Callianassa* spp.).

Across the reef, several major habitats or zones are distinguished. The most extreme environment of the reef itself is a rubble-covered portion of the Reef Flat (Figure 3D), formed from drifts of coral rubble deposited by waves during hurricanes. The coral pieces are irregular, riddled with boreholes, and encrusted with algae. The shallowest portion of the flat is emergent at lowest tides and repeatedly subjected to intense solar radiation, elevated temperatures, strong currents, salinity stress from rainwater, and desiccation. Echinoderms and other organisms that survive in this ecological house of horrors insinuate themselves deeply between chunks of rubble and in the sediment. At depths of 1–4 m (3–13 ft), the reef flat supports dense stands of turtle grass (*Thalassia testudinum* Banks ex König), manatee grass (*Syringodium filiforme* Kützing), and algae, but relatively few small sponges and soft corals. The dense carpets of seagrass can be pockmarked with sand-filled "blowouts" rimmed by seagrass rhizomes. These irregular pits, generally several meters in diameter and a half meter deep, are created during storms.

Seagrasses and calcareous algae also flourish in the deeper, sandy Back Reef zone (Figure 3C) at 5–8 m (16–26 ft) depths, which is protected from surge by the ramparts of the fore reef. Nearby, hard-bottom mounds known as patch reefs support dense aggregations of gorgonian corals (sea fans and sea whips), which can grow to 2 m (7 ft) tall, and a diverse assemblage of fire corals, brain corals, and staghorn coral.

The Sand Plain and Side Channels are characterized by larger patch reefs dominated by enormous coral heads, sponges, and gorgonians (sea fans, sea whips, and sea plumes), surrounded by expanses of coarse sand. Sea urchins, brittle stars, and basket stars inhabit the patch reefs, and sand dollars and burrowing sea urchins are among the dominant invertebrates on the sand plains.

The Fore Reef (Figure 3E), exposed to the heavy pounding of incoming waves, is a "spur-and-groove zone" formation that supports populations of sea urchins, brittle stars, and sea cucumbers. Most algae on the fore reef are encrusting or filamentous; fleshy forms are generally removed by herbivores. The spurs, growing perpendicular to the reef axis, are walls of limestone rock and living corals, up to 200 m (656 ft) long and 3–7 m (10–23 ft) high. Neighboring spurs are separated by channels (grooves) lined with sand and rubble. At Looe Key, spurs originated 4,000–6,500 years ago from patches of elkhorn coral (*Acropora palmata*) growing on a flat bed of sand and rubble or directly on a shelf of 120,000-year-old limestone. The actively growing *Acropora* infrastructure went into decline roughly 800 years ago. Today, the spurs are composed of a slow-growing coral veneer capping dead *Acropora* ridges. The shallow spurs are densely encrusted with bladed fire coral (*Millepora* spp.), pillowy mounds of soft, tan colonial anemones (*Palythoa* spp.), and encrusting stony coral (*Porites astreoides* Lamarck), which are exposed at low tides. Somewhat deeper, the spurs are shingled with lettuce coral (*Agaricia* spp.) and stand among piles of stick-shaped rubble from staghorn coral (*Acropora cervicornis* [Lamarck]) and elkhorn coral. Spurs in 7–8 m (23–26 ft) depths are dominated by monumental boulders of star coral (*Montastrea annularis*) and brain corals (*Diploria* spp.), reinforced and infilled by encrusting *P. astreoides,* finger coral (*Porites porites* [Pallas]), boring sponges, and coralline algae, and capped by staghorn coral and sometimes by elkhorn coral.

The Intermediate and Deep Reef seaward of the spur-and-groove formations occupies a 45–60° slope beginning at 10–30 m (33–98 ft) depth. It exhibits irregularly shaped, low-relief, spurlike features jutting from drifts of carbonate sand. Dominating the hard substrate in this habitat are shinglelike star coral and lettuce corals (*Montastrea annularis* and *Agaricia* spp.), demosponges, numerous soft corals, and crusts and clumps of macroalgae (*Lobophora* spp. and *Halimeda* spp.). On the deep reef, an occasional feather star (*Nemaster* spp.) may be observed, along with other elements of the echinoderm fauna. Below the deep reef, at a depth of 30–35 m (98–115 ft), the terrain is mostly a monotonous Deep Sand Bed (Figure 3F) interrupted with sparse rocky outcrops with few corals. This zone is transitional between the reef and a deep-water community that lies beyond the range of conventional diving gear.

GENERAL FEATURES OF THE PHYLUM ECHINODERMATA

The echinoderms (Greek: *echinos*, spiny; *derma*, skin), represented by the sea stars and their relatives (sea urchins, sea cucumbers, feather stars, and brittle stars), have come to symbolize the ocean realm (Figure 2). They are extraordinarily attractive and usually harmless and are, therefore, a pleasure to observe and study. For the same reasons, they are readily collected, and the dried skeletons of sea stars or sea urchins have graced curio collections for several centuries.

The phylum Echinodermata is an ancient lineage, not too distantly related to the phylum Chordata, the animals with backbones. There are approximately 16 extinct classes of echinoderms, with more than 13,000 fossil species now described; many of them seem bizarre in comparison with their present-day relatives. The group apparently had a long Precambrian history, for echinoderms were already fairly common in some habitats in the Early Cambrian, nearly 600 million years ago. Even those earliest forms had a well-developed skeleton and a diversity of body plans.

At least 6,500 species of echinoderms are alive today, and six living classes have been recognized. The sea stars (Class Asteroidea, approximately 1,800 living species) have five or more hollow arms radiating from a central body or disk. Brittle stars and basket stars (Class Ophiuroidea, approximately 2,000 living species) have five solid snakelike or branching arms and a small central disk. The stalked sea lilies and unstalked feather stars (Class Crinoidea, approximately 700 living species) have a central body and five or more long, featherlike arms. In sea urchins (Class Echinoidea, 900 living species), there are no arms; the body is equipped with movable spines of various lengths. "Regular" urchins are typically subspherical; "irregular" urchins include the ovoid heart urchins and the discoidal sand dollars. Sea cucumbers (Class Holothuroidea, approximately 1,200 living species) have a more or less cylindrical body, lacking arms and spines, and the mouth is surrounded by a ring of feeding tentacles. In the recently discovered sea daisies (Class Concentricycloidea, 2 living species), the body is discoidal and arms are lacking. These strange concentricycloids appear to live exclusively on pieces of wood on the deep sea floor.

What do all the living species of echinoderms have, or lack, that makes them distinctive? The echinoderms possess three unique and distinctive features: a body plan with five-part (pentamerous) symmetry, an internal calcite (calcium carbonate) skeleton, and a water-vascular system of fluid-filled vessels that are manifested externally as structures called tube feet. However, many other specializations, such as their nervous and circulatory systems and mode of development, separate the echinoderms from other animal groups. Casual examination first reveals the absence of a head and eyes, and closer inspection shows that a brain and heart are lacking as well. The sexes are usually separate, as determined by examination of gonadal structures, but it is generally impossible to distinguish a male from a female echinoderm based on outward appearance.

Five-part symmetry is evident in the five-armed sea stars and brittle stars, and in the five "petals" on the top of a sand dollar. It is not obvious in most sea cucumbers, which are often cylindrical and nearly featureless externally. Internally, however, five muscle bands and other structures confirm the pentamerous body plan of the sea cucumbers. This general organization is unique in the animal kingdom, and it has been the exclusive property of echinoderms throughout most of their long fossil history. Some unusual fossil echinoderms from Lower Paleozoic rocks, 500 million to 600 million years ago, show no trace of radial symmetry; asymmetry was evidently much less successful than the five-part body plan.

In all living echinoderms the body has a skeleton (test) composed of calcium carbonate in the mineral form called calcite. The skeleton is internal, formed within the dermis, and almost always invested with soft living tissue. The individual pieces of calcite (known as ossicles) may be quite large, 3 cm (an inch) or more across in some species, and may form a hard, rigid test, as in sand dollars. In contrast, most sea cucumbers have microscopic ossicles embedded in a relatively soft body wall. Linking the ossicles is an array of muscle and/or ligament that is uniquely specialized to regulate the stiffness of the body wall and to control the movements of its appendages. This unique construction, restricted to the phylum, permits lightly calcified sea cucumbers rapidly and reversably to transform from a rock hard to an almost liquid state; it endows heavily skeletonized sea stars with amazing pliability.

Another feature that distinguishes echinoderms is a water-vascular system composed of branching, fluid-filled vessels that give rise to specialized structures such as the remarkable tube feet. Tube feet on different parts of the body are variously modified for locomotion, adhesion to the substrate, respiration, burrowing, manipulating food, sensory perception, or performing a combination of tasks. Contraction and relaxation of internal, muscularized reservoirs alters pressure in the system and can cause extension and retraction of the tube feet, with all of these movements under precise nervous control.

Although the tube feet and other locomotory structures of echinoderms are capable of rapid movement, echinoderm locomotion is generally a slow process. For that reason, the best records of echinoderm behavior have been made using time-lapse cinematography. Methods of locomotion vary. Holothuroids can glide along the seafloor with the aid of their tube feet or move in a sluglike manner. Sea urchins, sea stars, and some brittle stars can also creep with the aid of their bands of tube feet. The sea urchins can use the leverage provided by their spines to aid in locomotion, and some brittle stars hop across the sea bottom by "breast-stroking" movements of their arms. The feather stars can pull themselves slowly ahead with their arms and reattach to the substratum with their cirri. Swimming behavior is known in many feather stars and in several species of deep-sea sea cucumbers and brittle stars.

Echinoderms are usually of modest size, most ranging up to 10 cm (4 in) in length or diameter. Very large species are not common, although some brittle stars can exceed 1 m (3 ft) in diameter, and some sea cucumbers can exceed 2 m (6 ft) in length. However, in some extinct sea lilies, the stalk exceeded 20 m (66 ft) in length. Some species are brilliantly colored and others are drab; depending on their pigmentation pattern and behavior, they may be conspicuous, but many echinoderms are camouflaged in their chosen habitats. In addition, many echinoderms are cryptic during the day and venture in the open only after dark, when visual predators are at a disadvantage. Predation on some echinoderms is infrequent, in part because they are inconspicuous, because of their protective calcite skeleton, and because many species contain toxins that repel predators. Other defensive structures include long, sharp spines and the poisonous pincer organs (pedicellariae) of sea urchins and sticky Cuvierian tubules discharged by some sea cucumbers. However, echinoderms compose a major food resource for certain fish and crustacean species, and, as noted below, some are relished by humans as well.

Echinoderms also survive predation through their amazing ability to regenerate lost body parts. If a predatory fish bites off the arm of a brittle star, sea star, or feather star, the lost parts can be replaced in a matter of a few weeks or months. Some echinoderms are capable of voluntarily severing parts of their body (a process called autotomy) when attacked. A brittle star seized by its arm tip can amputate the arm and escape; some sea cucumbers can eject their intestine and associated organs when disturbed. The lost structures can be regrown; sand dollars regrow spines and larger parts of the body that are lost to predators, and crinoids regenerate lost arms and even the visceral mass.

For some species of brittle stars, sea stars, and sea cucumbers, this power of regeneration is employed during asexual reproduction. In that process, an individual divides into two or more pieces, and each piece develops into a complete adult. However, this method of reproduction is much less widespread than sexual reproduction, in which eggs from a female echinoderm are fertilized by sperm from a male. The gametes are typically ejected into the seawater, where fertilization takes place. The fertilized egg usually develops into a free-swimming larval stage that lives in the plankton for several days or weeks. The nearly microscopic larvae look so very different from adult echinoderms that they were once regarded as a totally unrelated group of organisms, a group with considerable diversity. Each of the echinoderm classes has distinctive larvae, and the characteristics of the larvae are generally indicative of the various echinoderm families, genera, and species. Some echinoderm species complete the cycle of birth, reproduction, and death in a year or less, and other species live for several decades and reproduce many times.

The echinoderms are almost exclusively marine animals, and only a few species can survive in brackish waters. Aside from that limitation, they occupy a wide spectrum of oceanic habitats, ranging from tide pools to the greatest depths of the sea. Their habits and habitats are greatly varied. In rocky intertidal zones and coral reefs, sea stars and sea urchins are often conspicuous, but brittle stars inhabit crevices. Offshore, burrowing brittle stars and sea urchins (sand dollars and heart urchins) may form vast "beds" in soft sediments. In deep waters, sea lilies occur on both rocky and soft bottoms, and sea cucumbers can dominate on soft sediments in the abyss.

Where echinoderms occur in immense numbers, they can exert considerable influence on the ecosystem. Dense populations of sea urchins can denude vast tracts of seaweed and seagrass; sea stars can decimate coral reef communities and plunder beds of oysters and clams. On the other hand, other sea urchins and some sea cucumbers are responsible for the production and winnowing of vast quantities of the world's seafloor sediments. Certain sea cucumbers and sea urchins are prized as food, and tons of them are fished commercially every year.

OBSERVING, COLLECTING, AND PRESERVING ECHINODERMS

FINDING AND HANDLING

Many echinoderms, readily seen, cling to rocks, hard and soft corals, sponges, and live on sand. Cryptic and nocturnal species can be found under rocks and slabs of coral rubble. These shelters should be returned to their original position after you look beneath them, to avoid crushing animals and plants and minimize damage to their habitat. Degradation of natural marine areas is already bad enough because of human greed and callousness. Never collect more than you must, and always take proper care of the animals you gather.

To find certain species, a degree of "destructive sampling" may be required. However, rock smashing, excavation, and the use of toxins are only warranted when absolutely necessary for scientific research. In Florida, the Bahamas, and elsewhere, permission to collect echinoderms, or to carry out sampling activities that are potentially harmful to the environment, including the use of fishing gear such as trawls, must be obtained in advance from appropriate government agencies. Applications must be submitted, and written approval or licenses obtained. Fieldwork may be absolutely prohibited in some parks and environmental preserves.

Burrowing brittle stars, sea cucumbers, and sea urchins are often associated with the roots of seagrasses and soft sediments. They can be located by selectively digging, sieving sediment through a screen, or gingerly combing the sediment with one's fingers. Echinoderms may emerge from rock and reef crannies in response to the ichthyotoxins (fish poisons) used by fish collectors. Many species can be taken from great depths by dredges, trawls, grabs, or baited traps.

Most echinoderms can be manipulated with the naked hand—they do not bite! It may be easier to pick up very small specimens with tweezers. Sea urchins should be handled with appropriate caution (gloves may help). The spines of one species in this region are mildly toxic, and wounds caused by urchin spines can cause considerable pain and result in infection. Brittle stars and feather stars have a largely undeserved reputation for "going to pieces," but most hold up well if they are treated gently. They should be held near the disk and arm bases, not the arm tips. If sensitive, delicate species must be critically examined, they can be anesthetized as described below. After animals have been studied in the field or in the laboratory, they should be returned alive to the collecting site. If that is not possible, they should be preserved and permanently archived so that their usefulness can be extended.

COLLECTING

In the field, plastic bottles, buckets, or bags can be used to gather algae, sponges, and encrusted rubble that will be examined in the laboratory for small echinoderms. Scuba divers can hold echinoderms in rigid-walled plastic containers or plastic bags that can be tightly closed to make a protective "water balloon"; animals placed directly in a mesh dive bag usually come to the surface much the worse for wear. Specimens retained for identification and observation can be kept in a bucket or, even better, in an insulated cooler. They will stay in good condition with a bit of sediment (for burrowing species) or chunks of rubble in the container. However, they will die if kept in water fouled with sponges or other toxic animals, or left to overheat in the sun.

PRESERVING

Under special circumstances, it might be necessary to preserve echinoderms. For example, in environmental surveys and ecological monitoring programs specimens are preserved for accurate identification, and vouchers are deposited in museums for future reference. Professional photographers may collect specimens they have photographed and submit them to taxonomic specialists for identification.

Because any specimen, accompanied by an accurate collecting label, is more informative for identification than no specimen at all, opt for a basic instead of optimal preservation procedure as circumstances dictate. The basic technique produces specimens that are adequate for identification. The optimal procedure is more time-consuming but produces the best possible preserved specimens for scientific study. In either case, a label with the following information should *always* be added to the sample container as soon as feasible after the specimen has been collected: locality (as precisely as possible, with coordinates of latitude and longitude, if available), depth, date, name of collector, and if possible collection gear, preservative, vessel (e.g., name of dive boat), and notes on habitat and color of the living specimen. Use good-quality paper that will not shred when wet, and write the label with a soft lead pencil or an indelible india ink pen. No preservation technique is perfect, and specimens invariably discolor or fade because of the effects of preservative fluids. Therefore, a color photograph of the specimen taken underwater or on land, before preservation, can be extremely informative.

Basic Preservation Procedure

Place the specimens in a tightly sealed container of 70–80% alcohol. When plain ethanol is unavailable, vodka, rum, or isopropanol (rubbing alcohol) can be substituted. If alcohol is in short supply, large sea cucumbers can be eviscerated before preservation by making an incision in the body, removing the internal organs, and preserving only the body wall.

Formalin (preferably 5–10% buffered formalin) can be used as a preservative for most echinoderms, but sea cucumbers should always be stored in alcohol. Formalin is a dangerous chemical, and it should only be used by properly trained scientists. Furthermore, it is only suitable for short-term storage, because it is inherently acidic, and prolonged contact destroys the echinoderm skeleton and weakens specimens.

Optimal Preservation Procedure

Museum-quality specimens are prepared by following the five steps given below, but for applications such as histological or biochemical studies specialized techniques may be required. For example, specimens and tissues intended for DNA sequencing should be placed directly in ethanol ($\geq 80\%$); their exposure to water or formalin must be avoided.

1. CLEANING. Animals that are covered with sediment or debris should be gently agitated or rinsed with a stream of seawater.

2. RELAXATION. This step keeps specimens from breaking or contorting when they are preserved. Relaxation (anesthesia) is carried out in containers that are covered, to confine the specimens and to exclude light that disturbs them.

Echinoderms are commonly anesthetized with magnesium chloride or magnesium sulfate solutions (see "Chemical Solutions and Warnings" below). The effect of these chemicals is reversible to a point; individuals recover from anesthesia when returned to seawater. An alternative, simple method for tropical echinoderms involves placing animals in a container of seawater that is at room temperature and leaving them overnight in a refrigerator. In that time, they die in a "relaxed" posture. Special relaxing methods for sea cucumbers, large sea stars, feather stars, and basket stars are explained below.

Relaxation with magnesium chloride is a preferred treatment for specimens to be used for histological analysis or for scanning electron microscopy of soft tissues and is ideal for many echinoderms. Individuals are transferred from seawater directly into a container with a solution of magnesium chloride that is isotonic with seawater. Anesthesia is usually complete in 15 minutes.

It can be more convenient to use magnesium sulfate (Epsom salts) because it is available from any pharmacy and need not be weighed before use. Place specimens in a shallow container (such as a photographic tray) and barely cover them with seawater. Put 15–45 g (1–3 tablespoons) of Epsom salts in a corner of the container (not directly on the specimens). Every 5 minutes, gently tip the tray to mix the dissolving salts. If the specimens actively move after 20 minutes, sparingly add more salts (the amount of salts required is proportional to the volume of water in the tray).

Apodous sea cucumbers respond rapidly to magnesium chloride and magnesium sulfate, but other sea cucumbers may not relax, even after hours in magnesium salts. For refractory species, propylene phenoxytol (PPO) can be used. Place a few milliliters of concentrated PPO (see warning note below) in a small bottle of fresh water. Shake the bottle vigorously, then let the tiny drops of PPO settle to the bottom. Draw off a pipette (medicine dropper) of the supernatant liquid (the clear liquid on top) and add it to a container of seawater with the sea cucumber; add additional pipettes full of the mixture at 1- to 2-hour intervals until the specimen extends its feeding tentacles. Leave the specimen in the solution until it is completely unreactive to prodding, overnight if necessary.

Large sea stars, which would require an inordinate amount of magnesium salts for anesthetization, can be relaxed using fresh water. Place specimens in a bucket or a tray with just enough seawater to cover them and wait for them to extend their arms. At intervals of about 20–30 minutes, add fresh water to the container to dilute the seawater from full strength to 90%, 80%, 70%, and so on. Usually it is not necessary to dilute the seawater to less than 50% to bring about relaxation.

Feather stars and basket stars do not respond to standard anesthetics such as magnesium salts, but some species of feather stars relax after 10 minutes in calcium-free seawater followed by 10 minutes in approximately 0.1% MS-222 (a fish anesthetic). However, feather stars and basket stars can be preserved without relaxation, as described in the following section on preservation.

It is not necessary to relax sea urchins before preservation unless special preparation of tube feet and other soft tissues is required. Note that if an urchin with long, outspread spines does not fit in a narrow container, its spines may be induced to "fold up" if the *live* individual is balanced on the opening of the seawater-filled container and allowed to settle to the bottom.

3. PRESERVATION. Relaxation is complete when the tube feet, arms, or tentacles of an individual do not react to prodding; at that point, the specimen is placed in preservative. In general, the best preservative for echinoderm specimens is 70–80% ethanol, but formalin can sometimes be substituted, as specified below. Because specimens stiffen in preservative, relaxed sea stars, brittle stars, and sea cucumbers should be carefully stretched out in a tray of alcohol, oral side down, to harden them in a suitable position for study. To facilitate storage, a brittle star's disk and arms can be arranged in a U shape or "comet" form with arms flat and trailing behind the disk (see Figure 34). Pick up the disk of a relaxed brittle star between two fingers (or with tweezers); gently shake the specimen so that the arms straighten, before placing it in alcohol.

The tentacles of completely relaxed sea cucumbers will not contract when individuals are transferred to alcohol. Very large individuals, or animals with a thick body wall, are best preserved using alcohol injection. While the completely relaxed specimen is still in PPO solution, 95% alcohol is carefully injected through the anus or body wall of the sea cucumber with a hypodermic syringe. When the tentacles are fully extended, the anus is pressed closed for several minutes before transferring the specimen to 70–80% alcohol. If anesthetic is unavailable, a sea cucumber with extended tentacles can be quickly, tightly grasped with tweezers behind its tentacles and held in alcohol for several minutes before releasing it; its tentacles will remain extended.

Feather stars and basket stars, which have not been relaxed, are preserved as follows. Keep an individual in a darkened container of seawater until it extends its arms (nocturnal species may not fully extend their arms until night). Fill a tray with ethanol, preferably a 90–95% concentration. Lift a specimen out of seawater, supporting its arms so that they do not cover the central region of the body. Rapidly transfer it to the tray of alcohol oral side down and maintain just enough gentle pressure on the specimen to keep the arms extended for a minute or two, until the specimen stiffens. Sea urchins can be transferred from seawater to a wide-mouthed bottle of alcohol without any relaxation treatment, unless special preparations are required. Urchins also can be anesthetized by the slow addition of ethanol to their seawater, before they are transferred to preservative.

Most echinoderms can be preserved in 5–10% buffered formalin, except for sea cucumbers, which must be preserved in alcohol. However, small or delicate echinoderms left in formalin for long periods of time invariably deteriorate; they should be transferred to alcohol within several days. Initial preservation in formalin may be beneficial for large specimens of sea stars and sea urchins. However, they should not be stored in formalin for more than a few weeks before being transferred to alcohol or dried. If many specimens are to be processed in formalin, the solution may be re-used.

4. WASHING AND DRYING. The preservative solution in which specimens are first placed always becomes diluted, acidified, and contaminated with seawater salts or formalin and eventually damages wet-preserved or dried material. Therefore, before drying or transfer to fresh alcohol, specimens should be soaked overnight in distilled water to remove formalin and salts. After washing in distilled water, drain specimens and place them in clean, 70–80% ethanol. Large specimens (or large numbers of specimens) should be transferred through two changes of alcohol after they are washed, because the wash water they release dilutes the first preservative bath.

Alcohol preservation is the most satisfactory method for long-term storage of echinoderms because it inhibits the deterioration of the skeleton and soft tissue. However, echinoderm specimens may be dried for several reasons. Some must be examined dried or are too large to store in containers of alcohol. Also, formalin-preserved and dried specimens can retain their natural coloration for a long time if they are stored in the dark. A specimen that has been preserved in formalin can be dried after it is washed in water; specimens preserved in alcohol need not be washed. The specimen is placed on a support, such as a metal screen, in a warm, dry, well-aerated place (e.g., under a lamp but not in direct sunlight or an oven) until it is completely desiccated.

5. STORAGE. Preserved specimens may be stored in plastic or glass containers. For temporary storage, self-sealing plastic bags are useful for small specimens; animals with sharp spines can be kept in plastic bottles or metal paint cans. Before shipping, samples may be drained, leaving moistened specimens in a minimal volume of preservative. For long-term storage it is best to keep echinoderms in glass containers with tightly sealed plastic lids, in a volume of 70–80% ethanol at least twice their body volume. Never cushion wet or dried echinoderms with cotton, for the fibers become tangled in their spines.

CHEMICAL SOLUTIONS AND WARNINGS

Read and heed the cautionary instructions on chemical containers.

70% ethanol (=140-proof ethyl alcohol) is prepared by diluting a concentrated alcohol solution with *distilled* water. Mixing tapwater with concentrated alcohol can produce damaging precipitates. A 70% solution is easily made from concentrated 95% ethanol by pouring 70 ml of ethanol into a 100-ml graduated cylinder and then adding distilled water to the 95-ml mark.

5–10% buffered formalin MUST BE HANDLED WITH CAUTION and should only be used after professional training. Concentrated formaldehyde solution should be buffered with 2% borax (sodium tetraborate) powder. Allow the borax to dissolve before diluting the concentrated solution 1:9 to make a 10% formalin solution. Use 10% formalin for short-term treatment and 5% formalin for longer storage of large specimens. Sea cucumbers and delicate specimens such as crinoids should not be preserved in formaldehyde. *Note: Formaldehyde solutions and fumes are toxic. Use only in open, well-ventilated areas and wear gloves and goggles to avoid contact with the solution and fumes. Store formalin in a tightly sealed container.*

Epsom salts (magnesium sulfate), in crystal or granular form, is available off the shelf in pharmacies.

Isotonic magnesium chloride solution has the same ionic concentration as seawater and therefore does not irritate most marine invertebrates. It is made by dissolving 73 g of $MgCl_2 \cdot 6H_2O$ in 1 liter of distilled water. The solution can be used a number of times.

Propylene phenoxytol (PPO) is a potential carcinogen. It is advisable to wear gloves and exercise caution when using PPO. Preparation of PPO solution for the anesthetization of sea cucumbers is described above.

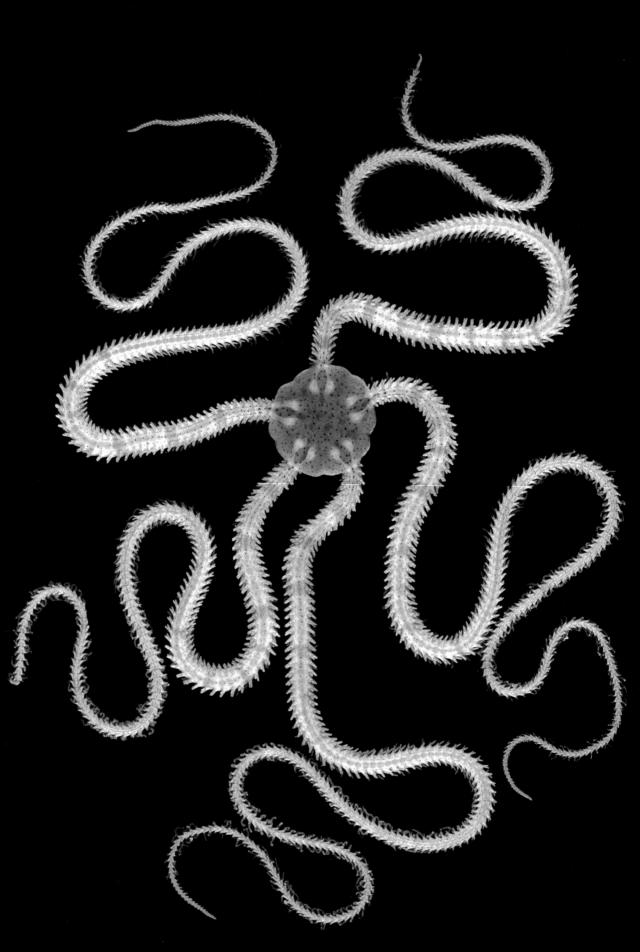

PHOTOGRAPHING ECHINODERMS

The phylum Echinodermata contains some of the most photogenic animals in the marine environment. Because they are plentiful and generally slow moving, especially in coral reef habitats, finding a suitable subject to photograph is not difficult. Even fast-moving echinoderms can be collected and afterwards photographed in an aquarium.

Although the techniques employed to photograph echinoderms in a studio or underwater are similar, underwater work requires that the photographer feel "at home" in the echinoderm's milieu. It also requires a considerably larger investment for scuba and submersible photographic gear. The essential equipment is listed below, and suggestions are offered to aid the reader in producing acceptable photographs with minimal experimentation. However, as with any hobby or sport, practice, study, and experience are essential to personal improvement. For the sake of brevity, it is assumed that the reader has some basic knowledge of photography and photographic gear.

STUDIO PHOTOGRAPHY

Certainly one can spend a great deal of money to photograph marine animals, but that need not be the case. To start with, the prospective photographer should consider purchasing the following equipment:

- 35 mm or larger format single-lens reflex camera equipped with a good-quality lens (preferably with 1:1 macro capability)
- high-grade, slow-speed film; ISO 25 or 64 (for color slides), ISO 25 to 100 (for color prints)
- portable flash units with sync cables (one pair)

- adjustable flash stands (one pair)
- inexpensive copy stand
- shutter cable release
- shallow glass aquarium (have five pieces of glass cut and cement them with clear silicone glue)
- two blocks of wood
- black velvet cloth
- Epsom salts or magnesium chloride solution

The items listed below can be helpful but they are not essential for superior photographs:

- film autowinder
- AC power supply for flash units
- electronic flash meter
- modeling light

The equipment is assembled as illustrated in Figure 4. The blocks of wood are used to support the aquarium above the velvet cloth. The velvet absorbs light passing through the aquarium, producing a uniform black background around the subject.

As indicated above, slow-speed films are preferred; they have a fine grain that produces sharp images. Because photographing specimens in a tank necessitates that the camera be positioned close to the subject, the lens will need to be stopped down (e.g., f/11 to f/22) to maximize the depth of field. Shooting slow-speed film with the camera lens stopped down requires a considerable amount of

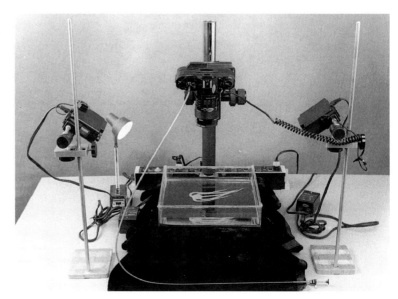

FIGURE 4. Equipment for photographing echinoderms. A camera, supported on a copy stand, is positioned above a specimen in an aquarium. A small lamp provides illumination for focusing, and the two strobes are triggered by the camera to expose the film.

light, making the placement of flash units a critical matter. Their proper angle and distance from the subject is determined by testing the system. This is accomplished by photographing a specimen with a test roll of film or by using an electronic flash meter to determine the target f-stop for the film type being used. The target f-stop can be altered by changing the power setting of the flash units or by moving them closer to, or farther away from, the subject. Caution must be used in placing the flashes so that reflected images of the firing strobes are not captured on the film.

When the system is ready for a shoot, fill the aquarium with enough seawater to cover the specimen. The seawater should be as clean as possible. Small floating or suspended particles will reflect light from the electronic flash units and mar the photograph. If you are photographing an animal that has been anesthetized (see the chapter "Observing, Collecting, and Preserving Echinoderms" for this technique), it is necessary to add some Epsom salts to the aquarium or keep the specimens in magnesium chloride solution; otherwise the subject will "wake up" and move about.

Before formatting the photo and focusing on the subject, check to see if glare from any room lighting is being reflected off the water onto the camera lens. Glare can be eliminated by turning off overhead lights and using a small modeling light (such as a desk lamp or flashlight) that is turned off after the subject has been brought into focus. A collar around the lens, made of black paper, is helpful in eliminating reflections from the camera body itself.

At this stage, the system is ready to go to work. However, one final recommendation—perhaps the most useful advice—is necessary. When considering the amount of money and time invested in setting up a photographic system, it should be obvious that film and developing costs are relatively cheap. Therefore, it is highly recommended that several frames be shot for each subject, varying the f-stop with each exposure. Using the following rule of thumb usually assures at least one properly exposed photograph for each subject photographed:

When photographing with slide film, shoot five exposures, one at the target f-stop and one at each of the two half stops above and below the target (e.g., target = f/16; take exposures at f/11, f/13, f/16, f/19, f/22). Color negative film allows more latitude in printing photographs, so generally only three exposures are necessary, one at the target f-stop and one at a full stop above and at a full stop below the target (e.g., target = f/16; take exposures at f/11, f/16, f/22).

UNDERWATER PHOTOGRAPHY

An especially rewarding way to photograph marine organisms is in the animals' natural habitat. Admittedly, working underwater adds another level of complexity to photography, but with a little experience the result can be dramatic pictures of a world that few people witness firsthand.

As in the case of studio photography, some specialized photographic equipment is required to get started, and this can be a costly proposition; not only must the photographer purchase an amphibious camera, or an underwater housing for a land camera, but a full complement of scuba gear is also essential. Besides the hardware, previous diving experience is invaluable, because the new underwater photographer will learn quickly that, when capturing a scene on film, little attention can be devoted to the surrounding environment. Currents or surge add to the photographer's difficulties in some settings. Furthermore, the diver must be willing to spend much of a dive looking through the camera, an experience that some find rather disconcerting.

When purchasing an underwater camera system, there are advantages and disadvantages to consider for each type of equipment. A popular system is an amphibious camera (such as Nikonos), which is compact and can be submerged without a separate protective housing. The major drawback of many versions is that the photographer must compose pictures through a viewfinder or with framing adapters that attach to the camera's lens. Focusing the lens and framing the picture require guesswork and experience, because the photographer does not see the subject through the camera lens and must compensate for the distortion of images viewed through air and water. Despite this drawback, the Nikonos line is extensive, with automatic exposure hardware, extension tubes for close-up photography, and underwater strobes. A single-lens reflex (SLR) model is also offered; it shows through the viewfinder precisely the image that is captured on film. Autofocus lenses and dedicated strobes designed for the SLR system can make underwater photography even more surefire.

The photographer-diver who already owns an SLR camera can purchase an excellent underwater housing for the price of an amphibious camera system. These housings are bulkier and heavier than amphibious cameras, and they can complicate entering the water, but once below the surface, their slight negative buoyancy allows them to be maneuvered with ease. They come with gears and levers to release the shutter, advance the film, and change the focus, f-stop, or film speed. Some housings accommodate motor drives that automatically advance the film. Moderately priced systems made of plastic (such as Ikelite) allow the diver to view all the workings of the camera. The greatest advantage to using an SLR in a protective housing is that focusing and formatting guesswork are eliminated because the subject is viewed through the camera's lens.

Regardless of the camera system selected, an amphibious flash unit (or two) is vital for underwater photography at depths greater than a few meters because seawater filters out colors of the visible spectrum. The farther the light must travel through the water column, the greater the filtration effect. Red light is eliminated by 6 m (20 ft) depth, orange light by 12 m (40 ft), and yellow and violet by 24 m (80 ft). The absorption of visible wavelengths makes objects appear bluish green or bluish gray as one descends deeper into the ocean.

A submersible strobe or flash unit can reveal the true colors of underwater subjects. These systems are designed to duplicate the visible spectrum of sunlight. Higher-priced units emit enough light to allow the use of slower-speed films, even with a lens stopped down to f/16 or f/22. Using two strobes, one higher powered than the other, can produce dramatic shadowing effects that enhance contrast. Strobes are also useful in "stopping action," because an active subject is illuminated for only a fraction of the time the camera shutter is open. Some strobes have a built-in modeling light that aids in focusing on the subject and aiming the flash. Lacking that, a small underwater light can be strapped to the strobe.

There are several suggestions to keep in mind when preparing for an underwater photo session. Find a reliable dive partner who understands that you will probably spend the dive in a limited area. Impress upon that partner the importance of staying clear of your work area so as not to stir up clouds of silt. Always check the camera and flash systems for leaks, testing them in fresh water before the dive. Clear, clean water is essential for producing good photographs underwater; even the finest equipment is no match for poor visibility caused by suspended sediment. Not only will suspended particles register as a cloud of white specks, they can also cause photographs to appear out of focus. Above all, it is far better to take several exposures (by varying the f-stop) of a few subjects than to take a single photograph of each of many subjects. In the former case, the photographer's efforts may yield six or seven outstanding shots per roll, but in the latter, it is quite possible that not a single picture will be satisfactory. Keep in mind that even the most successful photographers are pleased to produce just one or two excellent photographs per roll of film.

COMPOSITION AND SIGNIFICANCE OF SCIENTIFIC NAMES

In general, unlike birds, mammals, and flowers, echinoderms lack popular or common names. The few common names that have been applied to echinoderms convey little information (e.g., "long-spined urchin") or they are downright unwieldy or unattractive ("donkey-dung sea cucumber").

Every known echinoderm species has a scientific name, a two-word name derived from Latin or Greek. Throughout this book we have used only scientific names for the echinoderms. They may seem daunting, but as they become familiar they are easier to use, and their benefits become increasingly evident. Advantages include the fact that they are unambiguous (each name refers to just one species), and they are used and understood worldwide. In contrast, common names may apply to more than one species and may vary from one geographic region to another. Scientific names are often as descriptive and generally more informative than common names because they indicate the evolutionary relationships among species.

Biologists have devised a uniform system for naming animals and plants. Each species is given a two-word Latin or Greek name (a binominal), and the name is printed in italic type or underlined. The first name has an initial capital letter and is the genus name applied to a group of related species. The second part of the name is the species name. Thus, the sea urchins *Lytechinus variegatus* and *Lytechinus williamsi* are closely related species belonging to the same genus. Distinctive, geographically isolated entities may be regarded as subspecies. They are denoted by a trinomial, as in the case of *Echinometra lucunter lucunter* and *Echinometra lucunter polypora*.

A complex set of rules governs the formation and use of these names. It suffices here to say that generic names must be unique (the names of no two animal genera are identical), and that within a genus the species names must be different. Often the originator, the author of a name, is cited when the name is quoted, hence *Lytechinus williamsi* Chesher. The publication date of the first description may also be noted, for example, *Lytechinus williamsi* Chesher, 1968. A species may be transferred from its original genus to another for a variety of reasons; when this occurs, the author's name is placed in parentheses, hence *Lytechinus variegatus* (Leske). Genus and species names may call attention to a conspicuous character trait (*Lytechinus variegatus* varies greatly in color), or they may honor the discoverer (*Holothuria thomasi*), a colleague (*Holothuria rowei*), a spouse, or even the pet dog. When a name is being used frequently in a written discussion, the generic name may be abbreviated after its first use (e.g., *L. variegatus*).

Related genera are grouped into families, families into orders, orders into classes (e.g., Class Asteroidea: sea stars; Class Holothuroidea: sea cucumbers), and the classes of echinoderms constitute the phylum Echinodermata. Other animal phyla include the Mollusca (seashells) and Chordata (animals with backbones, e.g., fish, reptiles, mammals). This system, which has useful predictive value, is based on the physical similarities of the creatures that are classified together and on their ancestry (their genetic relatedness). For those reasons, the higher-order classification of an animal, such as its class, order, or family, is revealing of its overall body plan and the general features of its feeding and reproduction. Identification to genus and species means that an animal's appearance, and often its life history traits, can be described in greater detail. Individuals of the same species look alike and are ecologically similar, precisely because they are closely related genetically. However, there is always some variability in physical features (color, for example, or size) among the individuals of a species. Closely related species may also look alike, and they may differ only in subtle characteristics that can be discerned only by a scientific specialist.

SYSTEMATIC LIST OF THE ECHINODERMS
of the Florida Keys and the Bahama Islands
Found at Depths Less Than 30 m (98 ft)

Classes are arranged according to Smith (1984b). Orders and families are designated and arranged in accordance with recent studies, including Serafy (1979) for echinoids, Miller and Pawson (1984) for holothuroids, Clark and Downey (1992) for asteroids, Spencer and Wright (1966) for ophiuroids, and Meyer et al. (1978) for crinoids. Genera and species are arranged alphabetically in this listing.

CLASS CRINOIDEA
Order Comatulida
FAMILY COMASTERIDAE
Comactinia echinoptera (Müller, 1840)
Nemaster discoideus (Carpenter, 1888)
Nemaster rubiginosus (Pourtalès, 1869)
FAMILY COLOBOMETRIDAE
Analcidometra armata (Pourtalès, 1869)
FAMILY ANTEDONIDAE
Ctenantedon kinziei Meyer, 1972

CLASS ASTEROIDEA
Order Paxillosida
FAMILY LUIDIIDAE
Luidia alternata alternata (Say, 1825)
Luidia clathrata (Say, 1825)
Luidia senegalensis (Lamarck, 1816)
FAMILY ASTROPECTINIDAE
Astropecten articulatus (Say, 1825)
Astropecten duplicatus Gray, 1840

Order Valvatida
>FAMILY ASTERINIDAE
>>*Asterina folium* (Lütken, 1859)
>
>FAMILY OPHIDIASTERIDAE
>>*Copidaster lymani* A. H. Clark, 1948
>>*Linckia guildingii* Gray, 1840
>>*Ophidiaster guildingii* Gray, 1840
>
>FAMILY ASTEROPSEIDAE
>>*Poraniella echinulata* (Perrier, 1881)
>
>FAMILY OREASTERIDAE
>>*Oreaster reticulatus* (Linnaeus, 1758)

Order Spinulosida
>FAMILY ECHINASTERIDAE
>>*Echinaster* (*Othilia*) *echinophorus* (Lamarck, 1816)
>>*Echinaster* (*Othilia*) *sentus* (Say, 1825)

CLASS OPHIUROIDEA

Order Phrynophiurida
>FAMILY OPHIOMYXIDAE
>>*Ophioblenna antillensis* Lütken, 1859
>>*Ophiomyxa flaccida* (Say, 1825)
>
>FAMILY GORGONOCEPHALIDAE
>>*Asteroporpa annulata* Örsted & Lütken, 1856
>>*Astrophyton muricatum* (Lamarck, 1816)
>>*Schizostella bifurcata* A. H. Clark, 1952

Order Ophiurida
>FAMILY OPHIURIDAE
>>*Ophiolepis elegans* Lütken, 1859
>>*Ophiolepis gemma* Hendler & Turner, 1987
>>*Ophiolepis impressa* Lütken, 1859
>>*Ophiolepis paucispina* (Say, 1825)
>
>FAMILY OPHIOCOMIDAE
>>SUBFAMILY OPHIOCOMINAE
>>*Ophiocoma echinata* (Lamarck, 1816)
>>*Ophiocoma paucigranulata* Devaney, 1974
>>*Ophiocoma pumila* Lütken, 1859
>>*Ophiocoma wendtii* Müller & Troschel, 1842
>>*Ophiocomella ophiactoides* (H. L. Clark, 1901)
>>SUBFAMILY OPHIOPSILINAE
>>*Ophiopsila hartmeyeri* Koehler, 1913
>>*Ophiopsila riisei* Lütken, 1859
>>*Ophiopsila vittata* H. L. Clark, 1918
>
>FAMILY OPHIONEREIDIDAE
>>*Ophionereis olivacea* H. L. Clark, 1901

Ophionereis reticulata (Say, 1825)
Ophionereis squamulosa Koehler, 1914
Ophionereis vittata Hendler, 1995

FAMILY OPHIODERMATIDAE
Ophioderma appressum (Say, 1825)
Ophioderma brevicaudum Lütken, 1856
Ophioderma brevispinum (Say, 1825)
Ophioderma cinereum Müller & Troschel, 1842
Ophioderma guttatum Lütken, 1859
Ophioderma phoenium H. L. Clark, 1918
Ophioderma rubicundum Lütken, 1856
Ophioderma squamosissimum Lütken, 1856

FAMILY HEMIEURYALIDAE
Sigsbeia conifera Koehler, 1914

FAMILY OPHIACTIDAE
Hemipholis elongata (Say, 1825)
Ophiactis algicola H. L. Clark, 1933
Ophiactis quinqueradia Ljungman, 1871
Ophiactis rubropoda Singletary, 1973
Ophiactis savignyi (Müller & Troschel, 1842)

FAMILY AMPHIURIDAE
Amphiodia planispina (Martens, 1867)
Amphiodia pulchella (Lyman, 1869)
Amphiodia trychna H. L. Clark, 1918
Amphioplus coniortodes H. L. Clark, 1918
Amphioplus sepultus Hendler, 1995
Amphioplus thrombodes H. L. Clark, 1918
Amphipholis gracillima (Stimpson, 1852)
Amphipholis januarii Ljungman, 1867
Amphipholis squamata (Delle Chiaje, 1828)
Amphiura fibulata Koehler, 1913
Amphiura (Ophionema) intricata (Lütken, 1869)
Amphiura palmeri Lyman, 1882
Amphiura stimpsonii Lütken, 1859
Ophiocnida scabriuscula (Lütken, 1859)
Ophionephthys limicola Lütken, 1869
Ophiophragmus cubanus (A. H. Clark, 1917)
Ophiophragmus filograneus (Lyman, 1875)
Ophiophragmus pulcher H. L. Clark, 1918
Ophiophragmus riisei (Lütken, 1859)
Ophiophragmus septus (Lütken, 1859)
Ophiostigma isocanthum (Say, 1825)
Ophiostigma siva Hendler, 1995

FAMILY OPHIOTRICHIDAE
Ophiothrix angulata (Say, 1825)

Ophiothrix brachyactis H. L. Clark, 1915
Ophiothrix lineata Lyman, 1860
Ophiothrix orstedii Lütken, 1856
Ophiothrix suensonii Lütken, 1856

CLASS ECHINOIDEA
  Order Cidaroida
    FAMILY CIDARIDAE
      Eucidaris tribuloides (Lamarck, 1816)
  Order Diadematoida
    FAMILY DIADEMATIDAE
      Astropyga magnifica A. H. Clark, 1934
      Diadema antillarum (Philippi, 1845)
  Order Arbacioida
    FAMILY ARBACIIDAE
      Arbacia punctulata (Lamarck, 1816)
  Order Temnopleuroida
    FAMILY TOXOPNEUSTIDAE
      Lytechinus variegatus (Lamarck, 1816)
      Lytechinus williamsi Chesher, 1968
      Tripneustes ventricosus (Lamarck, 1816)
  Order Echinoida
    FAMILY ECHINOMETRIDAE
      Echinometra lucunter lucunter (Linnaeus, 1758)
      Echinometra viridis A. Agassiz, 1863
  Order Holectypoida
    FAMILY ECHINONEIDAE
      Echinoneus cyclostomus Leske, 1778
  Order Clypeasteroida
    FAMILY CLYPEASTERIDAE
      Clypeaster luetkeni Mortensen, 1948
      Clypeaster rosaceus (Linnaeus, 1758)
      Clypeaster subdepressus (Gray, 1825)
    FAMILY MELLITIDAE
      Encope aberrans Martens, 1867
      Encope michelini L. Agassiz, 1841
      Leodia sexiesperforata (Leske, 1778)
      Mellita isometra Harold & Telford, 1990
  Order Spatangoida
    FAMILY SCHIZASTERIDAE
      Moira atropos (Lamarck, 1816)
      Paraster doederleini Chesher, 1972
      Paraster floridiensis (Kier & Grant, 1965)
    FAMILY BRISSIDAE
      Brissopsis elongata elongata Mortensen, 1907

Brissus unicolor (Leske, 1778)
Meoma ventricosa ventricosa (Lamarck, 1816)
Plagiobrissus grandis (Gmelin, 1791)

CLASS HOLOTHUROIDEA
Order Dendrochirotida
FAMILY CUCUMARIIDAE
Duasmodactyla seguroensis (Deichmann, 1930)
Ocnus surinamensis (Semper, 1868)
Ocnus suspectus (Ludwig, 1874)
Thyonella gemmata (Pourtalès, 1851)
Thyonella pervicax (Théel, 1886)
FAMILY SCLERODACTYLIDAE
Euthyonidiella destichada (Deichmann, 1930)
Euthyonidiella trita (Sluiter, 1910)
Pseudothyone belli (Ludwig, 1886)
Sclerodactyla briareus (Lesueur, 1824)
FAMILY PHYLLOPHORIDAE
Neothyonidium parvum (Ludwig, 1881)
Phyllophorus (Urodemella) arenicola Pawson & Miller, 1992
Phyllophorus (Urodemella) occidentalis (Ludwig, 1875)
Stolus cognatus (Lampert, 1885)
Thyone deichmannae Madsen, 1941
Thyone pseudofusus Deichmann, 1930
Order Aspidochirotida
FAMILY STICHOPODIDAE
Astichopus multifidus (Sluiter, 1910)
Isostichopus badionotus (Selenka, 1867)
FAMILY HOLOTHURIIDAE
Actinopyga agassizi (Selenka, 1867)
Holothuria (Cystipus) cubana Ludwig, 1875
Holothuria (Halodeima) floridana Pourtalès, 1851
Holothuria (Halodeima) grisea Selenka, 1867
Holothuria (Halodeima) mexicana Ludwig, 1875
Holothuria (Platyperona) parvula (Selenka, 1867)
Holothuria (Platyperona) rowei Pawson & Gust, 1981
Holothuria (Selenkothuria) glaberrima Selenka, 1867
Holothuria (Semperothuria) surinamensis Ludwig, 1875
Holothuria (Theelothuria) princeps Selenka, 1867
Holothuria (Thymiosycia) arenicola Semper, 1868
Holothuria (Thymiosycia) impatiens (Forskål, 1775)

    Holothuria (*Thymiosycia*) *thomasi* Pawson &
     Caycedo, 1980
Order Molpadiida
  FAMILY CAUDINIDAE
    Paracaudina chilensis obesacauda (H. L. Clark, 1907)
Order Apodida
  FAMILY SYNAPTIDAE
   Epitomapta roseola (Verrill, 1873)
   Euapta lappa (Müller, 1850)
   Leptosynapta crassipatina H. L. Clark, 1924
   Leptosynapta multigranula H. L. Clark, 1924
   Leptosynapta parvipatina H. L. Clark, 1924
   Leptosynapta tenuis (Ayres, 1851)
   Protankyra ramiurna Heding, 1928
   Synaptula hydriformis (Lesueur, 1824)
  FAMILY CHIRIDOTIDAE
   Chiridota rotifera (Pourtalès, 1851)

CLASS CRINOIDEA
Feather Stars

Crinoids, with considerable justification, have been described as the most exquisite of all echinoderms. They are such archetypal inhabitants of tropical reefs that it may be forgotten that they occur from the Arctic to the Antartic and from shallow waters to the deep sea. Today, the group includes about 100 species of stalked sea lilies and about 600 species of unstalked feather stars, though more than 10 times that many fossil species have been described. The living crinoids are members of one subclass, the Articulata, and all the feather stars belong to the single order Comatulida.

The sea lilies, which were among the dominant biota of Paleozoic seas, now live only in water below 100 m (328 ft) depth; the feather stars range from the surface to great depths. On Caribbean coral reefs, only eight reasonably common species of feather stars are accessible to scuba observation. Four of these occur in the Florida Keys and the Bahama Islands area and are discussed in detail below; the other species are *Comactinia meridionalis meridionalis* (L. Agassiz), *Nemaster grandis* A. H. Clark, *Nemaster iowensis* (Springer), and *Tropiometra carinata* (Lamarck). Florida Keys and Bahama Islands feather stars are secretive for the most part, usually concealed in crevices in corals, with their arms extended for feeding. Most species prefer to feed at night, when chances of predation by fishes are decreased. In contrast to the low diversity in this region, the shallow reefs of the Indo-Pacific support a rich fauna of more than 120 species, many of them conspicuous and brilliantly colored, orange, yellow, red, black, and green, often in garish combinations.

FORM AND FUNCTION

The anatomical structures of feather stars are illustrated in Figure 5. Feather stars are fragile, "bony" animals that have a thin skin and a seeming minimum of soft parts. Their body diameter ranges from 3 to 50 mm (0.1–2 in), and they have arms 50–550 mm (2–22 in) long. The central body lies in a cuplike calyx, from which the cylindrical arms arise. In the sea lilies, the calyx caps a long stalk that supports the animal above the substrate. The feather star calyx rests upon a single discoidal to cylindrical plate, the centrodorsal. In most species, numerous (15–80) unbranched cirri radiate from the centrodorsal. The cirri are appendages made up of a series of ossicles, the last one of which is in the form of a recurved hook that opposes a spine on the penultimate segment. Using its cirri, a feather star can attach firmly to a suitable perch.

Depending upon the species, the number of arms can vary from five to 200 or more; most crinoids have at least 10 arms. Surprisingly, multibrachiate species develop from a five-armed juvenile by repeatedly growing back two arms in the place of arm branches that have been shed. The arms are long and flexible, made up of series of small cylindrical ossicles (brachials). Some connections between

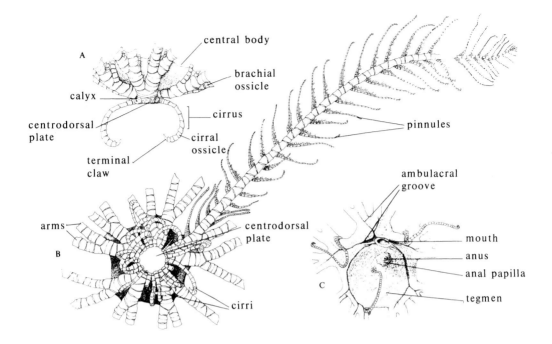

FIGURE 5. Major anatomical features of a feather star: A, central body in lateral view; B, body in aboral view, with all but one arm removed; C, central body in oral view, showing the visceral mass. Illustration by C. Messing.

brachials are purely ligamentary, lending rigidity to the arms. Most of the brachials are attached to one another by a ligament below a bony fulcral ridge and by paired muscle masses and ligaments above the ridge. Arm movements are powered by the contraction of the muscle masses pulling the arms upward, opposing the tension created by the ligaments. The ligaments serve to lock the extended arms in position.

In alternating series on each side of the arms are numerous pinnules, giving the arms a featherlike appearance. From the base to the tip of the arm are oral, then genital, and finally distal pinnules. Oral pinnules near the mouth are specialized for defense or for cleaning the oral surface. The genital pinnules carry the gonads. The majority of the pinnules are food-gathering organs. They are grooved, and the channel (ambulacrum) is bordered by tube feet. The groove of each pinnule feeds into an ambulacral furrow that runs down each arm to the mouth. Numerous microscopic tube feet are organized in groups of three associated with a protective lappet; the lappets alternate on opposite sides of the food groove.

The disk, also called the visceral mass, is covered by soft skin or tegmen. It surrounds the mouth, which lies at the meeting point of the ambulacral grooves, and an anus, the latter usually at the apex of a small muscular conical papilla. The mouth is central, except in the Comasteridae, the second largest family of feather stars, in which it is usually displaced to one side of the tegmen. The tegmen is perforated by 500–1,500 microscopic pores, probably serving in respiration, which lead into ciliated channels in the central body. Crinoids lack the voluminous body cavity characteristic of other echinoderms; their visceral mass is filled with connective tissue that is impregnated with microscopic deposits of calcium carbonate. The largest organs occupying the visceral mass are the intestine and the axial sinus. The soft body parts are supported by the calyx, composed of a large centrodorsal ossicle that bears the cirri, a nearly hidden series of five basal ossicles, and the five radial ossicles that bear the first brachials.

GENERAL BIOLOGY

Crinoids are conservative in their feeding habits; all subsist on small drifting organisms and organic particles. Since the advent of scuba observations, it has become clear that they are passive suspension feeders, extending their arms into the water to form a feeding fan that varies in configuration depending on the species and on environmental conditions. The orientation of the arms and pinnules can be adjusted to increase the effectiveness of particle capture. Drifting or swimming organisms are flicked rapidly by the long outer tube feet toward the food groove, where they are snared in mucus and then passed along the ambulacral grooves in the pinnules and arms to the mouth. They capture zooplankton (foraminiferans and actinopods, invertebrate larvae, and small crustaceans), phytoplankton (diatoms and unicellular algae), and particulate matter, mostly 0.05–0.40 mm (0.002–0.02 in) in size.

Many feather stars are rheophilic, preferring areas where currents reach 1–2 knots (0.5 m/second), carrying abundant food. Some species occupy exposed perches and others are cryptic. Many feed at night, and to take advantage of water flow they perch on elevated corals and sponges, extending their arms in parabolic, radial, or layered vertical feeding fans. Their arms are aligned perpendicular to the flow of water, with the food groove facing downcurrent. Reduced velocity, and vortices created by the pinnules, slow and direct particles toward the tube feet. In oscillatory currents the arms may form bidirectional fans, and the pinnules swivel rapidly to orient to the shifting current. Rheophobic species, which occupy low-velocity habitats, hold their arms in disordered arrays, with successive pinnules oriented at up to 90° angles to each other.

Feather stars can right themselves if overturned, and some can swim by thrashing their arms up and down in a coordinated fashion. Even species that cannot swim, including at least some stalked crinoids, can move by pulling themselves along the seafloor. Feather stars are relatively slow moving, and stalked crinoids even more so. Crinoids are subject to predation, mostly by fish, and they are capable of regenerating damaged arms. Often, however, fish crop only the soft visceral mass of feather stars, and the visceral mass of tropical species can be regenerated in less than a month. Perhaps more suprising, at least one species of stalked crinoid is capable of regenerating the entire crown—tegmen, arms, and all—from the base of the calyx atop the stalk. Perhaps because of their sedentary lifestyle, crinoids host a diverse assemblage of parasites and commensals, including gastropods, ophiuroids, clingfish, myzostome polychaetes, and crustaceans (shrimps, copepods, isopods, cirripedes). The colors of many of these associates precisely match those of their host species.

The gonads of crinoids are composed of masses of gametes that fill specialized cavities in the genital pinnules. For most feather stars the breeding season is restricted to a 1- to 2-month period in spring or summer. One of the most remarkable reproductive behaviors among the echinoderms is mass spawning, epitomized by *Oxycomanthus japonicus* (J. Müller). Each female releases around 2 million eggs during a spawning frenzy that lasts for 1 hour of 1 day each year, between 27 September and 9 October, at the first or last quarter of the moon. While spawning, many individuals whip their arms, dispersing gametes, but eggs often adhere to the arms and pinnules. Within a few days after fertilization, the uniformly ciliated embryo develops into a doliolaria larva encircled by several ciliated bands. It lacks a mouth and feeds, if at all, by absorbing dissolved nutrients. The doliolaria swims for a few days or less and then settles onto a suitable substratum, where it develops into a stalked stage called a "cystidean." Because the posterior end of the cystidean becomes the oral surface of the adult, its internal structures rotate through 90° during the metamorphosis to a "pentacrinoid" stage. The latter is a miniature stalked crinoid. It begins feeding about a week after settlement and, within a month, develops five arms and an anal opening. After some cirri have developed, the feather star separates from the top of the pentacrinoid stalk and attaches directly to the substratum. Relatively few feather stars have the capacity to brood their developing young; their eggs adhere to the genital pinnules or enter a brood pouch from which the doliolaria larvae or the stalked or stalkless juveniles emerge.

ECOLOGICAL AND COMMERCIAL IMPORTANCE

Crinoids have no commercial importance. Their ecological importance stems from their great abundance in preferred habitats, but the influence of crinoids on the ecology of marine communities has not been assessed. They are preyed upon by some fish, but do not seem to be a dietary mainstay of other animals.

TERMINOLOGY AND CONVENTIONS

The diameter of an individual is measured from the tip of an arm to the tip of the arm opposite. Arm length is measured from the boundary between the calyx and tegmen to the tip of the arm.

IMPORTANT REFERENCES

An extensive monograph of the world's crinoids was written by A. H. Clark during the period 1915 to 1967, the last part of which was coauthored by A. M. Clark. Background information on the structure and classification of the group is presented in the *Treatise on Invertebrate Paleontology Part T* (Moore and Teichert 1978). The *Challenger* Expedition reports, though out of date, also provide useful information and a profusion of superb illustrations (Carpenter 1884, 1888). Simms (1988) offered a contemporary overview of crinoid systematics. During the last 20 years, numerous papers on the systematics, distribution, and ecology of Caribbean-area feather stars have appeared (e.g., Meyer 1973a,b; Messing 1978; Meyer et al. 1978). The nutrition of the group was reviewed by Jangoux and Lawrence (1982), biotic interactions by Meyer and Ausich (1983), and reproduction was covered by Holland (1991).

FAMILY COMASTERIDAE

Comactinia echinoptera (Müller)
Figure 6

DESCRIPTION: This species has 10 arms and ranges from small and delicate in build to large and robust, often reaching a diameter of 30 cm (12 in). The four pairs of pinnules nearest the mouth on each arm have terminal comb teeth. The mouth lies near the edge of the tegmen, and the anal cone is more or less central. The arms tend to be unequal in size, those arising near the mouth usually being longer than the others. The arms are conspicuously thickened near the base.

In life, the cirri are white, the arms reddish brown flecked with yellow, and the pinnules are yellow. In preserved animals the arms become deep pink to lavender.

HABITAT: Coral reef and coral-algal bottoms, in rock crevices, and among corals.

DISTRIBUTION: South Carolina to southeastern Florida (not reported from the Florida Keys) and the Bahama Islands and through the West Indies; Caribbean coast of Central and South America to Brazil. DEPTH: 2–92 m (7–302 ft).

BIOLOGY: Messing and Dearborn (1990) noted that this species is widespread but few in number on shallow reefs in the West Indies. It is cryptic and is the only strictly nocturnal Caribbean comatulid (Macurda and Meyer 1977). Individuals remain concealed in crevices during the day and extend their longest arms to feed at night. The shorter arms and central body remain hidden. The arms form a loosely arranged monoplanar filtration fan at right angles to the direction of water movement (Macurda 1973, 1975; Meyer 1973a,b; Messing and Dearborn 1990). Macurda (1973) noted a negative

FIGURE 6. *Comactinia echinoptera*. Photo by D. Meyer.

phototaxic response when an individual was illuminated at night with a flashlight. A recent study of the axial coelom of the species described its blood-vessel and nephridium-like components and suggested that they function in excretion or nutrient transfer (Balser and Ruppert 1993).

REMARKS: Messing (1978) gave a careful and detailed description of this species, and he distinguished it from *Comactinia meridionalis*, with which it has frequently been confused. The latter occurs in shallow water along Florida's east coast and Central America, but has not been reported from the Florida Keys. In the Bahamas it is apparently restricted to deep (more than 100 m) water.

SELECTED REFERENCES: Müller 1840:92 [as *Comatula echinoptera*]; Meyer 1973a: 253, fig. 7; 1973b:118, fig. 5.4; Messing 1978:49, figs. 1–4; Messing and Dearborn 1990:23, fig. 10.

Genus *Nemaster*

It has been suggested that upwards of six species of *Nemaster*, *N. discoideus* (Carpenter), *N. grandis* A. H. Clark, *N. iowensis* (Springer), *N. mexicanensis* Tommasi, *N. rubiginosus* (Pourtalès), and one undescribed species, occur in the region. The divergent taxonomic conclusions of different authorities can be attributed to the diverse color patterns in this genus and variation in shape of the oral pinnules.

Because of striking morphological similarities between *N. iowensis*, *N. mexicanensis*, and *N. rubiginosus*, Meyer (1973b) and Meyer et al. (1978) synonymized these species, suppressing *N. iowensis* and *N. mexicanensis* as junior synonyms of *N. rubiginosus*. Later, Liddell and Ohlhorst (1982) used morphometric and chemical analyses to assay four nominal Caribbean species (*N. discoideus, N. grandis, N. iowensis,* and *N. rubiginosus*). They concluded that *N. rubiginosus* and *N. iowensis* represent a single, highly variable species and found substantial evidence of a new, closely related species. More recently, however, Hoggett and Rowe (1986), largely on the basis of oral pinnule morphology, determined that *N. rubiginosus* and *N. iowensis* were both valid species that belong in separate genera. They resurrected *N. iowensis* and created a new genus, *Davidaster,* for *N. rubiginosus* and *N. discoideus*. They also suggested that *N. iowensis* might be synonymous with *N. grandis*.

Only three of the nominal species in question (*N. discoideus, N. iowensis,* and *N. rubiginosus*) have been confirmed from waters of the Florida Keys and the Bahama Islands, and only these species are discussed below. In view of the taxonomic changes affecting the group, we take a conservative approach. The work of Meyer (1973a) and that of Liddell and Ohlhorst (1982), which support *N. rubiginosus* as a senior synonym of *N. iowensis,* are accepted here. Hoggett and Rowe's (1986) recognition of *N. iowensis* as a valid species and their referral of *N. discoideus* and *N. rubiginosus* to the new genus *Davidaster* require a further assessment on the basis of a more comprehensive systematic study.

FIGURE 7. *Nemaster discoideus.*

Nemaster discoideus (Carpenter)
Figure 7

DESCRIPTION: The slender, delicate arms of this species fragment easily when individuals are handled. The approximately 20 arms are nearly equal in length, 7–15 cm (3–6 in) long. Successive pinnules are directed at right angles to each other so that they stick out from the arm in four directions, not in one plane as in a feather. On the tegmen, an anal cone rises from the center, and the mouth opens near the margin. Two or three pairs of pinnules closest to the arm base terminate in distinctive comb-shaped structures.

N. discoideus has various color patterns; the arms are black to grayish green, light orange, yellow, or white. A distinctive characteristic, but one not seen in every individual, is the beaded appearance of the pinnules, which are usually white with contrasting black spots. Uniformly black individuals, and some with solid white pinnules tipped with black, have been reported. Frequently there is a solid, but more often a broken, black stripe on the ventral surface of each arm. In addition to differences in color, *N. discoideus* is a more delicate form than *N. rubiginosus*, usually making it possible to distinguish these similar species in situ.

HABITAT: Usually deeper coral reef zones including spur-and-groove and the fore reef slope.

DISTRIBUTION: Gulf of Mexico, southeastern Florida from Boca Raton to the Dry Tortugas, Bahama Islands, Turks and Caicos islands, Antillean Arc from the north coast of Cuba to Barbados and the Grenadines including Grand Cayman and Jamaica; Caribbean coast of Central and South America from the Yucatán to Colombia, Curaçao, and Bonaire (Meyer et al. 1978). DEPTH: 0.6–355 m (2–1,165 ft), but most common at 20–40 m (66–131 ft) (Macurda and Meyer 1977).

BIOLOGY: *N. discoideus* attaches to corals and sponges using short, curved cirri. The body is always well hidden within a crevice or under a ledge, and usually less than half of the arms are held out in a random fashion beyond the reef. They may be bent or slightly curled near the tips, with only a portion of each extended arm visible. This suspension-feeding posture is markedly different from that of many other feather stars, which form a circular or fan-shaped feeding array with their arms (see Macurda and Meyer 1983).

This is a rheophobic (current-avoiding) species and occurs in microhabitats with sluggish, multidirectional water flow. Presumably, the arrangement of its arms and pinnules is an adaptation for suspension feeding within the complex coral reef framework (Meyer 1973b). In contrast, circular or fan-shaped feeding arrays are advantageous to feather stars, such as *N. grandis*, that perch atop the reef in moderate to strong unidirectional currents. The tube feet of *N. discoideus* are extended day and night, and it is believed that the species feeds continuously (Macurda 1973; Meyer 1973b). Its stomach contents consist mostly of diatoms less than 485 μm in length (Liddell 1982). The species has a similar, but lower, oxygen consumption rate than other invertebrates of comparable size (Baumiller and LaBarbera 1989).

A recent study of the axial coelom of the species described its blood-vessel and nephridium-like components and suggested that they function in excretion or nutrient transfer (Balser and Ruppert 1993). Donovan (1993a) examined the microscopic structure and articulation of the claw-tipped cirri of this species and described possible adaptations for their role in attachment to the substrate.

At many localities, *N. discoideus* and *N.*

rubiginosus co-occur, sometimes almost side by side. However, *N. discoideus* often lives somewhat deeper on the reef and more deeply ensconced in the coral infrastructure than its more robust congener (Meyer 1973a; Meyer et al. 1978). On the western coast of Barbados, *N. discoideus* is the dominant feather star, with greatest average densities of approximately one individual per square meter at depths of 1–27 m (Liddell 1979), but at similar depths (17–26 m) off Discovery Bay, Jamaica, its mean density reaches only 0.02 individuals per square meter (Meyer 1973a). A shrimp commensal on individuals from Bonaire was reported by Llewellyn and Meyer (1991).

REMARKS: *N. discoideus* is similar in overall appearance to *N. rubiginosus*, but is smaller and more delicate. Its coloration is less bold than the bright orange to dark brown or black pigmentation patterns that prevail in *N. rubiginosus*. In addition, *N. discoideus* tends to have shorter and fewer (approximately 20 versus 20–35) arms than *N. rubiginosus*. The bathymetric ranges of both species overlap, but *N. discoideus* is generally deeper-dwelling than *N. rubiginosus*. *N. discoideus* usually occurs deeper in the reef infrastructure or near the reef base, but *N. rubiginosus* perches at margins or even near the reef crest in low-energy habitats.

SELECTED REFERENCES: Carpenter 1888:58 [as *Actinometra discoidea*]; Colin 1978:406, pl. on p. 405; Liddell 1979:2413, figs. 3f, 4e,f; Sefton and Webster 1986:92, pl. 152.

FIGURE 8. *Nemaster rubiginosus.*

Nemaster rubiginosus (Pourtalès)
Figure 8

DESCRIPTION: *Nemaster rubiginosus* typically has 20–40 arms, or more in very large individuals, which are almost always unequal, varying in length from 10 to 37 cm (4–15 in). As in *N. discoideus,* successive pinnules are arranged approximately perpendicular to each other. The mouth is near the disk margin, and the anus opens atop a central anal cone. Three or four pairs of oral pinnules closest to the disk terminate in comblike structures.

The color patterns of *N. rubiginosus* vary within local populations and between widely spaced populations. Individuals are most often a uniform bright orange with a black median stripe along the ventral surface of each arm. However, several other color varieties occur less frequently: (1) arms black, pinnules orange or black with orange tips; (2) arms and pinnules black; (3) arms orange, pinnules orange with yellow or white tips (Meyer 1973b).

HABITAT: On margins and crests of coral reef structures, usually in low-energy situations.

DISTRIBUTION: Western Gulf of Mexico, southeastern Florida from the Dry Tortugas to Key Largo, Bahamas, Turks and Caicos islands, Antillean Arc from Hispaniola to Barbados and the Grenadines including Grand Cayman and Jamaica; Caribbean coast of Central and South America from Belize to Bahia, Brazil (Meyer et al. 1978). DEPTH: 1–334 m (3–1,096 ft), but most common at 6–15 m (20–50 ft) (Meyer 1973a).

BIOLOGY: *N. rubiginosus* is the most frequently encountered shallow-water feather star throughout the tropical western Atlantic (Meyer 1973b), although at some localities it is greatly outnumbered by its congener, *N. discoideus* (Liddell 1979). Both species are moderately rheophobic (current avoiding) and cling to the reef infrastructure. However, *N. rubiginosus* occupies a higher position on the reef than does *N. discoideus* and is apparently more resistant to local current activity and wave surge. In Discovery Bay, Jamaica, densities of *N. rubiginosus* reach 0.6 individuals per square meter at depths of 13–14 m (Meyer 1973a), but at Barbados densities are only 0.16 individuals per square meter at depths of 21–24 m (Liddell 1979).

During the day, *N. rubiginosus* conceals the central portion of the body and many of its arms beneath its perch, but during hours of darkness some individuals advance from cover, thereby exposing their tegmen (Meyer 1973a). The arms used for suspension feeding protrude for approximately one-half of their length above the substratum. Its feeding behavior is similar to that described for *N. discoideus;* it appears to feed continuously, except in turbulent conditions (Macurda 1973; Meyer 1973b). Similarly to *N. discoideus,* its stomach contents consist chiefly of diatoms less than 475 µm in length (Liddell 1982). It has a similar, but somewhat lower oxygen consumption rate than other invertebrates of comparable size (Baumiller and LaBarbera 1989). A recent study of the axial coelom of the species described its blood-vessel and nephridium-like components and suggested that they function in excretion or nutrient transfer (Balser and Ruppert 1993). Donovan (1993a) examined the microscopic structure and articulation of the claw-tipped cirri of this species and described possible adaptations for their role in attachment to the substrate.

N. rubiginosus has a well-defined, synchronous annual reproductive cycle, unlike several other tropical crinoids that have more prolonged or variable cycles (Mladenov and Brady 1987). In Jamaica, the breeding season occurs in late fall and winter (October–March), and during that period the sex of mature adults can be determined by examination of the genital pinnules (those pinnules arising from the proximal half of the arms). In females, the pinnules are orange; in males they are white. The microscopic structure of the ovaries and eggs was described by Holland (1971).

Criales (1984) found two species of commensal shrimps, *Periclimenes bowmani* Chace and *Synalpheus townsendi* Coutière, associated with *N. rubiginosus* off the Santa Marta region of Colombia.

REMARKS: *N. rubiginosus* is similar to the preceding species, *N. discoideus,* in many ways, but the two can be distinguished with relative ease by coloration and body size, including arm length and thickness (see *Remarks* under *N. discoideus*). Also refer to the comments under Genus *Nemaster* for additional information concerning other taxa that have been confused with *N. rubiginosus*.

SELECTED REFERENCES: Pourtalès 1869: 356 [as *Antedon rubiginosa*]; Liddell 1979: 2413, fig. 4a–d; Macurda and Meyer 1983: 360, fig. 6; Sefton and Webster 1986:92, pl. 151.

FAMILY COLOBOMETRIDAE

Analcidometra armata (Pourtalès)
Figure 9

DESCRIPTION: This is the smallest feather star known from shallow depths in the western Atlantic, rarely exceeding 10 cm (4 in) diameter. It has only 10 short arms, usually of unequal length. The pinnules are aligned in two planar rows, one to either side of each arm. They vary in length, being shorter near the tegmen and longer near the arm tips. Each cirrus segment has a small spine on its underside. The mouth and adjacent anal cone are near the center of the tegmen.

A. armata has arms with alternating bands of white and red, white and lavender, or grayish green and red. Its pinnules and cirri also display a banded pattern.

HABITAT: The fore reef slope, in areas with slow to moderate water currents.

DISTRIBUTION: Widespread throughout the Caribbean Sea. It has been found near the Dry Tortugas (64 m [210 ft]), but is not recorded from shallow reefs off the Florida Keys. DEPTH: 3–148 m (10–486 ft), but usually found at 50–70 m (164–230 ft) (Meyer et al. 1978).

BIOLOGY: This feather star lives exposed, perched above the seafloor on gorgonian sea plumes (*Pseudopterogorgia* spp.) or sea whips such as *Ellisella barbadensis* (Duchassaing & Michelotti) (Meyer 1973a). There, it is in an advantageous position to intercept water-borne food particles. Meyer (1973a,b) found that off Jamaica and Panama *A. armata* forms a circular feeding filtration fan by fully extending its 10 arms, with the feeding fan aligned parallel to the branches of the host gorgonian.

Like several other species of feather stars, *A. armata* has the ability to swim, and it abandons its perch when disturbed. Individuals swim using coordinated upward and downward movements of three or four groups of arms, with each arm oscillating four to five times in 5 seconds (Macurda 1973). The feather star swims upward or sideways by orienting its oral surface in the direction of movement. Macurda found that

FIGURE 9. *Analcidometra armata*.

an individual in an aquarium could swim for 4 minutes, and that after a brief rest it could once more be stimulated to swim for 2–2.5 minutes.

Liddell (1979) noted that the distribution of *A. armata* off the west coast of Barbados was patchy and restricted to depths of 13.5–18 m (44–59 ft). At that locality it lives almost exclusively on *Pseudopterogorgia* spp.; from one to 20 individuals occur on a single gorgonian. Its stomach contents are predominantly diatoms, smaller on the average than the particles captured by other Caribbean feather stars (Liddell 1982). Meyer (1973a) found that one individual lived on the same gorgonian for approximately 5 years. A small shrimp, *Periclimenes* sp., is associated with this feather star. Its cryptic coloration blends perfectly with the host's coloration (Macurda and Meyer 1983); small size and an interrupted color pattern tend to conceal the feather star as well.

REMARKS: Before 1978, two nominal species of *Analcidometra* (*A. armata* and *A. caribbea* A. H. Clark) were thought to occur in the western Atlantic. After a comprehensive survey of tropical western Atlantic crinoids, Meyer et al. (1978) concluded that *A. caribbea* is a junior synonym of *A. armata*.

SELECTED REFERENCES: Pourtalès 1869: 356 [as *Antedon armata*]; Meyer et al. 1978: 417–418; Sefton and Webster 1986:92, fig. 153.

FIGURE 10. *Ctenantedon kinziei*. Photo by D. Meyer.

FAMILY ANTEDONIDAE

Ctenantedon kinziei Meyer
Figure 10

DESCRIPTION: This species is of delicate build; there are 10 arms, all of which are approximately equal in length. The animal reaches a diameter of about 20 cm (8 in). The modified oral pinnules near the mouth possess conspicuous terminal comb teeth, and in this respect this species resembles members of the family Comasteridae (for example, *Nemaster*). It differs from comasterids in having the mouth nearly central on the disk and the anus near the margin. There are approximately 50–90 cirri used for attachment to the substratum. In life, the arms are light brown, with numerous small black spots, and the pinnules are banded brown and white.

HABITAT: Within the infrastructure of coral reefs.

DISTRIBUTION: The Bahama Islands southward to Barbados; also Jamaica, Belize, Panama, Colombia, and Curaçao. DEPTH: 9–49 m (30–161 ft).

BIOLOGY: Meyer (1972) noted that this species remains well concealed in crevices, partly emerging at night, never extending its arms beyond the fabric of the reef. Liddell

(1979) found only four individuals at Barbados, two of which were feeding during the day, but he suggested that the overall numbers are much higher because most animals are concealed, feeding solely within cavities in the coral framework.

This species "hops" away when disturbed by a diver. It can swim actively, lashing alternate arms upward and downward, and may remain suspended in the water for several minutes (Meyer 1972; Meyer and Macurda 1977). Meyer also noted that it settles toward a substratum by flexing all the arms upward at the same time; this has the effect of pushing the animal downward. Along the edges of the ambulacra of this species are minute spherical structures called saccules; they are transparent in living individuals and pigmented in preserved specimens. These are of uncertain function, but may act as a "photoreceptive lens" (Holland 1967; Meyer 1972). Llewellyn and Meyer (1991) mentioned the presence of commensal shrimps on *C. kinziei*.

SELECTED REFERENCES: Meyer 1972:54, figs. 1–5; Liddell 1979:2413.

CLASS ASTEROIDEA
Sea Stars

CLASS ASTEROIDEA

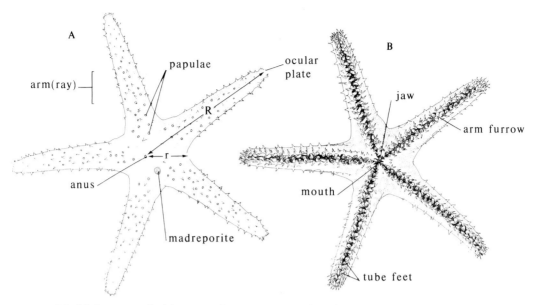

FIGURE 11. Major anatomical features of a sea star: *A,* in dorsal view; *B,* in ventral view. Illustration by C. Messing.

Sea stars (also known as starfish) are so familiar that their multirayed image is emblematic of the oceans. A robust skeleton, sedate behavior, and exquisite symmetry make sea stars highly attractive and woefully susceptible to curio collectors. Unfortunately, with only a dried specimen in hand it is difficult to understand how capably the complex system of skeletal plates and soft tissues affords a living sea star both remarkable rigidity and enormous flexibility. The body wall provides sufficient strength in some species to withstand the severe battering force of waves on rocky intertidal cliffs, yet sea stars can twist gracefully and right themselves when overturned. Faded sea star souvenirs also fail to convey the splendid living palette of pigments among different species: red, purple, orange, brown, white, gray, blue, and yellow.

Approximately 1,800 extant species of sea stars have been described, ranging from tiny ones, 1 cm (½ in) in diameter, to veritable giants that can amply cover the bottom of a large washtub more than a meter (3 ft) across. Compared with species-rich reef localities in the Red Sea and the Indo-West Pacific, the diversity of coral reef sea stars in Florida and the Bahama Islands is relatively low. In the Caribbean and the Gulf of Mexico, at depths less than 46 m (150 ft) at least 18 species of sea stars are known, and at depths up to 3,658 m (12,000 ft) another 160 species are reported.

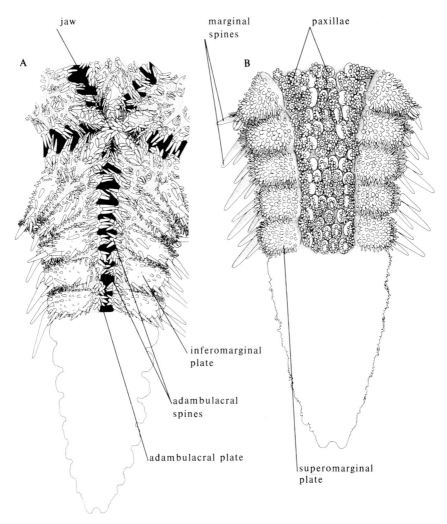

FIGURE 12. Anatomical features of a sea star arm and mouth region, in an astropectinid species; only the proximal part of an arm is shown in detail: A, in ventral view; B, in dorsal view. Illustration by C. Messing.

FORM AND FUNCTION

The general body form of an asteroid is stellate (star-shaped) with a central disk and symmetrical projecting arms (rays) (Figure 11). The rays typically number five, although species with up to 50 rays are known. The length of the arms of shallow-water species can be several times the diameter of the disk (Figure 20), or they may barely extend beyond the disk margin in species with a pentagonal shape (Figure 24); a few sea stars are nearly spherical.

The anatomical structures of sea stars are illustrated in Figures 11 and 12. Among the distinctive features on the upper surface of the disk is an interradially located madreporite, often of a contrasting color and usually intricately furrowed and perforated with microscopic pores; species with more than five arms, and asexually reproducing sea stars, may have numerous madreporites. The surface of a living sea star may look and feel "furry" because of abundant papulae, which are transparent, retractable evaginations of the body cavity, ciliated both inside and out. Singly or in groups, papulae protrude through microscopic pores between the plates of the disk and arms, providing a surface for respiratory gas exchange. Asteroids typically have an array of spines, tubercles, and granules attached to underlying skeletal plates. The upper surface of some sea stars is covered by paxillae, closely spaced plates with a mushroom-shaped central column crowned with radiating spinelets. Under moderate magnification, these paxillae look like a field of daisies. Small pincerlike pedicellariae, sometimes large enough to be noticeable with the naked eye, may occur on the top and bottom surface of the arms and disk. They differ from pedicellariae in the sea urchins, because they usually have only two valves and lack a stalk ossicle. In different species, sea star pedicellariae take a variety of shapes and sizes, from simple modified spines, clustered in a circlet, to highly specialized structures with two opposing hooklike valves. Their characteristic shapes are of great taxonomic importance. Pedicellariae act to rid areas around papulae of small organisms and debris and may also have a role in prey capture; the large pedicellariae or batteries of smaller pincers of some species can secure small fish or shrimplike crustaceans on which the sea star feeds. Some, but not all, sea stars have an anus near the center of the disk, often set off by a distinctive ring of spines or plates. Taxa lacking an anus eject waste material through the mouth.

On the lower surface of the arms, a double series of ambulacral ossicles are united by muscle and connective tissue in an inverted V-shape, the arcade of interconnected plates forming the central arm furrow (ambulacral groove). Two series of tube feet project from the furrow, one foot between each successive pair of ambulacral ossicles, and a terminal tube foot lies below the ocular (terminal) plate at the tip of the arm. It often carries an optic cushion, composed of cups of red-pigmented, light-sensitive cells, which in some species have associated lenses. Depending on the family of sea star, the remaining tube feet possess, or lack, suckered tips. Tube feet serve the sea star in locomotion, burrowing, and food gathering and function in respiration in some species. Sea stars move by the coordinated stepping motions of groups of tube feet that push the individual forward. In species with suckered tube feet, the suction generated by muscular retraction of the tube foot tip creates half the force of attachment, and mucous adhesion contributes the rest. It has been suggested that the pull of the sea star body against the tube feet enables starfish to sense the pull of gravity. The tube feet can be retracted into the arm furrow and the furrow then covered by movable spines of the ambulacrum.

In several families of sea stars, the double series of skeletal plates (the upper superomarginals and lower inferomarginals), aligned at the edges of the arm, are distinguishable from the plates composing the upper and lower surface of the arms. The "marginals" may bear erect spines, forming a conspicuous fringe that defines the perimeter of the sea star. Between the marginals and the plates at the top of the arm, sometimes arranged in a carinal series, there are numerous calcareous plates that form a reticulate (netlike), imbricate (overlapping), or tessellate (mosaic) meshwork. Connective tissue and series of circular and longitudinal muscles are responsible for a sea star's ability to change the shape of its arms.

A mouth, surrounded by five, multi-plate, triangular jaws, is situated ventrally at the center of the disk. Often the jaws are elevated and prominent, forming a conspicuous, composite star-shaped structure. The digestive tract organization varies considerably among different sea stars. It is generally better differentiated than that of other echinoderms, consisting in many species of a two-chambered stomach, intestine, rectal ceca, and paired pyloric ceca, secretory-absorptive organs that fill the large body cavity within each arm. The gonads also have a prominent place in the arms and may form large, paired tufts of tubules aligned with the pyloric ceca. Depending on the species, they may also form a series of smaller tufts connected by a shared gonoduct or opening independently through a series of gonoducts on the interradial edge of the arm.

GENERAL BIOLOGY

Sea stars are renowned for both their appetite and their diverse feeding strategies. Carnivorous species prey on sponges, shellfish, crabs, corals, worms, or other echinoderms; a few are cannibalistic. Some are scavengers, feeding on decaying fish and invertebrates. Still others are deposit feeders, filling their stomachs with mud from which they extract microscopic organisms and organic matter, and suspension feeders glean prey and food particles from the water. Intraoral feeders take live prey into their stomach, sometimes drastically distending or even rupturing their disk in the process. Intraoral digestion of resistant prey such as clams may take over a week. Extraoral feeders, which extrude their stomach from the disk, consume a greater variety of animal prey. In the case of species that attack mollusks such as oysters and mussels, their everted stomach can enter a gap just 0.1 mm wide (0.04 in). The opening is created by the persistent pull of the sea star's tube feet against the bivalve's retractor muscles, or the predator may take advantage of an imperfection in the seal of its prey's valves. In the case of coral-eaters such as the large tropical crown-of-thorns starfish of the Pacific, the everted stomach is applied to a patch of coral until the soft tissue is digested and only bleached coral skeleton remains. The impressive effectiveness of various sea stars as predators has led to the evolution of some drastic escape behaviors of their prey. Referred to as "fright responses," the reactions include leaping and burrowing in bivalves, somersaulting in gastropods, "mushrooming" in limpets, and swimming in sea cucumbers and sea anemones.

A general relationship between sea star systematics and their ecological adaptations was discerned by Blake (1983, 1987, 1990), and he has redefined three superorders and seven orders of asteroids. Blake characterized members of the Order Forcipulatida (Superorder Forcipulatacea) as offensive specialists on macroinvertebrates; their feeding mode and adaptations to wave-washed habitats hinge on their specialized three-element pedicellariae and their broad ambulacral furrows with quadriserial tube feet. In contrast, within the Superorder Valvatacea, the Valvatida are defensive specialists with heavily armored skeletons, feeding on particulate material or sessile prey. They tend to be the most common sea stars in shallow-water tropical habitats. The Paxillosida are lightly armored burrowers in soft sediments, feeding on vagile mollusks and echinoderms, and the Spinulosida (Superorder Spinulosacea) are defensive specialists that feed on organic deposits and colonial organisms.

The ability of sea stars to regenerate severed or voluntarily autotomized arms is well known. Their regenerative capability figures in the asexual reproduction of almost two dozen species that divide through the disk, producing clones of sea stars with identical genetic makeup. Another seven species are known to voluntarily pinch off one or more arms ("autotomous asexual reproduction") that subsequently regenerate a complete disk and arms. These asexually reproducing species tend to be of small size. Most sea stars reproduce sexually, and many have free-swimming larvae, but asexuality may occur at various stages in the life cycle of sexual species. Even the larvae of some sea stars pinch off body structures that are capable of growing into independent feeding larvae.

Some of the biochemical aspects of sexual reproduction are better understood for the asteroids than for other echinoderms. Their production of steroids has been followed during gametogenesis; chemical compounds that stimulate shedding of gametes, oocyte maturation, and meiosis have been characterized, and pheromones have been implicated in some spawning behaviors. In addition, the light regime (daylength) has been shown to control the reproductive cycle of certain species.

Spawning and external fertilization result in the production of free-living bipinnaria larvae. They characteristically have graceful larval processes (arms) that support sinuous ciliary bands used for locomotion and food gathering. In many species, the bipinnaria transforms into a brachiolaria larva that attaches during metamorphosis using three anterior adhesive arms that surround a central sucker. The early developmental stages of yolky nonfeeding larvae, and the embryos of brooding sea stars, may be a modified feeding larva or a "barrel-shaped" larva.

Asteroids occur at nearly all latitudes and depths: in tide pools, on rock jetties, in seagrass and kelp beds, beneath rock rubble, on sandy sediment, on coral reefs. They sometimes thrive under stressful conditions. In high-energy areas, such as wave-washed rocky shores, some species withstand the constant beating of the sea by clinging fast to the substrate with powerful sucker-tipped tube feet. Hardy species occur near the high-tide mark, resisting extended periods of desiccation by retreating to the moist crevices beneath rocks. In the deep sea, at depths to more than 9,100 m (30,000 ft), sea stars may be found on flat sandy sediment or associated with steep hard pavements.

ECOLOGICAL AND COMMERCIAL IMPORTANCE

Sea stars, especially the more prolific species, have an impact on marine communities of considerable ecological, and sometimes economic, importance. On the Pacific coast of the United States, *Pisaster ochraceus* (Brandt) can influence the species composition of the intertidal zone. The subtropical *Heliaster kubiniji* Xantus and the antarctic *Perknaster fuscus antarcticus* (Koehler) have similar roles as "keystone predators." The infamous coralivore, *Acanthaster planci* (Linnaeus), has inhabited Pacific coral reefs for the past 1–3 million years, but recent devastating infestations of "crown-of-thorns" have resulted in losses of over 90% of the corals from large reef tracts. For obvious reasons this sea star is regarded as "*the* major management problem in a number of coral-reef areas of the Indo-Pacific region" (Birkeland and Lucas 1990:202).

On North Atlantic coasts, the phenomenal depredation of *Asterias* species on oysters, mussels, and scallops is well known. Aggregations of these sea stars, sometimes numbering nearly 150,000 per hectare (2.5 acres), have been responsible for massive damage to commercial shellfisheries. Therefore, it seems fitting that the same *Asterias* species are themselves harvested as an ingredient in "fishmeal" for poultry. The major fishery is in Denmark, where up to 1,100 metric tons have been harvested annually in recent decades. Sea stars are also collected by biological supply companies for sale to schools, and as curios. The fishery for *Oreaster reticulatus* has resulted in the drastic decline of shallow-water populations (Sloan 1985).

TERMINOLOGY AND CONVENTIONS

The major radius (R) of a sea star is measured from the center of the disk to the tip of a ray; the minor radius (r) spans the center of the disk to the interradial margin (Figure 11). If identifications are being made from dried specimens, desiccated tube feet can be examined for the presence of suckers by rehydrating a piece of an arm in a solution of water and detergent. The structure of integument-covered plates can be examined by cleaning an arm, or dissected pedicellariae, with a dilute solution of household bleach, taking care to rinse the specimen in running water to terminate the cleaning process and remove residual bleach.

IMPORTANT REFERENCES

Knowledge of asteroids of this western Atlantic region dates to Perrier's (1884) report on the sea star fauna from the West Indies. Additional information was provided by Sladen (1889) in his *Challenger* report, which covered many tropical and subtropical species from the western Atlantic. Verrill (1915) published another notable account of the shallow and deep-water sea stars from Florida, the West Indies, and Brazil, and study of the region was further expanded by H. L. Clark (1933, 1941).

Two recent taxonomic references are invaluable: the monograph on western Atlantic sea stars by A. M. Clark and M. E. Downey (1992) and a preliminary paper by Downey (1973). A phylogeny of the Asteroidea was developed by Blake (1987). Information on asteroid feeding biology was reviewed by Sloan (1980) and Jangoux and Lawrence (1982), reproductive biology by Chia and Walker (1991), and various aspects of ecology by Birkeland (1989).

FAMILY LUIDIIDAE

Luidia alternata alternata (Say)
Figure 13

DESCRIPTION: Fully grown *Luidia a. alternata* commonly reach 20 cm (8 in) from arm tip to arm tip, although individuals up to 40 cm (12 in) are encountered occasionally. The species has five flat, straplike arms bordered by a fringe of slender, sharp, marginal spines. The rows of paxillae on the upper arm surface are irregularly arranged, giving the body a fragile appearance. The paxillae near the arm margins are larger than those along the midline of the arms. Several of the paxillae in the marginal rows bear a single long, pointed, erect spine surrounded by small spinelets.

The distinctively colored living individuals have a white or cream-colored dorsal surface, with mottling or bands of dark green, purple, brown, or black. The ventral surface is yellow, and the tube feet bright orange.

HABITAT: Sandy and muddy sediment.

DISTRIBUTION: Discontinuous between Cape Hatteras, North Carolina, and Buenos Aires, Argentina; known from several Carib-

FIGURE 13. *Luidia alternata alternata.*

bean Islands and from the coasts of Mississippi, Texas, and Mexico in the Gulf of Mexico, but not reported from the Bahama Islands or from south of the Yucatán Peninsula to Colombia. Off Florida, it occurs from Jacksonville to the Dry Tortugas and northward on the west coast to St. Petersburg. *L. alternata numidica* Koehler is reported from West Africa and the Cape Verde Islands. DEPTH: Generally 3–50 m (10–164 ft), but reported from low-tide mark to 200 m (656 ft).

BIOLOGY: *L. a. alternata* does not occur in numbers as great as *L. clathrata,* but it is frequently dredged or trawled from level bottoms on the continental shelf by scientific vessels and fishing boats. Schwartz and Porter (1977) reported that it is a common inhabitant of scallop beds off North Carolina, where it preys on small individuals of the sea star *Astropecten articulatus.* The species is fragile, and it readily autotomizes its arms when handled out of water.

The testes of males are milky white, and the pale salmon pink ovaries of females contain oocytes that reach 0.19 mm in diameter. Like other members of the genus, the species has a bipinnaria larval stage (Komatsu et al. 1991a,b).

SELECTED REFERENCES: Say, 1825:144–145 [as *Asterias alternata*]; Gray et al. 1968a: 138, fig. 6A,B; Zeiller 1974:103; A. M. Clark and Downey 1992:8, figs. 4b,c, 5d, 6f, 7a–g,q, 8a,b, pl. 1B.

FIGURE 14. *Luidia clathrata*.

Luidia clathrata (Say)
Figure 14

DESCRIPTION: This species has a small disk and five long, flat, straplike arms. Large individuals are 20–30 cm (8–11 in) in diameter, with arm lengths two to three times greater than the disk diameter. The upper surface of the body is covered with closely set plates; those at the margins of the arms are square to rectangular, arranged in regular longitudinal and transverse series. In the central portion of the arms, the plates are quadrangular, considerably smaller than the marginal plates, and not aligned in rows. Each arm is bordered by a fringe of short, erect spines, which decrease in length toward the arm tip. The ventral surface is covered with tiny spines, set on rectangular plates in parallel series. The tips of the tube feet are pointed, without suckers.

L. clathrata differs from other *Luidia* species in the Florida Keys and the Bahama Islands by the presence of a dark gray or black stripe on the dorsal midline of each arm. The stripe is usually wider and darker near the arm tips and may be bordered by one or more narrower stripes closer to the arm margins. The rest of the upper surface is usually gray, although light brown, rose, or salmon colors have been reported; the ventral surface is usually cream-colored.

HABITAT: Protected inshore areas such as bays and lagoons on sand or mud sediments; offshore on sand and shell hash.

DISTRIBUTION: New Jersey to southern

Brazil including Bermuda, the Gulf of Mexico, and most Caribbean Islands. In Florida it is reported from Jacksonville, Cape Canaveral, Fort Pierce, Miami, Charlotte Harbor, Tampa, and Cedar Key. Specimens from off Cape Sable, Key West, and the Dry Tortugas in the collections of the Florida Department of Natural Resources, St. Petersburg, confirm its presence in the Florida Keys. DEPTH: 0–100 m (0–328 ft), but most common less than 40 m (131 ft) deep.

BIOLOGY: Populations of *L. clathrata* may be quite dense, and it is frequently trawled by shrimpers. After storms, large numbers of dead specimens often wash onto sandy beaches. The species has been investigated by Lawrence and his colleagues for over a decade, possibly making it the most exhaustively investigated sea star in the western Atlantic (see, e.g., Lawrence 1973; Ellington and Lawrence 1974; Lawrence and Dehn 1979; McClintock and Lawrence 1981, 1982, 1984; McClintock et al. 1983, 1984; Watts and Lawrence 1986, 1990; Polson et al. 1993).

L. clathrata shows negative phototaxis and burrows to avoid light. It is capable of surviving relatively low salinities (to 14 ppt). Lawrence and others have characterized it as a highly mobile, soft-bottom predator and scavenger that feeds on the infauna and organic detritus. It is capable of detecting and following gradients of the metabolic chemicals produced by prey organisms (McClintock et al. 1984). Surface feeding is accomplished by ingesting sand and mud and then straining the soft sediment through oral spines that retain small prey such as mollusks, crustaceans, and brittle stars. While buried, it can move about with the aid of its tube feet and feed on detrital material by everting its stomach.

It can ingest larger prey than might be expected. Individuals can have entire specimens of *Mellita* sp. and *Moira atropos* in their stomach. The prey greatly exceeds in size the 2- to 3-mm diameter of the relaxed mouth opening of the sea star and creates a striking distortion of the sea star's disk (Blake 1982). On the Caribbean coast of Venezuela, where *L. clathrata* coexists with two other *Luidia* species, its diet consists of ophiuroids (35.3%), gastropods (24.4%), bivalves (19.7%), and other invertebrates (Penchaszadeh and Lera 1983). Schwartz and Porter (1977) reported that *L. clathrata* is an active predator on calico scallops (*Argopecten gibbus* [Linnaeus]) off North Carolina. In contrast, McClintock (1983) found that *L. clathrata* was unsuccessful in overcoming the swimming escape response of the bay scallop (*Argopecten irradians* [Lamarck]) in a laboratory setting. The gonads of this species resemble those of *L. alternata*, but its eggs are somewhat smaller. Its development, through metamorphosis, takes approximately 1 month and involves a large (2 mm long) bipinnaria larva (Komatsu et al. 1991b).

Like its close relatives, *L. alternata* and *L. senegalensis*, *L. clathrata* is the host for a small segmented worm, *Podarke obscura* Verrill, that resides among the tube feet in the ambulacral grooves (Ruppert and Fox 1988).

SELECTED REFERENCES: Say 1825:142 [as *Asterias clathrata*]; Downey 1973:22, pl. 1C,D; Zeiller 1974:103; Ruppert and Fox 1988:69, pl. A28; A. M. Clark and Downey 1992:13, figs. 4d, 5e–g, 6g,i, 8g, pl. 4B.

Luidia senegalensis (Lamarck)
Figure 15

DESCRIPTION: This species is unusual in having nine long and straplike arms. The dorsal surface is covered by close-set, square to irregular plates and is fringed with erect spines. Adult specimens can reach a total diameter of 30–40 cm (12–16 in); the individual arm lengths of most adults range from 12

CLASS ASTEROIDEA

FIGURE 15. *Luidia senegalensis*.

to 15 cm (5–6 in). In live or carefully preserved specimens, the disk appears circular in outline.

In live individuals collected near Fort Pierce, Florida, the marginal spines are white, the plates carrying these spines are yellow (especially so near the disk), the square plates at the margin of the arms are cream-colored and the irregular-shaped plates along the midline are dark gray. The dark midline zone is typical of most museum specimens. Ventrally, the plates and spines are white. The tube feet are transparent, with light brown tips.

HABITAT: Generally in calm lagoons, on sandy or muddy sediments. In Southwest Florida it is found on hard shell and quartz sand (Halpern 1970).

DISTRIBUTION: Reported from the Greater and Lesser Antilles and the coast of South America to southern Brazil. In Florida it occurs from Sebastian Inlet on the east coast (Turner, personal communication) to Bradenton on the west coast. Based on collections of the HBOI museum and the Florida Department of Natural Resources (St. Petersburg), the species is known in the Florida Keys at Key West and Content Key. Zeiller (1974:104) mentioned *L. senegalensis* occurring at the Bahama Islands, but this report is not confirmed elsewhere in the literature. Reports of *L. senegalensis* occurring off the West African coast also are dubious (Walenkamp 1976). DEPTH: Low-tide mark to 46 m (151 ft).

BIOLOGY: Komatsu et al. (1991a,b) described the development of the species through metamorphosis, the morphology of the brachiolaria larva, and settlement of the 0.5-mm juvenile. A study by Halpern (1970) of a population from Florida showed that settlement occurs late in the spring. The recently metamophosed individuals triple their overall diameter in 4 weeks, but their growth rate is drastically reduced during the winter months. Individuals of 15-cm diameter may be sexually mature. Halpern identified 20 species of mollusks (nine bivalves, 11 gastropods) from their stomach contents, but 71% of the individuals examined had one or more *Abra aequalis* (Say), a bivalve, in their stomachs. The second most frequently ingested mollusk, the snail *Nassarius vibex* (Say), was found in only 3.6% of the sea stars. Bivalves were the dominant prey of a Venezuelan population of this sea star; gastropods, *Astropecten* spp., irregular echinoids, and other invertebrates were of lesser importance (Penchaszadeh and Lera 1983). In Brazil, Monteira and Pardo (1994) found that the diet of this sea star usually consists of mollusks, especially bivalves, less often includes polychaetes and crustaceans, and incorporates a diversity of other components such as foraminiferans, tunicates, ophiuroids, and sediment.

Some individuals from the Indian River Lagoon, near the Fort Pierce Inlet, Florida, have a dark brown, commensal polychaete worm, *Podarke obscura*, living in the ambulacral groove (Ruppert and Fox 1988). Several of the 2.5-cm-long worms could be found on a single sea star (Miller, previously unpublished).

Tiffany (1978) documented a mass mortality of *L. senegalensis* at Captiva Island off the west-central Florida coast, where several hundred moribund specimens were found on a sandy beach during February 1977. The mortality was not attributed to stranding in rough seas. On sandy tidal flats in the Indian River Lagoon near Stuart, Florida, many decomposing specimens have been found, apparently stranded by a rapidly receding tide. They rapidly succumbed to desiccation and predation by marine birds (Miller, previously unpublished).

SELECTED REFERENCES: Lamarck 1816: 255 [as *Asterias senegalensis*]; Downey 1973:22, pl. 1A,B; Kaplan 1982:176, pl. 33, fig. 7; A. M. Clark and Downey 1992:21, pl. 4A.

FAMILY ASTROPECTINIDAE

Astropecten articulatus (Say)
Figure 16

DESCRIPTION: The smooth appearance of this sea star is a consequence of the closely set, granulose paxillae covering the disk and arms. The arms are moderately long, up to 9 cm (3.5 in), and gradually tapering. They are bordered by prominent, compact superomarginal plates, which protrude above the level of the paxillae. In many individuals, some of the marginal plates carry a short, erect, acute spine on their outer margin. These spines are generally lacking on marginals near the disk and are never found on the pair of basal marginals at the junction of adjacent arms. Surrounding each arm is a fringe of flat, acute spines arranged two to a plate. They are set at right angles to the long axis of the arm, and their length nearly equals the width of the marginal plates. Five elongate jaws, covered with tiny spines, can seal the stellate mouth opening.

This species can usually be recognized by its coloration. Dorsally the deep blue or purple paxillar region is framed by conspicuous white to orange marginal plates; ventrally the sea star is white or beige. Voss

CLASS ASTEROIDEA

FIGURE 16. *Astropecten articulatus*.

(1976) reported the color as light brown to dark purple or reddish brown.

HABITAT: Soft, sandy bottoms.

DISTRIBUTION: Chesapeake Bay, Virginia, to Colombia (South America), including the Bahama Islands, the Yucatán, Jamaica, Dominica, Puerto Rico, St. Thomas, Martinique, and the Lesser Antilles. In Florida, from Jacksonville to Tampa Bay, the Florida Keys, and the Dry Tortugas. DEPTH: Low-tide mark to 165 m (541 ft).

BIOLOGY: This species is common along the outer edge of the continental shelf and reportedly is "the most common sea-star in North Carolina, especially in Onslow and Raleigh Bays" at depths of 18–110 m (Gray et al. 1968a:144). Beddingfield and McClintock (1993) found that *A. articulatus* forages most intensively at dawn and dusk and adjusts its searching behavior according to the density of the prey. They suggested that the species can discriminate among prey items by contact chemoreception. However, Wells et al. (1961) characterized it as a voracious, nonselective predator. By examining the gut contents of 124 individuals, collected at depths between 7 and 13 m (24–42 ft) near Ocracoke Inlet, North Carolina, they identified 91 species of invertebrate prey. The predominant items were gastropod, scaphopod, and bivalve mollusks (73 species); other significant prey consisted of small crustaceans, juvenile sand dollars (*Mellita* sp.), and juvenile specimens of *A. articulatus*. A similar diet, with mollusks predominating, was reported for a population from Cuba (Espinosa 1982).

A. articulatus and *Luidia clathrata* both feed on calico scallops in scallop beds off North Carolina (Schwartz and Porter 1977). According to Ruppert and Fox (1988), both species move rapidly, for sea stars, reaching speeds of 75 cm/minute. *Luidia a. alternata* was also common in the beds, and although it did not feed on scallops, it devoured small individuals of *A. articulatus*. Polson et al. (1993) isolated intraskeletal matrix protein from this species. The molecular weights of the protein fractions, and the amino acid composition of one fraction, are similar but not identical to the matrix protein of *L. clathrata*.

SELECTED REFERENCES: Say 1825:141 [as *Asterias articulatus*]; Gray et al. 1968a:144, figs. 14, 15; Downey 1973:28, pl. 4, figs. C, D; Zeiller 1974:102; Ruppert and Fox 1988:69, pl. A28; A. M. Clark and Downey 1992:31, fig. 9a, pl. 5A,B.

Astropecten duplicatus Gray
Figure 17

DESCRIPTION: The moderate-sized adults of this species reach a diameter of 20 cm (8 in), although individuals of 10–15 cm (4–6 in) are more frequently seen. The five flat, narrow, and slightly tapering arms are two to

FIGURE 17. *Astropecten duplicatus.*

spines are often white, with an orange band at their base. The ventral surface is white to tan, with pale orange tube feet.

HABITAT: Soft sediment composed of sand or shell hash.

DISTRIBUTION: North Carolina to Brazil, including the Bahama Islands, Jamaica, Tobago, Trinidad, Puerto Rico, Dominica, St. Vincent, and St. Thomas. In the Gulf of Mexico, the species is found off Louisiana, Mississippi, Texas, and Mexico. Around Florida, it ranges from Jacksonville on the east coast to Tampa Bay on the west coast, and to the Florida Keys and the Dry Tortugas. DEPTH: 1–550 m (3–1,805 ft), but most common at 5–20 m (16–66 ft).

three times greater in length than the disk diameter. The edge of each arm is bordered by large, smooth, rectangular superomarginals that are set above the level of the paxillae covering the disk and most of the arm. One or two (sometimes up to five) marginal plates at the base of each arm carry a single (occasionally two or more) small, erect spine on their inner edge. Starting with the third, superomarginal plates along the length of the arm may be armed with a similar, but smaller, spine on their outer edge. Below these plates lies a marginal fringe of flat, erect, slightly curved spines with acute tips. The spines are arranged two to a plate. Ventrally, the five triangular jaws, covered with numerous spines, are prominent and distinct. Within the ambulacral groove are two rows of tube feet with pointed tips; suckers are lacking.

In life, *A. duplicatus* is quite variable in coloration and color pattern. Some individuals are simply a drab gray or light brown. Others have light brown on the central portion of the arms and the disk, and marginal plates mottled with light pink to dark reddish brown. The small erect spines on the superomarginal plates and the fringe of marginal

BIOLOGY: *A. duplicatus* burrows for protection and to reach its prey. Its diet consists predominantly of small bivalve and gastropod mollusks (Espinosa 1982). An individual slowly crawling over a sandy bottom at Looe Key, Florida, was stimulated to burrow by blocking its forward movement. After remaining motionless for 15–20 seconds, the sea star began to burrow by flexing its arm margins upward so that the marginal plates met over the center line of each arm. In this configuration, the paxillar region on the arms was concealed, and presumably the ambulacral grooves were pulled wide open, allowing the tube feet to be fully extended. Employing its tube feet, the animal gently rotated back and forth as it sank below the sediment surface, completely concealing itself beneath the sand in 2 minutes (Miller, previously unpublished).

REMARKS: *A. duplicatus* resembles *A. articulatus,* but in addition to the differences in color noted above, *A. duplicatus* has paxillae that are covered by tiny spines. The paxillae are not tightly arranged; often adjacent paxillae along the midline of the arm do not touch. In contrast, the paxillae of *A. articulatus* are covered with granules and set close together, giving the arms a smooth, uniform appearance. Furthermore, the stout, erect spine found on the first marginal plates at the base of each arm in *A. duplicatus* is lacking in *A. articulatus*.

SELECTED REFERENCES: Gray 1840:181; Gray et al. 1968a:144, fig. 16A,B; Downey 1973:29, pl. 6, figs. A, B; Zeiller 1974:102; A. M. Clark and Downey 1992:34, fig. 10d, pl. 8F,G.

FAMILY ASTERINIDAE

FIGURE 18. *Asterina folium*. Photo by W. Stearns, courtesy of P. Humann and W. Stearns.

Asterina folium (Lütken)
Figure 18

DESCRIPTION: This tiny species rarely exceeds 2.5 cm (1 in) in overall diameter. Generally, its shape is pentagonal, although large specimens may have short, bluntly rounded arms. The dorsal and ventral surfaces are flat, except for slight convexities on the radial areas of the disk. The dorsal surface is covered with small, overlapping scales, and the proximal edge of many scales in the radial area is notched for passage of a single papula. There are six longitudinal rows of papulae in each radius. The margin of the body is fringed by a series of short, webbed spines. On the oral surface, the interradial plates bear three to five small acute spines. The spines adjacent to the ambulacral groove are embedded in a stiff membrane.

Generally, small individuals of *A. folium* are mostly white, those of intermediate size are yellow or yellowish red, and the largest are blue to blue-green.

HABITAT: Coral reefs, particularly on the reef flat, and spur-and-groove zones, usually under coral rubble or rock, or in the reef framework.

DISTRIBUTION: Bermuda, the Florida Keys, the Dry Tortugas, the Bahama Islands, and most Caribbean islands, and the northern coast of South America, southward to Brazil. DEPTH: Low-tide mark to 15 m (49 ft).

REMARKS: Although *A. folium* is reported from numerous localities, it has never been found in large numbers, and it is always hidden beneath or within rubble, rock, or coral. At Bermuda, where it reportedly was once common (H. L. Clark 1942), it is now absent or extremely rare (Pawson, previously unpublished).

SELECTED REFERENCES: Lütken 1859a: 60 [as *Asteriscus folium*]; Verrill 1915:58, pl. 3, fig. 5, pl. 11, fig. 4, pl. 28, fig. 2 [as *Asterinides folium*]; Voss 1976:128, 129 (fig.); A. M. Clark and Downey 1992:182, figs. 31c,d,?e, 32f,?k, pl. 42I,J.

FAMILY OPHIDIASTERIDAE

FIGURE 19. *Copidaster lymani*.

Copidaster lymani A. H. Clark
Figure 19

DESCRIPTION: A moderate to large sea star, *Copidaster lymani* can attain a diameter of 31 cm (12 in) or more, but averages approximately 9 cm (3.5 in). The disk is small, little more than a junction point for the five radiating arms. The cylindrical arms, often unequal in length, taper little from disk to tip. They are constricted near their base, with the greatest diameter approximately midway along the length of each arm. The entire animal is covered with a thin skin that is slimy to the touch and conceals the underlying skeleton. However, the body has a pitted appearance to the naked eye, because of the numerous pedicellariae that occupy tiny oblong niches on both the upper and lower surfaces. The ambulacral margins carry a series of large, dagger-shaped spines that lie flat against the arms.

In living individuals, the upper surface is reddish tan to reddish orange, with a darker red mottling or irregular banding on the arms. Very dark specimens are almost uniformly blood red. The tiny pedicellariae and well-defined madreporite are a distinctly contrasting white. The lower surface is considerably lighter than the upper surface, usually a shade of orange. The transparent tube feet are light orange.

HABITAT: In the western Atlantic, this species is found in reef settings, usually beneath coral rock or rubble. At Ascension Island it has been found on a rock and sand bottom. In the Gulf of Guinea (eastern Atlantic), this species occurs on calcareous algal substrate.

DISTRIBUTION: In the western Atlantic the species is reported from Belize, Panama, and the Florida Keys. A juvenile specimen has been collected from the Bahama Islands (off San Salvador) (Miller, previously unpublished). The largest specimens collected to date are from Ascension Island in the mid-Atlantic. In the eastern Atlantic, *C. lymani* is reported off Guinea, Annobón Island, and Cape Verde. DEPTH: Less than 5 m (16 ft) in the Caribbean; elsewhere it occurs at 10–34 m (32–110 ft).

BIOLOGY: Like other ophidiasterid species, *C. lymani* is slow to respond to stimuli and remains immobile for extended periods. It is often found with *Ophidiaster guildingii*, a species that it closely resembles. It can be distinguished from *O. guildingii* by the presence of numerous pedicellariae and a slimy skin.

SELECTED REFERENCES: A. H. Clark 1948:55, figs. 1, 2; Miller 1984:195, fig. 1–3A; A. M. Clark and Downey 1992:270, pl. 66A–C.

Linckia guildingii Gray
Figures 20, 21

DESCRIPTION: The disk of this species is very small, and the slender arms are almost always of unequal lengths. They can number from four to seven, with the longest arms up to 22 cm (9 in) long, 9–10 times the diameter of the disk. Most individuals are considerably smaller than that and frequently have one arm markedly longer than the others.

The arms, viewed from above, appear cylindrical; the lower surface is flattened and somewhat concave along the ambulacral grooves. The width of the arms varies little from the disk to the distal tip. The upper surface of the arms and disk is constructed of small swollen plates that are covered with smooth granules. The plates are irregularly arranged, except along the lateral margin of the arms, where they form two regular longitudinal series. The papular areas lie in shallow depressions between the swollen plates of the upper and lateral surfaces of the arm. Larger specimens have groups of up to 30

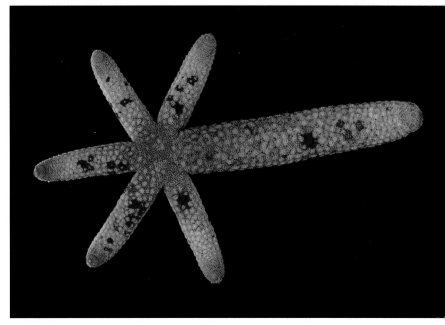

FIGURE 20. *Linckia guildingii*: a large, uniformly colored individual.

FIGURE 21. *Linckia guildingii*: a small, mottled individual with regenerating arms.

papular pores and two or more madreporites, one per interradius. Granules covering the lower surface increase in size toward the ambulacrum. The spines bordering the ambulacral grooves are actually flat granules with bluntly rounded tips.

The coloration of *L. guildingii* is highly variable and usually different in juvenile and adult individuals. Juveniles are mottled with shades of red, brown, violet, or purple; adults are usually a uniform reddish brown, yellowish brown, tan, or violet.

HABITAT: Usually on coral reef hard bottom; also reported from sandy beds between reefs.

DISTRIBUTION: *L. guildingii* is a circumtropical species. In the western Atlantic it occurs off Bermuda, from Florida to Brazil, from the Bahama Islands to Mexico, at numerous Caribbean islands, and from the reef complexes in the southwestern part of the Gulf of Mexico (Henkel 1982). DEPTH: Usually less than 1–2 m (less than 3–6 ft). However, off West Palm Beach, Florida, it is common at 18–27 m (60–90 ft) (Rouse, personal communication). An immature specimen in the HBOI museum was dredged from an *Oculina* coral reef at a depth of 99–110 m (325–361 ft) off Sebastian, Florida.

BIOLOGY: According to H. L. Clark (1933: 24), *L. guildingii* "is one of the commonest species in the West Indies," though it does not appear to be common anywhere along the Florida Keys (Miller, previously unpublished). In spite of its widespread distribution, very little has been published on its biology. It has been suggested that the species is a ciliary feeder and that it ingests the film of microorganisms adhering to hard surfaces; in captivity it will consume carrion (Anderson 1960, 1966; Yamaguchi 1975).

Regenerating specimens of *L. guildingii* with one long arm and several shorter arms are found frequently. These odd-looking forms are called "comets," for a resemblance to the streamlined shape of their celestial namesake. They arise from asexual reproduction, the progeny of adults that voluntarily detached their arms, and the unequal arm length noted in large specimens is a result of this process. The arm stubs form multi-rayed comets by sprouting a new set of arms along the fracture plane. An adult may autotomize one or more complete arms, and the full-grown "daughter" sea stars that eventually grow from the regenerating comets are genetically identical to the parent.

H. L. Clark (1933) suggested that variation in the coloration of *L. guildingii* is associated with a change in habitat that accompanies maturity and that sexual reproduction occurs only in full-grown specimens when asexual reproduction (autotomy and regeneration) ceases. Neither of these suggestions has been substantiated.

L. guildingii usually is cryptic, living beneath corals and coral rubble, but some individuals occur in the open. Despite its dense skeleton and thick skin, it is preyed upon by fish. Randall (1967) reported it from the stomach contents of the black margate, *Anisotremus surinamensis* (Bloch).

REMARKS: Small specimens of *L. guildingii* are sometimes confused with *Ophidiaster guildingii*, and the two species can be similarly pigmented and occupy the same habitats. However *O. guildingii* differs from *L. guildingii* in having arms equal (or nearly equal) in length, plates of the upper and lateral surfaces arranged in regular longitudinal series, less than 15 pores per papular area, and the outer row of spines along the ambulacral groove with pointed tips.

SELECTED REFERENCES: Gray 1840:285; Pawson 1986:526, pl. 174; Sefton and Webster 1986:96, pl. 161; A. M. Clark and Downey 1992:275, fig. 42c,d, pl. 67C,D.

FIGURE 22. *Ophidiaster guildingii.*

Ophidiaster guildingii Gray
Figure 22

DESCRIPTION: This species has a small disk and five narrow arms that taper only near the tip. It never attains a large size, and mature specimens rarely exceed 10 cm (4 in) in diameter. The arms are of equal length and three to four times the disk diameter. On the upper and lateral margins of each arm, seven longitudinal rows of flat plates alternate with eight rows of papular pore areas. The plates are covered with extremely fine granules, around 200 of which cover plates near the disk. The pore areas have 15 or fewer perforations. At the tip of each arm there is a distinctly raised ocular plate that carries several blunt granules. The ambulacral grooves are bordered by two rows of spines, an outer row of longer, bluntly pointed spines and an inner row of shorter, flat spines with truncate tips. The outer row of spines lies flat against the lower arm surface of preserved specimens, giving each ambulacrum a fringed appearance.

Like *Linckia guildingii,* the coloration of *O. guildingii* is variable and can change as an individual grows. According to H. L. Clark (1933), young individuals are dull purplish red variegated with a lighter shade; mature specimens range from pale yellow to orange and scarlet to brownish red, more or less blotched with blue, purple, maroon, or brown. The overall coloration and the pigmentation of the papulae may differ. Uniformly pigmented specimens are rare. Living individuals have a conspicuous, bright red optic cushion at the tip of each arm.

HABITAT: Coral reef zones with coral rock or rubble.

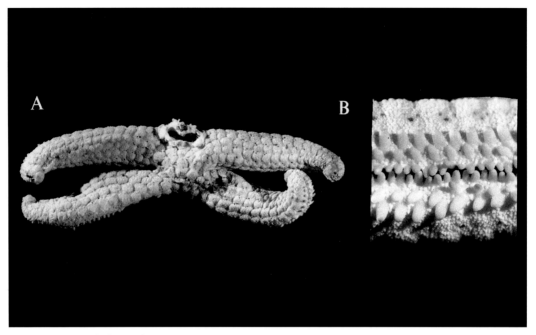

FIGURE 23. *Ophidiaster bayeri*: A, dorsal view (USNM E7176, holotype); B, ventral view of arm showing two series of subambulacral spines bordering the ambulacral groove. Photos by G. Hendler.

DISTRIBUTION: The Florida Keys, the Dry Tortugas, many of the islands of the Greater and Lesser Antilles, Belize (Miller, previously unpublished), and off Brazil. DEPTH: Usually less than 6 m (less than 20 ft). There is a questionable record of *O. guildingii* at 70–87 m (230–285 ft) on the Flower Gardens reefs in the northwestern Gulf of Mexico (Dubois 1975). Henkel (1982) reported it from Enmedio Reef off Veracruz, Mexico.

BIOLOGY: *O. guildingii* is a cryptic, slow-moving species that clings to the underside of coral rock or rubble in reef flat areas. Mortensen (1917) found that *O. guildingii* collected off Tobago had large eggs during the latter part of April.

REMARKS: This species resembles *Copidaster lymani* and young *Linckia guildingii*, two sea stars with which it sometimes occurs. The distinguishing marks of those species are discussed above. Another related species, *Ophidiaster bayeri* A. H. Clark, 1948, was described on the basis of one specimen collected from the outer reefs off Key Largo, Florida (see Figure 23A,B). It is distinguished from *O. guildingii* by (1) a cover of very coarse granules (about 120 on a median plate at the base of the arm, about one-half the number in *O. guildingii*; (2) the presence of two (rather than one) series of subambulacral spines, the outer of which is larger, broad, and somewhat flattened; and (3) the absence of pedicellariae. A. M. Clark and Downey (1992) suggested that this sea star might be a subspecies of the Pacific congener *Ophidiaster trychnus* Fisher.

SELECTED REFERENCES: Gray 1840:284; Downey 1973:68, pl. 28, figs. C, D; Hess 1978:26, fig. 7; A. M. Clark and Downey 1992:281, fig. 44c,d, pl. 69A,B.

FIGURE 24. *Poraniella echinulata.*

FAMILY ASTEROPSEIDAE

Poraniella echinulata (Perrier)
Figure 24

DESCRIPTION: This is one of the smallest sea stars in the tropical western Atlantic. Specimens with an overall diameter greater than 3 cm (1 in) are rare. Individuals are star-shaped, with a broad disk and five short arms. The arms are relatively wide, flat, and very thin, tapering to blunt tips. Surrounding the center of the disk, the 10 largest aboral plates form a double circlet. The five papular areas on the disk and arms are oval in outline and swollen, and between adjacent papular areas the disk is sunken or slightly concave. In live animals, the upper surface is covered with a thick, fleshy tissue that obscures some features of the skeleton noted above. The margin of the body is marked by two series of large, swollen superomarginal and inferomarginal plates. The inferomarginals each carry a bundle of spines. On the lower surface, the plates between the marginals and the ambulacral grooves also carry tiny sharp spines. At the apex of each of the five jaws surrounding the mouth there is a fanlike array of long, flat spines.

The upper surface of *P. echinulata* is a bright orange-red to blood red, variegated with white. In some individuals, the white pigment forms a pentagon at the center of the disk and a distinct stripe along the midline of each arm. The distal tips of the arms are mottled black and white, the fleshy papulae are pigmented pale orange, and the madreporite is light tan. The lower surface is uniformly orange-red except for the white tips of the spines at the jaws' tips. The tube feet are transparent and colorless. Specimens preserved in alcohol rapidly lose their pigmentation.

HABITAT: On hard substrates, beneath coral rubble and rock in shallow reef habitats.

DISTRIBUTION: Florida, the Bahama Islands, Cuba, Yucatán, Panama, Belize, and Barbados. Specimens collected off Carrie Bow Cay, Belize, represent the shallowest depth records to date (Miller, previously unpublished). DEPTH: 3–309 m (10–1,014 ft).

BIOLOGY: This species has never been found in abundance, but population density of such a cryptic species is difficult to judge. Despite striking coloration, individuals of this species are difficult to discern among the brightly colored sponges, foraminiferans, and bryozoans to which they cling.

P. echinulata is most often found in fore reef localities, indicating that its preferred habitat may be on the deep reef, beyond scuba depths. Specimens were collected from vertical rock walls in the Bahama Islands at depths near 150 m (492 ft) using a research submersible (Miller, previously unpublished).

SELECTED REFERENCES: Perrier 1881:17 [as *Marginaster echinulatus*]; Verrill 1915: 70, pl. 7, figs. 1–1a, pl. 15, figs. 5–5b [as *Poraniella regularis*]; Downey 1973:81, pl. 36, figs. C, D [as *Poraniella regularis*]; A. M. Clark and Downey 1992:290, pl. 46A,B.

FAMILY OREASTERIDAE

Oreaster reticulatus (Linnaeus)
Figures 25, 26

DESCRIPTION: *Oreaster reticulatus* is probably the most widely known and most easily identified member of the Caribbean marine biota. Individuals are robust and can reach a diameter of 50 cm (20 in). The massive central disk is inflated and supports five short, slightly tapering arms. Most individuals are stellate, but some with a nearly pentagonal outline have been reported. The disk of juveniles is generally less inflated than in adults.

The upper surface is invested with thick, heavy plates forming a reticulate pattern; the central plates are arranged in a circlet. The plates carry numerous prominent tubercles with bluntly rounded tips, and a continuous series of tuberculate plates demarcates the lateral margin. The lower surface is flat, except for a shallow concavity near the mouth. It is covered with granulose plates arranged in chevrons in each interradius. The granules covering the central portion of these plates are larger than those at the plate margins. The ambulacral grooves are covered by a double series of large, flat, blunt-tipped spines.

The color patterns of *O. reticulatus* are extremely variable, even among individuals in the same population. The upper surface of juveniles is usually mottled green (often olive green), brown, tan, and gray. In contrast, the upper surface of adults is yellow, brown, or orange, with the large erect tubercles dis-

FIGURE 25. *Oreaster reticulatus:* an adult individual.

FIGURE 26. *Oreaster reticulatus:* a juvenile individual.

tinctly lighter or darker than the disk and arms. The lower surface of both juvenile and adult specimens is beige or cream.

HABITAT: *O. reticulatus* prefers the shallow, quiet waters of reef flats, lagoons, and mangrove channels. It generally occurs in beds of *Thalassia testudinum, Halodule wrightii* Ascherson, and *Syringodium filiforme* and on the sand flats associated with those seagrasses.

DISTRIBUTION: On both sides of the Atlantic, ranging from North Carolina (Cape Hatteras) and Bermuda (occasional) south to Florida, the Bahama Islands, and Brazil; also from the Cape Verde Islands off western Africa. DEPTH: 1–37 m (3–120 ft).

BIOLOGY: In clear water on calm days, specimens of this stout-bodied species can be seen from a boat while motoring over grass beds and sand patches. The large size, sluggish nature, and accessibility of *O. reticulatus* make it especially vulnerable to human exploitation. Dried, sometimes garishly painted, carcasses of this species have been sold as souvenirs in seaside curio shops for many years, and because of overcollecting, *O. reticulatus* is now considered rare in some areas where it was once common.

This species is among the most thoroughly studied Caribbean echinoderms. The structure of its skeleton was illustrated by Agassiz (1877). A wealth of information on its ecology (feeding behavior, predation, homing and foraging movements, habitat utilization, population structure) is provided by Scheibling (1980a,b,c,d, 1981, 1982a,b) and others (Thomas 1960; Anderson 1978). *O. reticulatus* feeds primarily on microorganisms and the particulate matter associated with sand, seagrass, and algal substrates (Scheibling 1982a). It also has the ability to graze on algae or prey upon echinoderms such as *Tripneustes ventricosus, Meoma v. ventricosa*, even other *O. reticulatus,* and a variety of sponges. The expansive stomach is everted to cover an item of food, and digestion occurs outside the body (extraoral feeding).

Scheibling (1980c,d) found that *O. reticulatus* reaches greatest population densities in areas with the seagrass *Halodule wrightii* and occurs in lowest numbers on bare sand plains. He also calculated that it can move at speeds of 12–33 cm/minute, with greatest speeds reached on sandy bottoms.

The triton gastropod *Charonia variegata* (Lamarck) is the only confirmed predator of adult *O. reticulatus. Charonia,* with its low feeding rate and small numbers, is unlikely to seriously limit the abundance of *Oreaster* (Scheibling 1980a,b). Keller's (1976) experiments with caged enclosures suggest that juvenile *O. reticulatus* are consumed by fish. There is support for this hypothesis in Randall's (1967) report that *O. reticulatus* is eaten by the queen triggerfish (*Balistes vetula* Linnaeus) and the spotted trunkfish (*Lactophrys bicaudalis* [Linnaeus]).

SELECTED REFERENCES: Linnaeus 1758: 661 [as *Asterias reticulata*]; Downey 1973: 60, pl. 24, figs. A, B; Colin 1978:416; Pawson 1986:525, pl. 174, pl. 14, fig. 13; Sefton and Webster 1986:94, pl. 160; A. M. Clark and Downey 1992:293, pl. 72C,D.

FAMILY ECHINASTERIDAE

Echinaster (Othilia) echinophorus
(Lamarck)
Figure 27

DESCRIPTION: This moderately small species can attain a diameter of approximately 7 cm (2.8 in). It has a small disk and five arms that taper little, except near their tips. They are subcylindrical, flattened ventrally, and composed of widely spaced plates covered by a relatively thin skin. The upper surface of each arm bears one to two irregular rows of conspicuous, erect, thornlike spines. There are six to nine spines per row, each spine measuring 2–3 mm (0.08–0.12 in) in length. Along the side of the arm there is a straight row of spines similar to those on the upper surface. On the lower surface, the ambulacral grooves are bordered on both sides by three series of spines: the spines in the middle and outer rows are twice and four times as large, respectively, as those in the inner row. The ventral margin of each arm carries another series of spines, more numerous and slightly smaller than those on the upper surface.

The uniform bright red or crimson color of *E. echinophorus* is fairly consistent, but a specimen from Belize was dark reddish orange when alive (Miller, previously unpublished).

HABITAT: Usually associated with hard subtrates. Found on reef fringing mangrove cays in Belize (Hendler, previously unpublished) and among mangrove roots in Puerto Rico (B. Cutress, personal communication).

DISTRIBUTION: Florida Keys (Shark Key, Key Largo), Puerto Rico, Jamaica, and Nicaragua (A. M. Clark and Downey 1992),

FIGURE 27. *Echinaster (Othilia) echinophorus.*

Belize (Miller and Hendler, previously unpublished), and off Lee Stocking Island, Bahama Islands (Pitts, personal communication). DEPTH: Usually in shallow water, although A. M. Clark and Downey (1992) reported it to 55 m (180 ft).

BIOLOGY: The species has been reported in greatest abundance off Brazil. Its relative scarcity at other localities perhaps indicates that more northerly habitats are less than optimal for the species. Individuals generally live beneath coral rubble and rock and have been found clinging to mangrove roots.

REMARKS: *E. echinophorus* is similar to the far more common Florida Keys sea star *E. sentus*. It is distinguished from the latter by its fewer but longer and more conspicuous arm spines and its uniform color. *E. sentus* has a body wall and spines with contrasting colors.

SELECTED REFERENCES: Lamarck 1816: 560 [as *Asterias spinosa*]; Downey 1973: 86, pl. 39, figs. C, D [as *Echinaster echinophorus*]; Zeiller 1974:105 [as *Echinaster sentus*]; Kaplan 1982:175, pl. 32(8) [as *Echinaster sentus*]; A. M. Clark 1987:70; A. M. Clark and Downey 1992:367, pls. 89A,B,F, 90F–H.

Echinaster (Othilia) sentus (Say)
Figures 28, 29

DESCRIPTION: This species attains a moderate size, with a maximum overall diameter of 18 cm (7 in) and arms approximately twice as long as the disk diameter. The five arms are stout, slightly tapering to a bluntly rounded tip, and usually equal in length. They are nearly circular in cross section, except for a flattened area adjacent to the ambulacral grooves. The upper and lower surfaces are covered with a thick opaque skin, making it difficult to see the underlying plates. Some of the plates bear sharp conical spines 1.5–2 mm (0.06–0.08 in) long, which are more prominent in dried than in live specimens. The spines are often arranged in regular longitudinal rows (approximately 15–20 spines per row) on the upper surface, but they are scattered in some individuals. Those bordering the ambulacral grooves are ar-

FIGURE 28. *Echinaster (Othilia) sentus:* an adult individual.

ranged in three rows: an inner row of short spines, a middle row of intermediate height, and an outer row of spines similar to the long spines on the upper plates.

E. sentus is reported to be deep red, reddish brown, dark purple, pale violet, yellow brown, or purple. However, color can be altered drastically by preservation, and it is not clear which previous records were based on living specimens. Color patterns of living juvenile *E. sentus*, 2.5–5.5 cm (1–2 in) in diameter, illustrated in Figure 29, and a large adult (12 cm [5 in] in diameter) shown in Figure 28, depict the change in color as an individual matures. The upper surface has a white ground color and contrasting orange-brown spines and violet papulae. The tube feet are orange. When large preserved specimens are dried, the upper surface becomes dark reddish brown, the papulae appear black, and numerous dermal glands form dark brown dots on the skin; these are especially noticeable on the lower surface near the ambulacral grooves.

HABITAT: On sand or on a mixture of sand and coarse shell hash, in seagrass beds, and along rocky coastlines. Juveniles have been found off Big Pine Key, Florida Keys, at 3 m (10 ft) depth on uniformly mixed gravel and clay sediment adjacent to a mangrove island (L. Cameron, personal communication).

DISTRIBUTION: The geographic range is problematic because reports of western Atlantic *Echinaster* species have so often been based on misidentified specimens. However, its occurrence is confirmed from the Bahama Islands and Florida (Biscayne Bay to Key West). DEPTH: Shallow intertidal to 13 m (42 ft).

BIOLOGY: Lane Cameron (personal communication) found that a food source for *E. sentus*, near Big Pine Key, Florida, is the fire sponge, *Tedania ignis* (Duchassaing & Michelotti). He also noted that the large size and bright orange color of *E. sentus* eggs makes them visible to scuba divers. The newly hatched juveniles are almost 0.2 cm in diameter and are abundant during May and June. Juveniles appear to avoid bright light, residing beneath drift algae or in the shade cast by mangrove trees.

REMARKS: A similar species, *Echinaster (Othilia) echinophorus,* has also been verified from the Florida Keys (A. M. Clark 1987; A. M. Clark and Downey 1992), but only a few specimens have been collected from this area. Differences between the species are discussed in the *Remarks* section for *E. echinophorus*. The variability of *Echinaster* species makes their identification decidedly difficult. Aside from the two species discussed above, other *Echinaster (Othilia)* that occur in shallow water off the Gulf and Atlantic coasts of Florida include *E. graminicola* Campbell & Turner, *E. guyanensis* A. M. Clark, *E. paucispinus* A. M. Clark, *E. serpentarius* Müller & Troschel, and *E. spinulosus* Verrill (A. M. Clark and Downey 1992).

FIGURE 29. *Echinaster (Othilia) sentus:* juvenile individuals.

SELECTED REFERENCES: Say 1825:143 [as *Asterias sentus*]; Downey 1973:89, pl. 42, figs. A, B [as *Echinaster* species C]; Hess 1978:32, fig. 10; A. M. Clark 1987:70; A. M. Clark and Downey 1992:377, figs. 56a, 58a, pl. 91A,B.

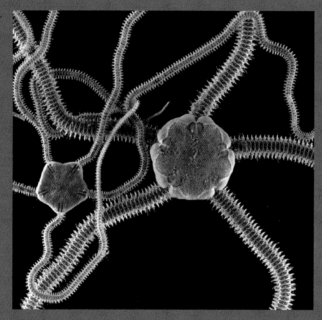

CLASS OPHIUROIDEA
Brittle Stars, Basket Stars

The brittle stars and basket stars are graceful, agile relatives of the sluggish sea stars. Approximately 2,000 species have been described from around the world. On Caribbean reefs, coral-associated brittle stars occur in densities of 20–40 individuals per square meter; burrowing species living in soft sediment can be 10 or 100 times more numerous. In the Florida Keys, brittle stars can constitute 27–52% of macrofauna beneath rubble; "after scleractinian corals and octocorals, ophiuroids are perhaps the most abundant and characteristic macroinvertebrate of Florida coral reefs" (Kissling and Taylor 1977: 226). The use of scuba has enriched our appreciation of the shallow-water Caribbean brittle star fauna. We identified 37 species just at Looe Key, Florida, and over 44 species have been reported near a single cay on the Belize Barrier Reef.

FORM AND FUNCTION

Despite a superficial similarity in form, this group exhibits a considerable range of body plans, from smooth or spiny brittle stars to the so-called basket stars with branching arms. At one extreme in size are commensal species a few millimeters in diameter that cling to the tiny spines of sand dollars, and at the other are basket stars with intricately arborescent arms that can span over a meter.

The anatomical structures of brittle stars are depicted in Figures 30–32. The disk and arms of a brittle star are well demarcated. They are protected by series of integument-covered ossicles called shields, plates, scales, spines, and granules. Minute, fingerlike tube feet emerge beneath each arm joint. Some species employ tube feet for locomotion, but the majority move by muscular exertions of pairs of arms that flex and extend to push the disk ahead; several species can swim. The adhesive properties of the tube feet enable many brittle stars to scale smooth vertical surfaces and to catch and manipulate particles of food.

Within the disk are a capacious stomach, series of gonads, and sack-shaped bursae. The latter structures, which open beside the base of the arms, are used in respiration and reproduction. In many species, paired radial shields are evident on the dorsal surface of the disk near the base of the arm. Conspicuous structures below the disk include five oral shields, at least one of which (the madreporite) may be enlarged and perforate.

Many accounts state that brittle stars move their arms only in a horizontal plane. However, the arms are composed of a series of flexible joints, and nearly every species is capable of a degree of vertical bending or coiling. At the core of the arm joints are the vertebrae, ossicles resembling bones of the human spine, linked by muscles and connective tissues. Modified arm ossicles are joined in a ring beneath the disk, forming five triangular jaws that frame the centrally placed mouth. Attached to the jaws are ossicles called oral

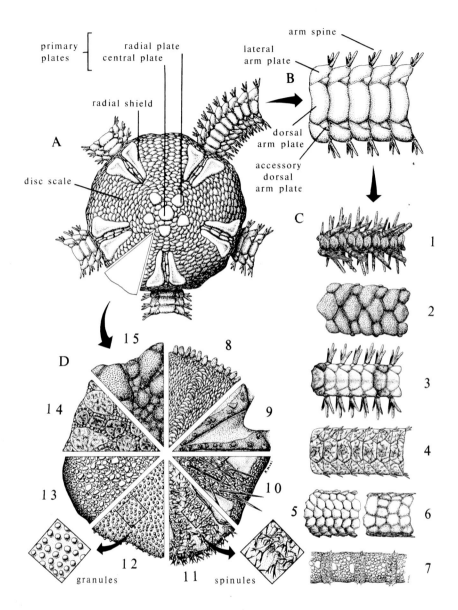

FIGURE 30. Major anatomical features of a brittle star: *A*, the dorsal surface of the disk; *B*, the dorsal surface of an arm; *C*, typical patterns of arm plates and spines including the (1) asymmetrical arm joints in *Ophiocoma echinata*, (2) dorsal arm plates, accessory dorsal arm plates, and lateral arm plates in *Sigsbeia conifera*, (3) *Ophionereis reticulata*, and (4) *Ophiolepis elegans*, (5) regularly fragmented dorsal arm plates of *Ophioderma squamosissimum*, (6) irregularly fragmented dorsal arm plates of *Ophioderma cinereum*, (7) rings of hooks encircling the distal arm joints of *Astrophyton muricatum*; *D*, typical coverings of the dorsal surface of the disk, with sectors representing the (8) imbricating scales and fence papillae in *Ophiophragmus pulcher*, (9) thick integument with tubercles in *Astrophyton muricatum*, (10) long spines in *Ophiothrix suensonii*, (11) trifid and bifid spinules in *Ophiothrix angulata*, (12) rounded granules in *Ophiocoma wendtii*, (13) flattened granules in *Ophioderma squamosissimum*, (14) small scales surrounding large scales in *Ophiolepis elegans*, (15) thickened irregular scales in *Sigsbeia conifera*. Illustration by R. Mooi.

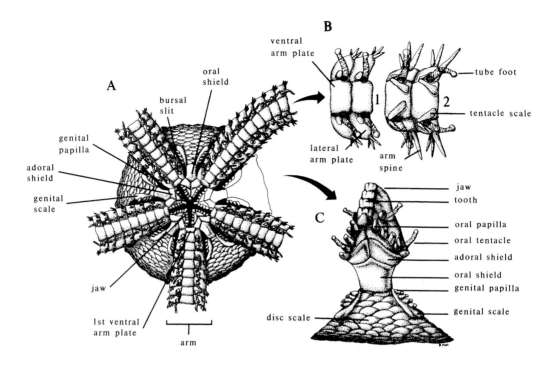

FIGURE 31. Major anatomical features of a brittle star: A, the ventral surface of the disk (note that the ventral interradius at 8 o'clock has four bursal slits, typical of *Ophioderma* species); B, the ventral surface of an arm of (1) a typical brittle star with small tentacle scales beside the tube feet, and (2) an *Ophiopsila* species with elongated, ciliated tentacle scales; C, the jaw structures that project into the mouth, and the ventral interradius of the disk. Illustration by R. Mooi.

papillae, dental papillae, and teeth (Figures 31, 32). The arm joints beneath and beyond the edge of the disk have protective lateral arm plates that bear series of arm spines; the joints are capped above and below by dorsal and ventral arm plates. The arm spines may be obvious or inconspicuous, and they, and associated arm hooks and tentacle scales, serve in feeding and defense. At the tip of the arm is a cylindrical terminal plate, through which the distalmost tube foot protrudes. New joints are added to the arm at the inner edge of the terminal plate; thus the youngest joints are those nearest the arm tip. As the disk grows, new plates arise between older ones, and the disk covers the older, proximal arm joints.

GENERAL BIOLOGY

Brittle stars owe their name to a notorious capability for voluntarily breaking their arms. This seemingly self-defeating reflex is actually an advantage when an individual is attacked by a predator, a common predicament for brittle stars. For some brittle star species, their talent for defensive self-mutilation extends to an ability to cast off the disk, thereby discarding the stomach, gonads, and other tissues.

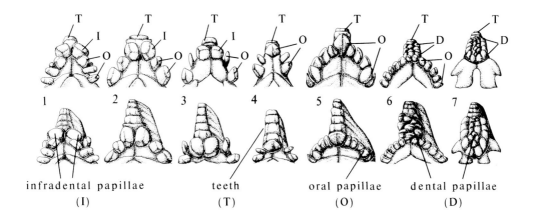

FIGURE 32. Brittle star jaw structures showing configurations of the dental papillae (D), infradental papillae (I), oral papillae (O), and teeth (T) that are characteristic of different families and genera. Upper row: ventral view of the jaw; lower row: the proximal edge of the jaw, showing the tooth row that projects into the mouth. (1) *Amphioplus thrombodes,* (2) *Ophiophragmus pulcher,* and (3) *Amphiura kinbergi* with paired infradental papillae at the apex of the jaw; (4) *Ophiactis savignyi* with two oral papillae on each side of the jaw, and (5) *Ophiolepis elegans* with several pairs of oral papillae; (6) *Ophiocoma echinata* with a cluster of dental papillae and a series of oral papillae, (7) *Ophiothrix suensonii* with a large cluster of dental papillae, oral papillae lacking. Illustration by R. Mooi.

That process is all the more remarkable because these individuals can completely regenerate the broken arms and regrow the disk and associated viscera, usually in 2 weeks to 2 months. Other brittle star defenses include distastefulness and luminescence, speedy escape, and the shadow response, a stop-motion reflex triggered by passing predators.

The reproductive organs (testes and ovaries) consist of clusters of gamete-lined tubules that are housed within the disk. Spawning brittle stars assume a push-up posture, raising their disks off the bottom while pumping a stream of gametes from the bursal slits. Because they can be attractive and susceptible to predators, it is no wonder that brittle stars tend to spawn at night, when predation is reduced or less effective. Spawned oocytes meet spermatozoa in midstream and join; the fertilized eggs are wafted off in the water currents and give rise to planktonic larvae.

The larvae develop for a few days to over a month before completing metamorphosis, the terminal transformation to a juvenile brittle star. Some species have large, yolky eggs and nonfeeding, barrel-shaped vitellaria larvae that swim using rings of cilia; they metamorphose in less than a week. Others produce hundreds of thousands of tiny eggs that develop into more elaborate, microscopic long-lived larvae. These are beautiful, translucent ophiopluteus larvae, with an intricate crystalline supporting skeleton and with sinuous ciliary bands that are used for food gathering and locomotion.

A minority of brittle star species are hermaphroditic, either sequentially (transforming from one sex to another) or simultaneously. There is a curious correlation between hermaphroditism and this mode of development: all brittle star hermaphrodites (and a few species with separate sexes) brood their young. That is, the embryos are held in the bursae of the parent until they have developed into miniature crawl-away juveniles. Another small group of brittle stars is capable of asexual reproduction by a process called fission, splitting across the disk to produce a pair of half-brittle stars. The halves heal, regenerate missing parts, and split again when they reach an appropriate size.

The feeding behaviors of brittle stars, like their reproductive habits, are remarkably diverse. They obtain some nutrients without lifting an arm, by the uptake of dissolved compounds through their skin. But generally, particulate food is conveyed to the mouth by the tube feet and muscular arms. Some brittle stars can coil their arms quickly, particularly the slender tip of the arm, and capture small active organisms, even fish. Many species bend the arm in broad loops, like an elephant's trunk, to convey larger chunks of scavenged material to the mouth. The coordinated movements of the hundreds of tube feet under the arm can collect and relay small particles from the surface of the sediment, snatch particles floating past in the water, or glean material adhering to the hooked or sticky, mucus-coated spines on the arm. The feet are capable of sensing extremely dilute concentrations of chemicals such as amino acids and vitamins, enabling individuals to detect and respond to food, even at a distance.

Most species remain concealed during the day, perhaps hiding from diurnal predators. Brittle stars lack eyes, but it is thought that nerves beneath microscopic transparent nodules in the skeletal plates are photosensitive. Furthermore, the discovery that some brittle stars can change color suggests how the amount of light striking the nerves is controlled. The pigment-containing cells cover the transparent nodules during the day (serving as "sunglasses") and uncover them at night. The adjustment enables brittle stars to detect crevices in sunlight or moonlight, an ability that is essential for their escape from predators and competitors. The colors of most species are fixed, although pigmentation patterns can change, sometimes dramatically, as individuals grow. Color patterns of an individual are as specific as a fingerprint, and sometimes blatant intraspecific variations are exhibited.

The most frequently cited classification of ophiuroids distinguishes two orders in the class, the Phrynophiurida (slimy-skinned brittle stars and basket stars) and Ophiurida (all others), divided into five or six subordinal groupings and roughly 17 families. Higher-order classification is based primarily on the shapes of internal structures such as genital, oral, and dental plates and features of the vertebrae. The major ecological and morphological differences among groups of brittle stars are readily recognized at the family level.

Brittle stars may show greater fidelity in body plan to the substratum they occupy than to phylogenetic affinity or distribution. For example, species in the family Amphiuridae that live in soft sediments have adaptations for burrowing such as soft disks and long delicate arms, which would jeopardize their survival in an exposed habitat. Burrowing *Ophiopsila* species, in the family Ophiocomidae, are remarkably similar to amphiurids. Species in several families, which are epizoic on soft coral and sponges, have hooked spines used to cling to their host and structural and behavioral defenses against predation.

ECOLOGICAL AND COMMERCIAL IMPORTANCE

Brittle stars have never been an important item of human commerce, although at least during the eighteenth century, people of Indonesia cooked and ate basket star roe. Brittle stars and basket stars are not often harvested for the souvenir trade because much of their fragile beauty disappears when they are dried. They may have an important role in the ecology of marine communities, but the extent of the role seldom has been tested and their interactions with other organisms have been sporadically documented (Ambrose 1993). Despite their refractory skeleton, brittle stars figure in the diets of commercially important crabs, shrimps, and fish, including at least 33 species of Caribbean reef fish, and they are consumed by hermit crabs, mantis shrimps, sea stars, and other brittle stars.

TERMINOLOGY AND CONVENTIONS

The colors noted in the species accounts describe living brittle stars; pigmentation and color patterns are altered or removed by preservative chemicals. Arm structures are described for the proximal one-third of the arm, unless otherwise noted. Toward the tip of the arm, the relative dimensions of arm plates change, and the arm spines and tentacle papillae diminish in number. The number of arm spines and tentacle scales refers to the number on each *side* of an arm joint on a portion of the arm near the disk. For example, "this species has three arm spines" means that near the disk there are six arm spines on each joint (three on each lateral arm plate). Similarly, the number of oral papillae cited for a species refers to the number on one side of a jaw.

Disk diameter is measured from the distal edge of a pair of radial shields to a point on the opposite edge of the disk. Arm length is the distance from the disk edge (the base of an arm) to the tip of the longest arm.

IMPORTANT REFERENCES

Brittle stars are treated in most textbooks of invertebrate zoology, in a book for the layman (A. M. Clark 1977), and in a thorough but outdated scientific treatise (Hyman 1955) on the Echinodermata. Two indispensible taxonomic monographs are H. L. Clark's (1915) catalog, which lists all the (then) known species of brittle stars, and Fell's (1960) keys to the families and genera of brittle stars worldwide. Accounts of higher-level classification, based on internal anatomy, appear in Matsumoto (1917), Murakami (1963), and Spencer and Wright (1966). Recent topical reviews address brittle star feeding (Jangoux and Lawrence 1982), reproduction (Hendler 1991), and their ecology on coral reefs (Birkeland 1989).

Contributions to the identification of brittle stars of this region include the general works of H. L. Clark (1918, 1933). Keys and taxonomic information for the family Amphiuridae appear in Thomas (1962b, 1963, 1964, 1965a,b, 1966) and A. M. Clark (1970). References are available to the genera *Ophiocoma* (Devaney 1970, 1974b) and *Ophionereis* (A. M. Clark 1953; Thomas 1973), and A. M. Clark (1967b) provided a taxonomic overview of the genus *Ophiothrix*. Ziesenhenne (1955) presented a key to the genus *Ophioderma*. It omits *Ophioderma devaneyi* Hendler & Miller, a large, reddish, deep-water species that occurs off Florida, and recently described members of the Caribbean fauna such as *O. anitae* Hotchkiss and *O. ensiferum* Hendler & Miller. The Caribbean *Ophiolepis* species are keyed in Hendler (1988b). Existing keys (H. L. Clark 1918, 1933) to the genera *Ophiactis* and *Ophiopsila* are of limited value.

KEY TO THE FAMILIES OF BRITTLE STARS[1]

1a. The disk has soft, slick skin covering the radial shields (Figures 33, 34). There are no long, thin spines on the disk; dental papillae lacking ..... **Ophiomyxidae**
 b. The disk bears scales, spines, tubercles, or granules (Figure 30D) 2
2a. The arms branch dichotomously and/or bear rings of microscopic hooks especially near the arm tip (Figure 30C-7) .... **Gorgonocephalidae**
 b. The arms are unbranched and only bear spines 3
3a. There is a pair of blocklike infradental papillae at the tip of the jaw. Oral papillae are present on the sides of the jaw (Figure 32-1,2,3). Accessory dorsal arm plates are lacking .... **Amphiuridae**[2]

b. There is a cluster of dental papillae at the tip of the jaw (Figure 32-6,7). Oral papillae are either present or lacking 4

c. There is usually a single oral papilla or tooth at the tip of the jaw (Figure 32-4,5); in the cases where two papillae are present, they are not blocklike. Oral papillae are present on the sides of the jaw; dental papillae are lacking 5

4a. Oral papillae are lacking (Figure 32-7), and the disk bears spines, spinules, or stumps (Figure 30D-10,11). The dorsal arm spines are usually prickly and glassy; tiny ventral arm spines may be hook-shaped .... **Ophiotrichidae**

b. Oral papillae are present on the side of the jaw (Figure 32-6), and the disk bears rounded microscopic granules (Figure 32-6). Arm spines are smooth and solid .... **Ophiocomidae** (*Ophiocoma* and *Ophiocomella* species)

c. Oral papillae are present on the side of the jaw, and the disk is not granulated. Inner tentacle scales are very long and slender (Figure 31B-2) .... **Ophiocomidae** (*Ophiopsila* species)

5a. The disk bears round, or faceted, microscopic granules (Figure 30D-12,13). There are four bursal slits in each interradius (Figure 31A) .... **Ophiodermatidae**

b. The disk is covered by scales and may also bear small, scattered spines. There are two bursal slits per interradius 6

6a. The disk scales are small, thin, rounded, imbricating and mostly of uniform size 7

b. The disk scales are of two contrasting sizes and/or irregular in shape 8

7a. The disk is smooth and covered by fine, imbricating scales. Accessory dorsal arm plates are present (Figure 30C-3). The radial shields are small. There is a continuous series of oral papillae at the margin of the jaw (Figure 32-5) .... **Ophionereididae**

b. The disk is covered by small, imbricating scales and may bear small, scattered spines. Accessory dorsal arm plates are lacking. The oral papillae are usually inconspicuous and do not form a continuous series along the margin of the jaw (Figure 32-4). Some species are fissiparous; individuals have six or fewer arms .... **Ophiactidae**

8a. The disk is covered by groups of small scales surrounding larger scales (Figure 30D-14). Pairs of tentacle scales, composing an operculum, cover the tentacle pores. Adjacent dorsal arm plates are in contact (Figure 30C-4); adjacent ventral arm plates are in contact .... **Ophiuridae**

b. The disk scales are variously shaped and wedged into 10 narrow zones between the large radial shields (Figure 30D-15). The tentacle scales are single. Dorsal arm plates at the middle of the arm are separated by lateral arm plates (Figure 30C-2). Adjacent ventral arm plates are separated by tissue-filled gaps. The arms are often coiled .... **Hemieuryalidae**

[1] Applicable only to the shallow-water fauna of the Florida Keys, the Bahama Islands, and much of the Caribbean. Based on the features of living animals. A microscope is required to see some important structures.

[2] Small Ophiocomidae and Ophionereididae having only two or a few small dental papillae may key out as Amphiuridae (see Figure 32). See key to genera under the family Amphiuridae and Figures 100–106.

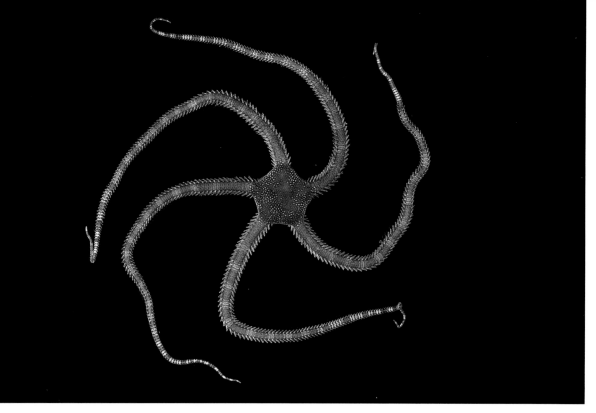

FIGURE 33. *Ophioblenna antillensis*.

FAMILY OPHIOMYXIDAE

Ophioblenna antillensis Lütken
Figure 33

DESCRIPTION: The largest known specimen is approximately 23 mm (0.9 in) in disk diameter, with arms 116 mm (4.6 in) long. The soft, smooth disk integument of the species is the basis of its scientific name, which translates as "slimy snake of the Antilles." The arms have a thin skin through which the broad, unfragmented dorsal arm plates and slender thorny arm spines are evident even with the naked eye. Additional characteristics that distinguish this species from *Ophiomyxa flaccida* are paired, ovoid tentacle scales and smooth-edged teeth and oral papillae. Embedded in the flesh of the disk are microscopic, pointed ossicles that can be seen on dried specimens.

The disk of *O. antillensis* is usually brown or reddish brown dorsally, sometimes spotted or otherwise variegated; the ventral surface of the disk always is marbled with a contrasting cream color and very dark brown or black. The arms are banded with light and dark brown and sometimes also with purple; those of small individuals are purplish, as are the regenerating arm tips of adults. The disk of juveniles has a radiating purple pattern on a cream-colored background (Hendler, previously unpublished).

HABITAT: Under rubble on shallow reef flats, beneath dead coral slabs in the turbulent spur-and-groove zone, and in the interstices of corals on the fore reef slope.

DISTRIBUTION: The Bahama Islands (Hendler, previously unpublished), Puerto Rico, St. Thomas, Belize, and Panama. DEPTH: 1–24 m (3–79 ft).

BIOLOGY: This brittle star is infrequently collected, and individuals are few in number when encountered. Some individuals harbor a commensal ostracod, *Pontocypria hendleri* Maddocks (Maddocks 1987).

SELECTED REFERENCES: Lütken 1859b: 239, pl. 4, figs. 4a–d; H. L. Clark 1901a:249, pl. 15, figs. 1–4 [as *Ophialcaea glabra*]; Hotchkiss 1982:391, Figure 171a,b [as *Ophiomitrella glabra*].

Ophiomyxa flaccida (Say)
Figure 34

DESCRIPTION: Individuals grow to 26 mm (1 in) in disk diameter with arms 130 mm (4.3 in) long. The disk and arms are both covered by soft, smooth, slick skin (Byrne and Hendler 1988), hence its scientific name, loosely translated as "soft mucus-snake." Minute serrations on the arm spines give the arms a rough texture and provide another tactile clue for identification. The teeth and oral papillae of this species have a transparent, serrated edge, visible under the microscope. There is a thick collar of connective tissue reinforced with microscopic scales, rather than discrete tentacle scales, at the base of the tube feet.

The species is green, yellow, orange, red, or brown. The disk is often flecked or mottled, and the arms usually are banded with lighter and darker shades of the ground color or cream color. At a given locality, individuals may all be a single color, or a variety of colors may be represented. Many juveniles from the Belizean reef slope are reddish with a white spot on the disk, the same color as juvenile *Ophioderma rubicundum* in the same habitat (Hendler, previously unpublished). Whether the similarity is evidence of camouflage, mimicry, or is mere coincidence is not yet known.

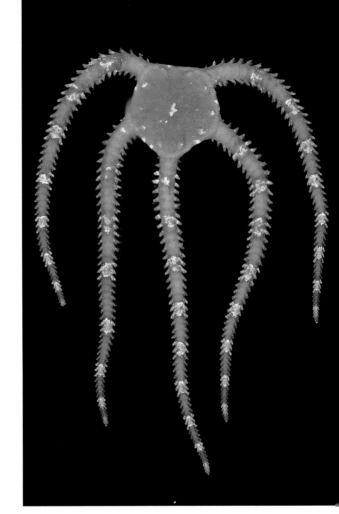

FIGURE 34. *Ophiomyxa flaccida*.

HABITAT: All reef habitats and zones, and seagrass beds; associated with many substrates, even in the holes manufactured by rock-boring mollusks.

DISTRIBUTION: Bermuda, the Florida Keys, the Dry Tortugas, the Bahama Islands, Cuba, Jamaica, Haiti, Puerto Rico, the Virgin Islands, the Leeward and Windward Islands, Barbados, Trinidad, Tobago, Curaçao, Belize, Isla de Providencia, and the mainland of Central and South America from Belize to Brazil. DEPTH: Less than 100 m (less than 328 ft). Records for greater depths, such as one for St. Paul's Rocks in the equatorial mid-Atlantic (Edwards and Lubbock 1983), may be based on occurrences of a deep-water species, *O. tumida* Lyman.

BIOLOGY: Very small *O. flaccida* reside in clumps of algae, and increasingly larger ones occur in coral and under rubble slabs (Hendler and Littman 1986). Individuals sometimes occur in water so shallow that they are exposed and die during extremely low tides (Glynn 1968). H. L. Clark (1901b) described the species as a rapidly moving brittle star, which skillfully eludes the collector by retreating into rocky interstices and by autotomizing its arms. Actually, it is neither particularly sluggish nor the most agile species. Sides (1987) found that almost half of the arms of *O. flaccida* that she sampled were regenerating and attributed the high proportion to slow regeneration. The arms of *Ophiomyxa* stiffen in response to disturbance, and their rigidity and rough spines presumably deter predators (Byrne and Hendler 1988).

Individuals are active nocturnally and emerge from crevices shortly before sunset. They crawl about for part of the night, sometimes appearing to engulf algae or other material with their open mouth. Large items of food are collected by looping of the arms, and stomach contents include sponge, algae, and detritus (Byrne and Hendler 1988; Hendler, previously unpublished; Sides, personal communication).

In Panama, males are slightly smaller and less common than females. There, between March and July, females that brood embryos have large, yolky, negatively buoyant eggs. Other females with more numerous, smaller, positively buoyant eggs do not brood (Hendler 1979a). Evidently the species has two modes of reproduction, or *O. flaccida* may have an undescribed Caribbean congener.

Infrequently, the arms of individuals from Looe Key, Florida, had an attached ostracod associate, probably *Pontocypria hendleri* (Maddocks, personal communication).

SELECTED REFERENCES: Say 1825:151 [as *Ophiura flaccida*]; Lütken 1859b:240, pl. 5, fig. 1a–d; Lyman 1865:178, pl. 2, figs. 6, 18, 19; H. L. Clark 1919:69, pl. 1, figs. 1, 2.

FAMILY GORGONOCEPHALIDAE

Asteroporpa annulata Örsted & Lütken
Figure 35

DESCRIPTION: The disk can reach 22 mm (1.2 in) in diameter, and the arms attain 180 mm (7 in) in length. The arms are not branched and are not clearly demarcated from the disk. The central body is thickened in a hump at the juncture of the disk and each arm base. The arms are annulated; raised ridges bearing microscopic hooks alternate with depressed rings paved with flattened granules. The alternating sequence of raised hooks and sunken granules is repeated on the disk humps, concealing the radial shields; large flattened granules occupy the center of the disk. The arm spines arise from the lower surface of the arm and are hook-shaped.

The disk has a "bull's eye" pattern of brown and tan rings, sometimes tinted yellow or pale green at the center. The rings continue, as a bold series of dark and light alternating bands, for the full length of the arm on the dorsal side. The bottom of the arm is light tan, sometimes with a median brown stripe.

HABITAT: Associated with hard and soft corals and with the stems of stalked crinoids,

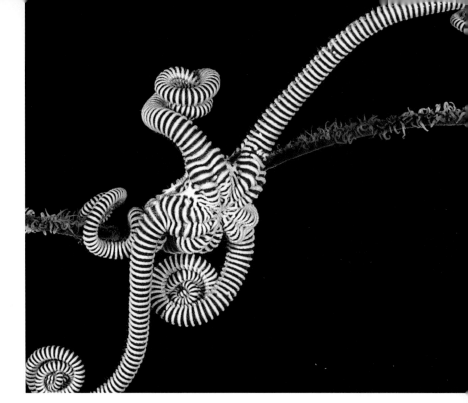

FIGURE 35. *Asteroporpa annulata*.

on hard substrate on the continental shelf and upper continental slope; also found on offshore *Oculina* coral mounds and on shipwrecks.

DISTRIBUTION: Bermuda, North Carolina, and the Gulf of Mexico, through the Bahama Islands and the Caribbean to Brazil. DEPTH: 37–305 m (121–1,001 ft); collected by scuba off North Carolina (Littman, personal communication).

BIOLOGY: Individuals are slow-moving and may be more active nocturnally than during the day. They cling, with one or more prehensile arms, to gorgonians, corals, and crinoid stems. The dorsal surface faces into the current and up to five arms are extended to feed, sometimes bent in a sinusoidal pattern. They ingest copepods and other microscopic plankters, passing prey to their mouth with their smooth, pointed tube feet. An unidentified siphonostome copepod lives in the brittle star's stomach. A 20% incidence of regenerated arms found in one population is an indication that these deep-water animals are attacked by predators (Hendler and Miller 1984a; Woodley and Emson 1988). In the Gulf of Mexico, the spawning of *A. annulata* appears to be seasonal. The size of its eggs (0.16 mm) suggests an abbreviated mode of development. Juveniles are rarely collected, but are occasionally found clinging to adults (McClintock et al. 1993).

SELECTED REFERENCES: Lütken 1856: 17; 1859b:254, pl. 5, figs. 4a–d, 5a,b [as *Asteroporpa annulata* and *Asteroporpa affinis*]; Hendler and Miller 1984a:449, fig. 1a–h. A. H. Clark's (1948) key to *Asteroporpa* species is not reliable.

Astrophyton muricatum (Lamarck)
Figures 30C-7, 30D-9, 36

DESCRIPTION: The meshlike complexity of the arms is the basis of the common name "basket star" for this and related species. The largest individuals have disks 70 mm in diameter and branched arms that span over

FIGURE 36. *Astrophyton muricatum*.

1 m (more than 3 ft) from tip to tip (Tommasi 1970). The disk and arms are covered with flattened, closely crowded granules. Microscopic hooks occur in small groups near the base of the arm, and a hook-bearing ridge straddles the dorsal surface of each distal arm segment. There are two to three arm spines on arm joints beyond the second bifurcation. Prominent, granule-free tubercles are irregularly distributed on the radial shields and on the dorsal surface of the proximal part of the arm. Each of the five arms gives rise to two thick, stubby locomotory branches and two long, slender feeding branches. Outer branches are unequal, with thick central forks and thin lateral forks arising on alternate sides of the arm. The complexity of *A. muricatum* has inspired some impressive calculations. Specimens reportedly may have 35 series of arm bifurcations (Döderlein 1911), more than 81,000 arm joints (Boone 1933), and over 10,000 terminal arm branches (Lyman 1877).

Juveniles (less than 1.5 mm [0.06 in] disk diameter) with relatively few arm branches are pink (Wolfe 1978). Small basket stars have banded arms and have yellow and white tubercles set in reddish and brown regions on the disk; such an individual is illustrated here. Large individuals are a homogeneous black, brown, light chocolate brown, yellowish brown, bright orange-yellow, tan, gray, green, or dirty white. Verrill's (1899a) observation that its color matches the host gorgonian is incorrect (Hendler, previously unpublished).

HABITAT: Reef habitats deeper than the reef flat and hard prominences on mud and seagrass flats; associated with various species of gorgonians, stony corals, fire corals, sponges, and on rocky substrate.

DISTRIBUTION: North Carolina to the Florida Keys and the Gulf of Mexico, the Bahama Islands (Hendler, previously unpublished), Cuba, Jamaica, Haiti, Puerto Rico, the Virgin Islands, the Windward and Leeward Islands, Barbados, Curaçao, and the mainland coasts of Central and South America to Brazil. Reported from Tenerife based on a specimen in the Berlin Museum (Döderlein 1911). DEPTH: 2–70 m (5–230 ft); a record from 508 m (1,668 ft) (H. L. Clark 1915) seems unlikely.

BIOLOGY: Small individuals cling to gorgonians; large basket stars hide in crevices by day, ascend to an elevated perch at dusk, and descend along the same path at dawn. The daytime residence may be shared with other invertebrates; one adult was found beneath a large brain coral with an *Ophioderma guttatum,* and juveniles have been found on gorgonians with *Schizostella bifurcata* (Hendler, previously unpublished). The feeding arm branches are extended in a parabolic array in moderately strong water currents, their tips directed into the current; the locomotor branches grip the substratum and protect the disk (Hendler and Meyer 1982). The arms rapidly coil toward the disk when illuminated or jostled.

The notion that *A. muricatum* feeds on gorgonian polyps is incorrect (Pearson 1937; Voss and Voss 1955). It is a nocturnal suspension feeder that captures planktonic prey in coiled terminal arm branches, probably securing plankters with the microscopic hooks girdling the arm (Fricke 1968; Macurda 1976; Wolfe 1982). Adults can capture items such as small fish, but 90% of their stomach contents consist of tiny copepods (Davis 1966; Fricke 1968; Wolfe 1978, 1982). The feeding branches periodically transfer plankters to the mouth (Hendler 1982a), and by morning the prey are completely digested (Wolfe 1978). Presumably, the large basket stars are rarely preyed upon, judging from the rarity of broken, regenerating arms (Wolfe 1978). It is unclear whether their nocturnal habit is a defensive adaptation or is tied to the cyclical availability of their prey.

Individuals are estimated to live over 7 years (Wolfe 1982), and an individual can occupy the same "perch" for 1–2 years (Wolfe 1982). Basket stars 22–33 mm in disk diameter are sexually mature; females have large eggs, up to 0.44 mm in diameter, throughout the year (Wolfe 1978). Their mode of development is not known, but Fricke (1968) suggested that the species has a bottom-dwelling larva.

A red, white, and yellow-brown commensal shrimp, *Periclimenes perryae* Chace, lives exclusively on *A. muricatum* (Davis 1966; Fricke 1968; Colin 1978; Criales 1984). Other shrimps (*Synalpheus* sp.), copepods, amphipods, isopods, mollusks, ophiuroids, and an internal myzostome worm are also associated with *Astrophyton* (Boone 1928, 1933; Fricke 1968). *P. perryae,* a juvenile *Branchiosyllis* sp. polychaete, and a parasitic lichomolgid copepod living in the stomach, *Critomolgus astrophyticus* Humes & Stock (Humes and Stock 1973), were all noted on *A. muricatum* at Looe Key, Florida (Hendler, previously unpublished). High numbers of the same copepod were found in the stomach of an unhealthy basket star (Williams and Wolfe-Waters 1990).

REMARKS: A "variety" of questionable systematic standing, *A. muricatum caraibica,* was described on the basis of a few, apparently juvenile, *A. muricatum* specimens (Döderlein 1911; Tortonese 1939).

SELECTED REFERENCES: Lamarck 1816: 538 [as *Euryale muricatum*]; Döderlein 1911:52, 108, fig. 11, pl. 5, fig. 1; 53, 108, pl. 5, fig. 4 [as *Astrophyton muricatum caraibica*]; Humann 1992:286–287 (figs.).

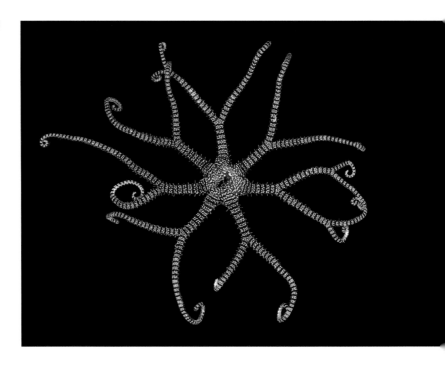

FIGURE 37. *Schizostella bifurcata*.

Schizostella bifurcata A. H. Clark
Figure 37

DESCRIPTION: This diminutive reddish brown or pinkish brown basket star is typically 4 mm (0.2 in) in disk diameter, with seven short arms that branch near the middle and sometimes again near the tip. The arm bases are swollen and poorly differentiated from the disk; proximal rows of arm hooks occur on top of the disk. Rings of elevated white granules border both sides of each darkly pigmented hook-bearing annulus on the dorsal surface of the arm. Two hook-shaped, ventrally placed arm spines are present on most arm segments. Scattered scales are embedded in the integument covering the ventral surface of the disk and arms.

S. bifurcata has multiple madreporites, distinguishing it from *A. muricatum*, which has a single madreporite. *Astrophyton* is further distinguished because its arms are clearly demarcated from the disk, and they lack ridges of hooks on the basal branch (though very small individuals may have one to two hooks per segment).

HABITAT: Fore reef walls bordering the edge of the continental slope (Hendler, previously unpublished); associated with gorgonians.

DISTRIBUTION: Elliott Key and Key Biscayne, Florida; the Cayman Islands, Cozumel, Puerto Rico, Barbados, Belize, and Colombia. DEPTH: 12–46 m (40–150 ft) (Hendler, previously unpublished).

BIOLOGY: This species lives on various species of gorgonians. During the day, individuals wrap their arms around soft coral branches (A. H. Clark 1952; Sefton 1987). At night the arms are extended in a suspension-feeding posture similar to that of *Astrophyton muricatum* (Hendler and Sefton, previously unpublished). As A. H. Clark (1952) speculated, *S. bifurcata* is fissiparous; it reproduces asexually by dividing across the disk. As a consequence, individuals usually have several large arms and several smaller, regenerating arms (Hendler, previously unpublished).

REMARKS: The variability in arm number, granule arrangement, and arm branching patterns of different populations cast doubt on the recognition of *S. bifurcata* and *S. bayeri* A. H. Clark as separate species (A. H. Clark 1952; Hendler, previously unpublished).

SELECTED REFERENCES: A. H. Clark 1952:452, pl. 40 [and as *S. bayeri*]; Sefton 1987:134–135 [as *Asteroporpa annulata*]; Humann 1992:284, 285 (fig.).

FAMILY OPHIURIDAE

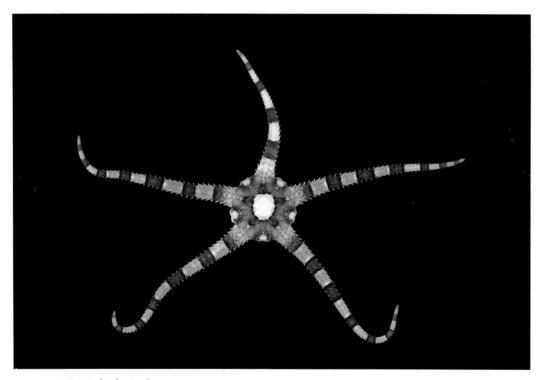

FIGURE 38. *Ophiolepis elegans*.

Ophiolepis elegans Lütken
Figures 30C-4, 30D-14, 32-5, 38

DESCRIPTION: This moderate-sized species reaches a disk diameter of 20 mm (0.8 in) and has short arms, two to three times the disk diameter. The disk is flat and smooth, and in large individuals the tilelike dorsal scales of the disk are visible to the naked eye. One column of scales lies between each pair of arms, and each major disk scale is framed by a row of microscopic scales. There are usually swollen scales at the edge of the disk, and at the outer tip of each of the paired radial shields a trio of scales is arranged in a moustache shape. The accessory dorsal arm plates are conspicuous. There are four to six arm spines. The tube feet are small, smooth, and can completely retract into the tentacle pores, where they are covered by paired, opercular tentacle scales. The disk is colored, and the arms are banded, with contrasting shades of cream color, gray, dusky green, and brown.

HABITAT: Coastal and estuarine inlets and bays, on firm shell and sand bottom or muddy sand; also sandy habitats around reefs, seagrass beds, and mangroves.

DISTRIBUTION: The Bahama Islands, North Carolina to the Dry Tortugas and the Gulf coast to Mexico, Texas offshore reefs, Cuba, Jamaica, St. Thomas, St. Martin, St. Barthélemy, Antigua, Martinique, Barbados, Trinidad, Mosquito Bank, and the coast of Central and South America from Mexico to French Guiana. Records for the coast of Africa are incorrect (Madsen 1970). DEPTH: 1–92 m (3–302 ft).

BIOLOGY: Near Florida Bay, Tabb and Manning (1961) collected 1,020 specimens in a 5-minute drag with a 2-meter otter trawl, but lower densities are often the rule. In Florida, *O. elegans* spawns from early May to late September or October. Females have as many as 16,500 greenish brown eggs, 0.25 mm in diameter. They develop into yolky, vitellaria larvae that metamorphose in 3 days. Even after several months the tiny, slow-moving juveniles have only five arm joints (Stancyk 1973). Warén (1980, 1984) described a gastropod, *Ersilia stancyki* Warén, parasitic on *O. elegans*. Up to 10 of the 2-mm-long brownish snails may attach to the disk and arm bases of a single host.

SELECTED REFERENCES: Lütken 1859b: 207; Lyman 1865:58, pl. 2, fig. 5; Boone 1933:122, pl. 74A–C, pl. 75, fig. A, pl. 76, fig. A; Madsen 1970:236, fig. 48.

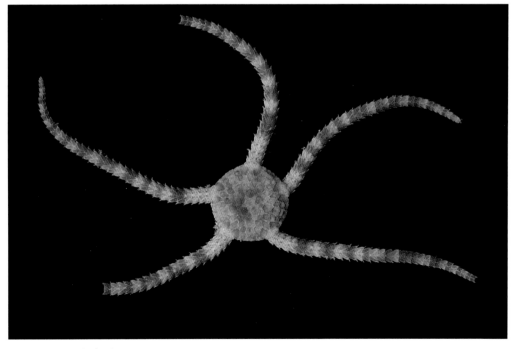

FIGURE 39. *Ophiolepis gemma*.

Ophiolepis gemma Hendler & Turner
Figure 39

DESCRIPTION: The disk diameter of this diminutive species reaches 6.2 mm (0.2 in) and the slender arms grow to 29.5 mm (1.2 in) in length. Large scales, each framed by a row of tiny scales, radiate from the center of the disk, three columns between each pair of arms. Many of the large scales are polygonal, with angled, sharply defined edges, and when moist their edges reflect light. A trio of scales in a moustache-shaped array are situated at the outer tip of each pair of radial shields. The proximal third of each arm joint is slightly constricted; the dorsal arm plates are generally separated by lateral arm plates. Accessory dorsal arm plates are present in large specimens, restricted to the first few arm joints. The lateral arm plates bear up to three tapered arm spines, and the dorsal distal arm spines are hook-shaped. As is typical for the genus, the tube feet are small and smooth; they retract into tentacle pores that are sealed by an opercular pair of tentacle scales.

The disk is pale reddish brown, changing to gray near the center; the arms are banded reddish and white. The thin disk scales reveal the dark coloration of the underlying viscera. Many of the disk and arm scales are marked with tiny orange or reddish brown spots and patches; the radial shields may have large dark spots at the distal corners. Small individuals are relatively pale, with a gray disk and whitish arms.

HABITAT: Usually on steep fore reef slopes; in clumps of calcareous algae and in sediment-filled crevices among algae, corals, and sponges.

DISTRIBUTION: Southwest Florida shelf, Belize, Barbados. The species is reported here for the first time from the Bahama Islands (Crooked Island). DEPTH: 3–139 m (10–456 ft).

BIOLOGY: What little is known about the life history of this small species was described by Hendler and Turner (1987). It often occurs with the much larger *O. impressa* and other reef-slope brittle stars such as *Ophioderma rubicundum* and *Ophiocoma wendtii*. Based on the incidence of regenerating arms, individuals seem frequently to be injured by predators or physical stress. Some specimens have abnormally enlarged ossicles on the arms or oral frame, but it is not known whether these swellings are induced by a parasite.

The species broods its young, and individuals as small as 3.3 mm in disk diameter may have young in their bursae. As many as five brooded young, all at the same stage of development, have been found in an interradius of one individual.

SELECTED REFERENCES: Hendler and Turner 1987:1, figs. 1–4, 5a–e,h, 8; Hendler 1988b:269.

Ophiolepis impressa Lütken
Figure 40

DESCRIPTION: An individual of 18 mm (0.7 in) disk diameter, with 70-mm (2.8-in) arms, approaches the maximum size for the species. On the dorsal side of the disk, rows of large, thick scales are bordered by small, irregularly shaped and arranged scales. Distal to the pairs of radial shields there are three scales arranged in a moustache shape. The four or five small arm spines, about the length of an arm joint, project from the side of the arm; the topmost is distinctly shorter and thinner than the ventral spines. The lateral edges of the dorsal arm plates are straight and diverging. The accessory dorsal arm plates, distal to the outer corners of the dorsal arm plate, are nearly microscopic and no larger than the topmost arm spine. The small, smooth tube feet are whitish. They can

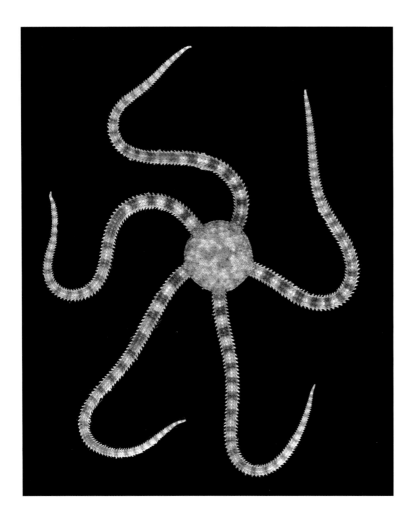

FIGURE 40. *Ophiolepis impressa*.

completely retract into the tentacle pores, which are protected by paired, opercular tentacle scales. *O. impressa* is reddish, yellowish, or greenish brown, the disk variegated with cream color and patches of brown, and the arms banded with gray and white.

HABITAT: All coral reef zones, but most often in shallow water; on sand under coral or coral rubble, and in algae.

DISTRIBUTION: The Bahama Islands, the Florida Keys, the Dry Tortugas, Texas, Cuba, Jamaica, Puerto Rico, St. Thomas, St. Croix, Guadeloupe, Barbados, Curaçao, Belize, Panama, Colombia, Venezuela, and Brazil. DEPTH: Intertidal to 24 m (79 ft); reported as deep as 549 m (1,800 ft) (Lyman 1883), but records beyond 156 m (512 ft) are probably based on misidentified specimens of *Ophiolepis ailsae* Hendler & Turner (Hendler and Turner 1987).

BIOLOGY: This species is one of the most abundant brittle stars in quantitative samples of reef rubble (Kissling and Taylor 1977; Lewis and Bray 1983). Individuals are usually immobile, but show some nocturnal activity (Sides, personal communication). It remains hidden even at night, at most exposing the tips of one or two arms, but it can be readily found under coral slabs on sand, frequently with other brittle star species (Henkel 1982; Sides 1985 and personal communication). At Looe Key, Florida, it occurred in branching corals with *Ophiothrix angulata* and under coral slabs with *Ophionereis reticulata*. Experiments have shown that the latter species is competitively superior, displacing *O. impressa* from shelters in several days (Sides 1985). This might be expected, considering the relatively small size of the tube feet of *O. impressa* and its sluggish rate of locomotion. The species is a deposit feeder and may use its arms and tube feet to gather particles above and below the surface of the sediment. Its stomach contents consist of algae, feces, and mucus-bound sediment (Sides, personal communication).

This species had the lowest value of arm autotomy among eight shallow-water brittle stars investigated; its rate of arm regeneration was relatively slow as well (Sides 1982, 1987). Aronson (1988) found that *O. impressa* is moderately palatable to some fish. However, its robust, heavily calcified arms must, to some degree, protect it from predators. Hendler and Turner (1987) pointed out that the minute hooked spines near the arm tips are more likely used for feeding than for defense. Individuals in the field are often covered with mucus-bound silt, but the function and significance of the coating of sediment are unknown. Small specimens coil their arms on top of the disk, in the same manner, and presumably to the same ends, as adult *O. paucispina* (see below). The precise mode of reproduction of this common species is not known, but based on the diameter of the yolky eggs (0.2 mm), it may have abbreviated larval development (Hendler 1979a).

SELECTED REFERENCES: Lütken 1859b: 203, pl. 2, fig. 3a,b; Lyman 1865:64, fig. 4 [as *Ophiozona impressa*]; Tortonese 1934: 45, pl. 2, fig. 3, pl. 8, figs. 54–55; Devaney 1974b:144, figs. 13–15.

Ophiolepis paucispina (Say)
Figure 41

DESCRIPTION: *O. paucispina* is a "pygmy" species. It is best identified under magnification, because it resembles the juveniles of larger *Ophiolepis* species and the adults of *O. gemma*. A typical individual of 4.5 mm (0.2 in) disk diameter has arms 9 mm (0.4 in) long, but it may reach 7 mm (0.3 in) in disk diameter with arms 22 mm (0.9 in) long. From the center of the disk, columns of large, regularly arranged scales radiate outward. These thick-edged major scales are separated from each other by continuous rows of small, close-set scales, imparting a symmetrical, faceted appearance to the disk. There is a trio of scales, in a moustache-shaped array, at the outer ends of the pairs of radial shields, as in other *Ophiolepis* species. The arm joints are swollen distally. The two conical arm spines are much shorter than an arm joint. The sides of dorsal arm plates are truncated, and an accessory dorsal arm plate, at least as wide as the dorsal arm spine length, fills the gap between the dorsal and lateral arm plates on most arm joints. The small, smooth tube feet can be retracted into the tentacle pores and completely covered by paired, opercular tentacle scales.

The colors of *O. paucispina* "harmonize well with the sand" (H. L. Clark 1933:74). The white, brownish, or bluish gray disk may have a few gray, greenish, or brown spots, and the pale arms often have several brown or gray bands.

HABITAT: Shallow, sandy reef flats, mangrove, lagoonal, and seagrass environments; under coral rubble on sand, in calcareous algae such as *Halimeda*, and among plant debris.

DISTRIBUTION: Bermuda, the Bahama Islands, the Florida Keys and the Dry Tortugas, Texas, Jamaica, Haiti, Puerto Rico, St.

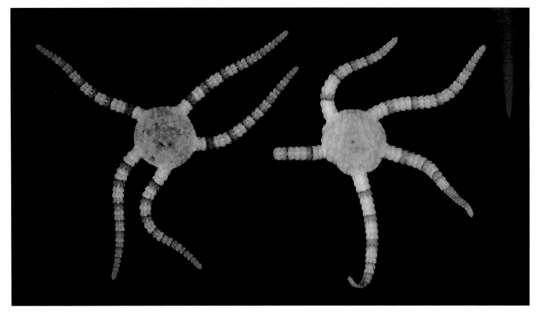

FIGURE 41. *Ophiolepis paucispina*.

Thomas, St. Barthélemy, Antigua, Barbados, Tobago (Trinidad?), Curaçao, Aruba, Mexico, Belize, Panama, Colombia, and Brazil, including Isla da Trindade. Off Africa, it is reported from the Canary Islands to south of the equator, but there are no reliable records from Ascension Island (Pawson 1978) and other mid-Atlantic islands. DEPTH: 1–37 m (2–120 ft).

BIOLOGY: This species has been described as extremely secretive, but it may be "fairly common, after one has learned where and how to find it" (H. L. Clark 1942:379). Hendler and Littman (1986) counted more than 100 individuals per liter of algae, and Lewis and Bray (1983) estimated that it was the tenth most abundant brittle star species on Barbados reefs.

It is an ovoviviparous, simultaneous hermaphrodite, with pink yolky eggs that reach 0.4 mm in diameter. Brooding individuals, with up to 41 embryos, were found during most of the year in Panama (Hendler 1979b; Byrne 1988, 1989, 1991). More than one "clutch" is brooded at a time, and the embryos within a parent are at different stages of development. Byrne (1988, 1989) has shown that oogenesis proceeds continuously, and she described the origin and deposition of egg yolk in the species.

O. paucispina (and small *O. impressa*) coil the arms over the disk when disturbed (Hendler and Turner 1987). This behavior may be analogous to the reflexive "balling" reaction (Emson and Wilkie 1982) of *Amphipholis squamata*. It "freezes" when its shelter is dislodged, and with arms coiled it quickly sinks like a pebble (Hendler 1979b, previously unpublished; contra H. L. Clark 1942).

REMARKS: Madsen (1970) discussed differences between eastern and western Atlantic specimens.

SELECTED REFERENCES: Say 1825:149 [as *Ophiura paucispina*]; Lütken 1859b:204, pl. 2, fig. 2a,b; Koehler 1914b:177, pl. 9, fig. 14.

FAMILY OPHIOCOMIDAE, SUBFAMILY OPHIOCOMINAE

Ophiocoma echinata (Lamarck)
Figures 30C-1, 32-6, 42, 43

DESCRIPTION: Specimens of this robust species can measure 32 mm (1.3 in) in disk diameter with arms 150 mm (5.9 in) long. Like other *Ophiocoma* and *Ophiocomella* species, it has a granule-covered disk, and the jaws bear a cluster of dental papillae and series of oral papillae. Two characteristics readily distinguish the species from Caribbean congeners: white tube feet and thickened or bulbous dorsal arm spines. The dorsal arm spines may be longer or shorter than adjacent spines near the disk, and they are the longest spine near the arm tip. There are two tentacle scales on arm joints beyond the disk, three spines on the first and second arm joints, and different numbers of arm spines on opposite sides of arm joints beyond the disk.

The disk usually is irregularly, and sometimes boldly, patterned (Figure 43), with a combination of black, brown, and gray; the arms are variegated or banded with the same hues. Juveniles are almost entirely black, with white-tipped radial shields and several white bands on the arms; they first develop disk granules at 2.3 mm (0.1 in) disk diameter.

HABITAT: All reef zones, seagrass beds, and mangroves; particularly abundant in rubble.

DISTRIBUTION: Bermuda, the Bahama Islands, Florida (Fort Pierce to the Florida Keys and the Dry Tortugas), Cuba, the Cayman Islands, Jamaica, Haiti, Puerto Rico, the Virgin Islands, the Leeward and Windward Islands, Barbados, Tobago, Isla La Tortuga, the Netherlands Antilles, Mexico, and the coast of Central and South America to Brazil. There

FIGURE 42. *Ophiocoma echinata*.

FIGURE 43. *Ophiocoma echinata*: detail of the disk of a boldly patterned individual.

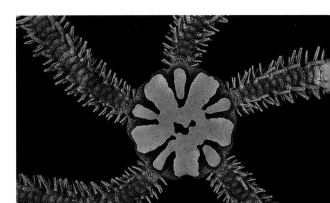

are questionable records for Liberia (Koehler 1907) and Ascension Island (Koehler 1914a). Reports of this species from Africa are incorrect (Devaney 1970). DEPTH: Intertidal to 24 m (79 ft).

BIOLOGY: This brittle star is a prominent denizen of coral rubble habitat throughout the Caribbean (Grave 1898b; H. L. Clark 1901a, 1933; Kissling and Taylor 1977; Bray 1981; Lewis and Bray 1983; Aronson 1993). Most individuals live on the back reef (Hendler and Peck 1988). Some in shallow intertidal areas can be killed by elevated temperatures during midday low tides (Glynn 1968). The juveniles occur in filamentous algae with *Ophiostigma isocanthum*, *Amphipholis squamata*, *Ophionereis squamulosa*, and *Amphiodia pulchella*, at Looe Key, Florida; a similar suite of alga-dwelling species occurs in Belize (Hendler and Littman 1986; Hendler, previously unpublished).

In 1921, A. H. Clark (1921b:44) wrote that *O. echinata* was referred to as "The Sea Scorpion," just as it had been in an account of Barbados from 1750. He stated that in 1903 it was the "most abundant and conspicuous form of animal life" under rubble; even now it occurs in densities up to about 30 individuals per square meter off Barbados (Bray 1981; Lewis and Bray 1983). There are similar long-term records of its abundance in Jamaica (see Grave 1898b; Fontaine 1953).

The permanence of populations may reflect this brittle star's physical durability. Sides estimated that a hurricane, which destroyed a swath of Jamaican reef, caused only a 10–17% level of arm loss in *O. echinata* (Woodley et al. 1981). Aronson (1993) found that the density of backreef populations was not altered by hurricanes. Individuals with relatively short arms are found in turbulent areas (Bray 1981), but the arms may have been shortened by breakage and autotomy. The species is eaten by some fish predators (Randall 1967; Reinthal et al. 1984; Aronson 1988), and a 35% incidence of regeneration indicates that arm cropping can cause a severe energetic drain (Sides 1982, 1987). An arm of *O. echinata*, broken at the disk, takes about 10 months to regenerate fully (Sullivan 1988; but see Stockard 1909).

This species has a pattern of diurnal activity and color change similar to that of its Caribbean congeners (Hendler 1984b; Sides and Woodley 1985; Hendler and Byrne 1987); adults are considerably paler at night than during the day. *O. echinata* rarely migrates from shelter and often occurs beneath the same slabs as *O. pumila*, *O. wendtii*, and *Ophioderma appressum* (Hendler, previously unpublished), but aggressive defense and competiton for burrow space between *O. echinata* and *O. wendtii* have been reported (Sides 1985).

This brittle star's arm tips are rarely seen during the day, but feeding individuals often expose several arms between dusk and dawn. The arms curve upward and are moved from side to side, intercepting plant and animal materials and detritus. The stomach contents mostly consist of sand and pieces of fleshy algae (Bray 1975; Hendler and Meyer 1982). Some rapacious individuals feed on eggs from damselfish nesting sites (Itzkowitz and Koch 1991).

Within a population, individual *O. echinata* have asynchronous breeding cycles (Mladenov 1983). The spawning season may shift from the rainy season (in Panama) to earlier, warmer months at more northerly island localities (Grave 1898a; Mortensen 1931; Hendler 1979a; Mladenov 1983; van Veghel 1993). The dark reddish ovaries of an individual 22 mm (0.9 in) in disk diameter contain an estimated 888,000 eggs, each about 0.07 mm in diameter (Schoener 1972; Mladenov 1983). Spermatozoan ultrastructure, fertilization (Hylander and Summers 1975), and the prickly fertilization membrane (Grave 1898a; Mortensen 1921; Devaney 1970) of the species have been described.

The ophiopluteus larva has been reared to a six-armed stage, but development through metamorphosis is as yet undescribed (Mortensen 1921).

Branchiosyllis exilis (Gravier), a polychaete, is sometimes associated with *O. echinata* (Hendler and Meyer 1982). *Ophiopsyllus reductus* Stock, Humes & Gooding, a cancerillid copepod, infests *O. echinata* and *Ophiocomella ophiactoides* (Stock et al. 1963). The "association" mentioned between *O. echinata* and the sea urchins *Echinometra viridis* and *E. l. lucunter* by Nutting (1919) is probably coincidental, and the story that exposure to *O. echinata* is harmful to octopus (Burrage 1964) seems without substance. Other accounts of this relatively well-studied species deal with its arm vertebrae microstructure (Bray 1985), lethal temperature limits (Singletary 1971), respiration (Petersen 1976), and locomotion (Cowles 1910).

SELECTED REFERENCES: Lamarck 1816: 543 [as *Ophiura echinata*]; A. H. Clark, 1939a:450, pl. 54, fig. 4; Hendler 1984b: 380, fig. 1E,F.

Ophiocoma paucigranulata Devaney
Figure 44

DESCRIPTION: Individuals can attain a disk diameter of 20 mm (0.8 in); one that is 17 mm (0.7 in) in disk diameter has arms 83 mm (3.3 in) long. The species has thin, tapering, cylindrical dorsal arm spines that are equal in length to several arm joints, similar to those of *O. wendtii*. Both species, typical of Ophiocominae, have dental and oral papillae and disk granules. However, in *O. paucigranulata* the distal portion of each small radial shield is free of scales and granules; the ventral surface of the disk is free of granules, whereas the interradii of *O. wendtii* bear a V-shaped region covered with granules. The dorsal arm plates of *O. paucigranulata* are symmetrical and bear the same number of spines on both sides of each arm joint, in contrast to the asymmetrical arrangement in *O. wendtii* and *O. echinata*. There are two tentacle scales on arm joints beyond the edge of the disk, three arm spines on the first and second arm joints, and the ventral arm spines are slightly compressed.

The disk is a dark brownish black, and

FIGURE 44. *Ophiocoma paucigranulata*.

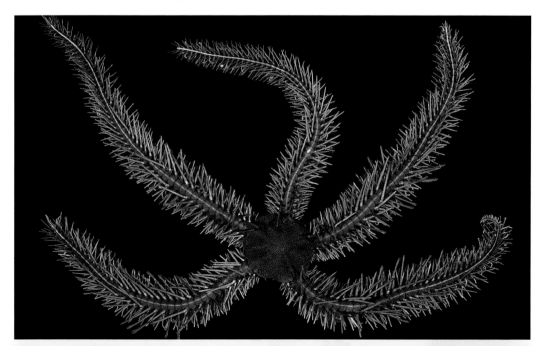

the arms have a thin, pale stripe that is most apparent near the arm tip. Juveniles, as well as adults, can be distinguished readily from O. *wendtii* of similar size by their striped arms and by the number of spines on the first arm joint. The tube feet are pale yellowish, orange, or tan rather than red as in O. *wendtii*.

HABITAT: Shallow back reef to the fore reef slope, but considerably more abundant in deeper reef zones; in the interstices of rubble, corals, and algae.

DISTRIBUTION: Previously reported only from Belize. A juvenile specimen has been collected at Looe Key, Florida; adults have been found in the Bahama Islands (Hendler, previously unpublished) and photographed off the Cayman Islands (Humann, personal communication). DEPTH: Less than 1–24 m (less than 1–79 ft); rarely occurring in shallow water, the species may be part of the Caribbean "deep reef fauna" (Hendler and Peck 1988).

BIOLOGY: The juveniles live in clumps of algae, larger individuals occur in branching corals, and the largest in rubble (Hendler and Littman 1986). Adults occupy spacious cavities, often living in the same crevices as O. *wendtii* and O. *echinata* (Devaney 1974b; Hendler 1984b). This species is active nocturnally, extending its arms from concealment at dusk and withdrawing them at dawn. The arms are raised off the substratum with at least the distal 1 cm of the arm curled upward, probably a suspension-feeding attitude. The species reacts to light by quickly withdrawing from illumination. It undergoes reversible diurnal color change in the same manner as O. *wendtii*. During the day it is brownish black with an indistinct light brown stripe on the arm, and at night it turns a dark gray, the arm stripe becoming a contrasting pale gray or white. O. *paucigranulata* is preyed upon by fish. It "locks" its long spines defensively and produces mucus for feeding and defense, similarly to O. *wendtii* (Hendler 1984b; Hendler and Byrne 1987).

SELECTED REFERENCES: Devaney 1974b:132, figs. 5–8; Hendler 1984b:380, figs. 1c,d.

Ophiocoma pumila Lütken
Figure 45

DESCRIPTION: A large individual, 17 mm (0.7 in) in disk diameter, has arms almost 140 mm (5.5 in) long. This species is the smallest ("pumila" means dwarf), most delicate, and most pallid Caribbean *Ophiocoma*. Like its congeners, it has jaws bearing both dental and oral papillae, and its disk is granule-covered. In this species the largest, longest arm spine is one of the middle arm spines. The species is further distinguished by elongate granules along the edge of the disk, single tentacle scales (except on a few proximal arm joints), and three arm spines on the first and second arm joints. Some individuals have club-shaped, integument-covered ventral arm spines on one or a few proximal arm joints; these are not, however, a consistent and reliable taxonomic characteristic (Hendler, previously unpublished). The tube feet are nearly transparent.

The disk of O. *pumila* is variegated with a combination of brown, reddish brown, tan, yellow, green, gray, or white; its arms are banded. Small individuals often are boldly colored with green and reddish pigmentation, and the arm tips and regenerating arms of adults are frequently green.

HABITAT: All reef zones, but more common in shallow habitats than on the fore reef slope.

BRITTLE STARS AND BASKET STARS

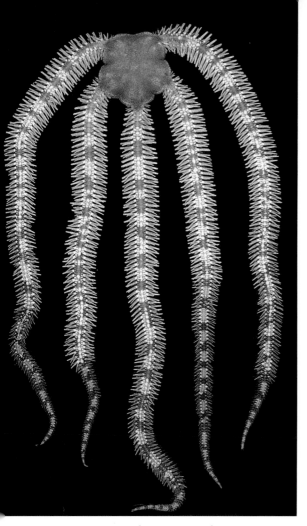

FIGURE 45. *Ophiocoma pumila*.

DISTRIBUTION: Bermuda, the Bahama Islands, Florida (Fort Pierce to the Florida Keys), the Dry Tortugas, Cuba, Jamaica, Pedro Bank, Haiti, Puerto Rico, the Virgin Islands, the Leeward and Windward Islands, Barbados, Tobago, Isla La Tortuga, the Netherlands Antilles, and the coasts of Central and South America to Brazil. Also known from the Azores, the Cape Verde Islands, Gold Coast, and São Tomé (Gulf of Guinea) in Africa. Some of these records are open to doubt because *O. pumila* has so often been confused with *Ophiocomella ophiactoides*. DEPTH: Usually intertidal to 24 m (80 ft), but reported from 368 m (1,206 ft) (e.g., Koehler 1914a; Madsen 1970).

BIOLOGY: This species is the first or second most abundant brittle star on many shallow Caribbean reefs (Kissling and Taylor 1977; Lewis and Bray 1983; Hendler and Peck 1988; Aronson 1993). Large individuals occupy crannies in rock and beneath corals or rubble resting on sand, and smaller individuals often live in branching coral. The very smallest *O. pumila* live in clumps of algae with a variety of other small brittle stars (Hendler and Littman 1986). Their density may exceed 100 per liter of algae (Emson et al. 1985a; Hendler and Littman 1986). Adults inhabit lower, shorter burrows than *O. wendtii* and *O. echinata* (Sides and Woodley 1985), and they can alter crevice dimensions, competitively excluding *Ophionereis reticulata* and *Ophiolepis impressa* (Sides 1985). The color patterns, particularly of the juveniles, may conceal them from predators (H. L. Clark 1933), but that possibility has not been tested.

O. pumila is the least light-sensitive of the Caribbean *Ophiocoma* species, but conceals itself in crevices (Hendler 1984b; Hendler and Byrne 1987). Between dusk and dawn, individuals extend two or three arms from crevices; the arm tip may be held nearly vertically to suspension feed or kept in contact with the sediment to deposit feed (Hendler 1984b; Sides and Woodley 1985). Its stomach contents consist of calcareous particles and pieces of fleshy algae. Microscopic, meshlike ossicles are embedded in the wall of the stomach (Irimura 1991).

In Panama, *O. pumila* spawns year-round (Hendler 1979a). Its pinkish red eggs are about 0.07 mm in diameter (Jordan 1908; Devaney 1970; Mladenov 1985a). Mladenov (1985a) discovered that after 79 days the four-armed ophiopluteus transforms into an armless, secondary larval stage. Thus, the long-lived larva is capable of long-distance dispersal, possibly accounting for the distribution of the species on both sides of the Atlantic.

Arm-cropping predators, including fish and crabs, are a serious hazard to O. *pumila* (Sides 1982, 1987), and the whole brittle star, rather than just the arm tips, is consumed by wrasses, parrotfish, and mojarras (Hendler 1984b). At a given time, almost 35% of the arms in a population are regenerating, growing at a rate of 1–2 cm per month (Morgulis 1909; Sides 1982, 1987). Its vulnerability may be a correlate of fewer highly acidic mucous glands compared with other species (Hendler 1984b; Sides 1987; Aronson 1988). The species may be less susceptible to environmental stress than to predation. Aronson (1993) found that the densities of back reef populations were not diminished by hurricanes. A scaleworm, *Hermenia verruculosa* Grube, is sometimes associated with O. *pumila* but is not an obligate commensal (Devaney 1974b; Pettibone 1993).

REMARKS: Minor morphological differences distinguish individuals living in eastern and western Atlantic waters (A. M. Clark 1955; Madsen 1970).

SELECTED REFERENCES: Lütken 1859b: 248, pl. 4, figs. 5a–c; Parslow and Clark 1963:37, fig. 11g,h; Hendler 1984b:380, fig. 1G,H.

Ophiocoma wendtii Müller & Troschel
Figures 30D-12, 46

DESCRIPTION: This species grows to 35 mm (1.4 in) in disk diameter and has arms longer than 176 mm (6.9 in). The most dorsal arm spines are the longest, usually equal in length to four to five arm joints. They are thin, irregularly cylindrical, and similar in dimensions to the dorsal arm spines of O. *paucigranulata*, but usually markedly thickened at the tip. As noted under O. *paucigranulata*, the disk of O. *wendtii* has a relatively dense cover of granules. It has a single tentacle scale

FIGURE 46. *Ophiocoma wendtii*.

on the joints beyond the disk, two arm spines on the first and three on the second arm joints, and alternating numbers of spines on successive arm joints.

The species has a characteristic black to reddish brown coloration and banded arm tips. Its red tube feet impart a reddish hue to the ventral surface of the arms. A radiating black pattern on the disk, and black bands on the arms, are more pronounced in small and pale individuals than in large, dark ones.

Juveniles are so unlike the adults that they may be mistaken for amphiurids (H. L. Clark 1922; A. H. Clark 1939a) or ophiacanthids (H. L. Clark 1922). They have only a pair of papillae, resembling amphiurid infradental papillae, at the tip of the jaw. As the individual grows, these are incorporated into a cluster of dental papillae, and flanked by oral papillae. The very smallest specimens have scattered, short spines on the disk; at 1–4 mm disk diameter they have a homogeneous black disk and uniformly orange or salmon-colored arms; those 2–7 mm (0.1–0.3 in) have only scales on the disk; disk granules and arm banding appear in larger individuals (Hendler, previously unpublished).

HABITAT: All coral reef zones, mangroves, and seagrass beds; beneath rubble slabs and rocks, in coral colonies, in algae, and under sponges.

DISTRIBUTION: Bermuda, the Bahama Islands, the Florida Keys, the Dry Tortugas, Texas offshore reefs, Cuba, Jamaica, Haiti, Puerto Rico, the Virgin Islands, the Leeward and Windward Islands, Barbados, Tobago, the Netherlands Antilles, Mexico, and coasts of Central and South America to Brazil. DEPTH: Usually less than 1–27 m (less than 1–89 ft); also reported from 384 m (1,260 ft) (Verrill 1899b; Koehler 1914a).

BIOLOGY: *O. wendtii* is one of the most abundant brittle stars on coral reefs of Florida and the Caribbean (Milliman 1969; Kissling and Taylor 1977; Lewis and Bray 1983; Hendler and Peck 1988). Nearly as common as *O. echinata*, both species often occur together beneath coral, rubble, or sponges (H. L. Clark 1933; Fontaine 1953), frequently living alongside other brittle star species.

Its behavior and ecology were described by Hendler (1984b) and Sides and Woodley (1985). *O. wendtii* tends to occupy higher, longer crevices beneath coral and rubble than *O. echinata* and *O. pumila*. Individuals cling to the upper side of the shelter during the day, protrude two to three arms from the burrow at dusk, and withdraw them at dawn. The extended arms capture floating particles, algal epiphytes, and benthic "scuzz" (Cowles 1910), and the presence of food pellets on the arms during the day suggests that they are suspension or deposit feeding within the crevice (Hendler, previously unpublished). They can occupy the same shelter for many months and may exhibit diurnal homing activity and aggressively defend burrows (Sides 1985), but their behavioral repertoire can vary markedly among populations. Their long, erect arm spines can lock defensively, preventing the removal of an individual from its crevice, and they secrete a sticky mucus. The locomotion of *O. wendtii*, its "righting response," and its ability to climb on glass were described by Cowles (1910).

The color patterns of individual *O. wendtii* are unvarying, but individuals undergo a cyclical change from brownish by day to a grayish brown with distinct black arm bands and disk markings at night. This species is more light-sensitive than other Caribbean congeners (Cowles 1910; Hendler 1984b), and its sensitivity is related to the daily color-change process (Hendler 1984b; Hendler and Byrne 1987; Cobb and Hendler 1990). Photosensitivity in this, and other *Ophiocoma* species, is mediated by expanded peripheral trabeculae (EPT) that overlie sensory cells located within the external ossicles of the arm.

O. wendtii is preyed upon by at least three species of West Indian reef fish (Randall 1967; Hendler 1984b). Arm-cropping by predators takes a toll, and 30% of the arms in a population may be regenerating at a given time (Stockard 1909; Sides 1982, 1987).

H. L. Clark (1933) noted that *O. wendtii* and *O. echinata* live together, yet never hybridize. He suggested that staggered breeding seasons prevent interbreeding, noting that in Jamaica *O. wendtii* breeds in May or earlier, and *O. echinata* does not spawn until July (Grave 1898a,b). However, in Panama both species spawn in September and October, and there are variations in spawning periods among the different populations of *O. wendtii* (Hendler 1979a). Thus, there must be barriers to hybridization other than temporal isolation. The ultrastructure of spermatozoa and fertilization of the species have been investigated (Hylander and Summers 1975), but the barriers to hybridization among sympatric *Ophiocoma* species remain unknown.

REMARKS: *O. wendtii* is not found outside the western Atlantic region, but the name of an Indo-Pacific relative, *Ophiocoma riisei*, has mistakenly been applied to this species (Devaney 1970).

SELECTED REFERENCES: Müller and Troschel 1842:99; Lütken 1859b:245, pl. 4, figs. 6a–d; A. H. Clark 1939a:450, pl. 54, fig. 5 [as *Ophiocoma riisei*]; Hendler 1984b: 380, fig. 1A–B [as *O. wendti*].

Ophiocomella ophiactoides
(H. L. Clark)
Figure 47

DESCRIPTION: This is a small six-armed species, usually no more than 5 mm (0.2 in) in disk diameter, with arms 19 mm (0.7 in) long. It has a granule-covered disk and the combination of dental and oral papillae typical of *Ophiocoma* species. It closely resembles *Ophiocoma pumila*, and it was long thought to be the young, asexually reproducing stage of the latter species (A. H. Clark 1939b; Parslow and Clark 1963; Devaney 1970). Several features have been suggested to distinguish them (Parslow and A. M. Clark 1963; A. M. Clark 1976a,b; Devaney 1970), but most reliable is the number of arms. Six-armed *O. ophiactoides* and five-armed *O. pumila* are typical. For example, only one six-armed *O. pumila* was found among more than 200 individuals examined from Looe Key, Florida (Hendler, previously unpublished). Individuals are a cream color or yellowish brown, variegated with reddish brown and green; the arms are banded with dark and pale green and reddish brown. The animal illustrated here is somewhat darker and more purple than typical.

O. pumila, unlike *O. ophiactoides*, has middle arm spines that are much larger than the dorsal and ventral arm spines, and the increasing length of its arm spines to the twelfth arm joint is more marked than in *O. ophiactoides* (Parslow and Clark 1963). Some differences between the species may be a result of repeated fission and regeneration of the disk and arms in *O. ophiactoides*, but that is not the case for a microscopic internal canal in its jaw (Devaney 1970), nor its distinctive larval stage.

HABITAT: All reef zones, except perhaps the fore reef slope (Hendler and Peck 1988); in rubble, coral, and algae.

DISTRIBUTION: Bermuda, the Dry Tortugas, the Florida Keys, the Bahama Islands (Hendler, previously unpublished), Jamaica, Puerto Rico, St. Martin, St. Barthélemy, Barbados, Tobago, Bonaire, Curaçao, Belize, and Venezuela. DEPTH: Less than 1–18 m (less than 1–60 ft).

BIOLOGY: *O. ophiactoides* is abundant, but seldom seen, because it is small, cryptically colored, and repelled by light (Emson et al. 1985a). In algal turf it may occur in excess of 100 individuals per liter, along with a fairly predictable suite of other small brittle stars, including fissiparous *Ophiactis* and *Ophiostigma* species (Henkel 1982; Emson et al. 1985a; Hendler and Littman 1986; Mladenov and Emson 1988).

Wilkie et al. (1984) and Mladenov et al. (1983) described the process of asexual reproduction in *O. ophiactoides*, suggesting that fission, which involves the weakening of connective-tissue bonds, is activated by the

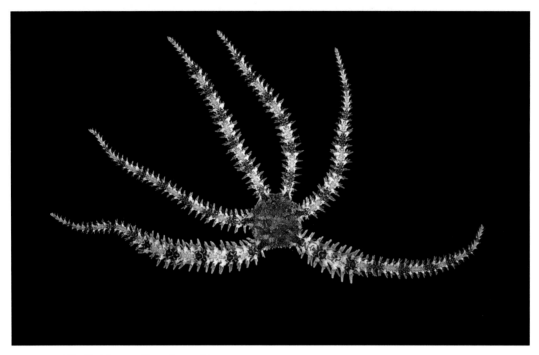

FIGURE 47. *Ophiocomella ophiactoides*.

nervous system. According to Mladenov and Emson (1984, 1988), gonads are present in some recently split individuals, indicating that sexual and asexual reproduction could occur concurrently. Only large individuals are capable of sexual reproduction; small individuals reproduce asexually. Asexual reproduction, the predominant reproductive mode, may occur year-round and results in large, genetically homogeneous clones (Hendler and Littman 1986; Mladenov and Emson 1988). Based on an estimated 3-month interval between fissions, one individual could give rise to 15 brittle stars in a year (Mladenov et al. 1983).

Females have a maximum of 46 pink ovaries, and over 7,000 tiny eggs (0.08 mm in diameter). Males tend to be smaller than the females (Mladenov and Emson 1984, 1988). Spawning individuals lift their disks off the bottom in a posture typical of spawning brittle stars (Hendler and Meyer 1982). The ophiopluteus larva develops four pairs of larval arms after 26 days (Mladenov and Emson 1984, 1988).

A copepod, *Ophiopsyllus reductus* Stock, Humes & Gooding, is a host-specific, external parasite of *O. ophiactoides*. It can visibly damage and suppress reproduction of the brittle star (Emson et al. 1985b; Emson and Mladenov 1987b).

REMARKS: It has been suggested, on the basis of some morphological and genetic evidence, that the genus *Ophiocomella* should be merged with *Ophiocoma* (Parslow and Clark 1963; Mladenov and Emson 1990).

SELECTED REFERENCES: H. L. Clark 1901a:249, pl. 15, figs. 5–8 [as *Ophiacantha ophiactoides*]; H. L. Clark 1918:265, pl. 7, fig. 5 [as *Ophiacantha oligacantha*]; A. H. Clark 1939a:451, pl. 54, fig. 3 [as *Ophiocoma pumila*]; Parslow and A. M. Clark 1963:37, fig. 11a–f; Devaney 1970: fig. 43.

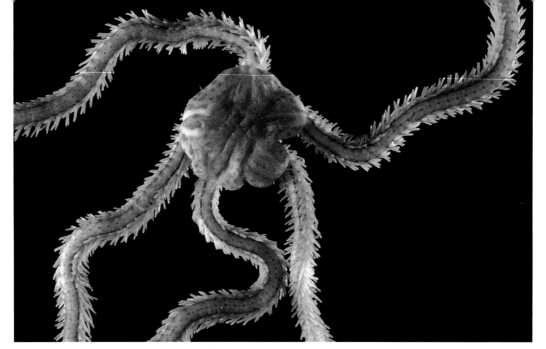

FIGURE 48. *Ophiopsila hartmeyeri*. Photo by G. Hendler.

FAMILY OPHIOCOMIDAE, SUBFAMILY OPHIOPSILINAE

Ophiopsila hartmeyeri Koehler
Figure 48

DESCRIPTION: This brittle star has the small disk, long arms, and elongate, heavily ciliated inner tentacle scale typical of *Ophiopsila* species. An individual of 6 mm (0.2 in) disk diameter has arms about 60 mm (2.4 in) long. *O. hartmeyeri* has both oral and dental papillae, like all Ophiocomidae, but its disk lacks the granules found in *Ophiocoma* species. It differs from other *Ophiopsila* species of this region by having up to eight flattened, blunt arm spines. The top two to three spearhead-shaped spines have a rounded tip; they are decidedly broader than the slender, delicate, more ventral spines.

O. hartmeyeri has been described as having a reddish or orange-yellow disk, but a living individual from Cozumel had a dark gray disk with scattered deep reddish brown patches and white radial shields. The pale tan arms may have dark median and lateral stripes and irregularly spaced, darkly pigmented joints. A discontinuous stripe on the lateral arm plates consists of series of small black spots. There may be pairs of dark patches on the ventral arm plates and beside the oral plates.

HABITAT: The fore reef, and offshore; in sandy sediment.

DISTRIBUTION: Pompano, Florida (based on LACM specimens), the Florida Keys, St. Thomas, Montserrat, Barbados, Cozumel, Venezuela, and Brazil. DEPTH: 12–161 m (40–528 ft).

BIOLOGY: Based on observations made at Paraiso Reef, Cozumel, *O. hartmeyeri* and *O. riisei* may occur at the same locality, but the former species burrows in soft sediment, and the latter in hard substrates (Fenner, personal communication).

SELECTED REFERENCES: Koehler 1913: 368, pl. 21, figs. 7, 8; de Roa 1967:293, fig. 23.

Ophiopsila riisei Lütken
Figure 49

DESCRIPTION: A large specimen measures 12 mm (0.5 in) in disk diameter and has arms 165 mm (6.5 in) long. This species, like other ophiocomids, has both dental papillae and oral papillae; its specialized inner tentacle scale is typical of the genus *Ophiopsila*. *O. riisei* differs from its congeners in having up to seven thick, flattened arm spines. The two most dorsal spines are almost ellipsoidal, short and broad, with a very blunt tip. The more ventral spines are relatively slender; the most ventral spine is longest. Thick integument obscures the disk scales, except for the radial shields and a group of small scales proximal to them.

The disk is blotched and the arms are irregularly banded with cream color and shades of gray, reddish, yellowish, maroon, or purplish brown. The arm bands may be considerably more pronounced than in the individual depicted here. There are black spots on the dorsal and ventral sides of the disk and arms, and generally there is a small black spot on the lateral arm plate near the base of each arm spine.

HABITAT: From the reef crest to the fore reef slope, in coral and rubble; also found in mangrove peat banks and in clumps of algae.

DISTRIBUTION: The Bahama Islands, Florida (Palm Beach to the Florida Keys), the Dry Tortugas, Cuba, Haiti, Puerto Rico, St. Thomas, St. Barthélemy, Tortola, Barbados, and the coasts of Central and South America

FIGURE 49. *Ophiopsila riisei*.

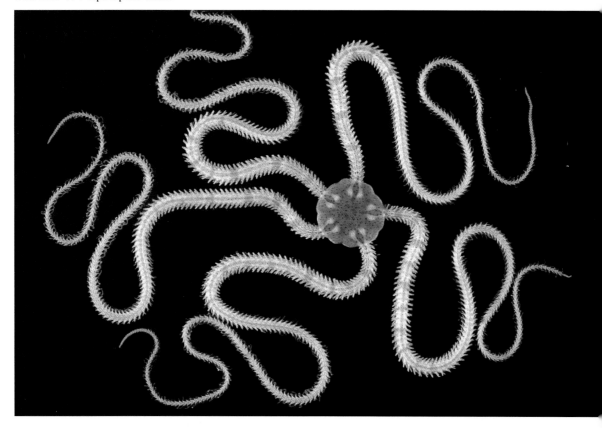

to Brazil. DEPTH: Reportedly less than 1–366 m (less than 1–1,200 ft), but about 91 m (300 ft) is a confirmed lower limit.

BIOLOGY: In this region, *O. riisei* is the most common *Ophiopsila* species living in hard substrates. At night, individuals hold several arms stiffly extended from small crevices and may give corals a hirsute appearance; they remain hidden during the day. After dark, they attract attention with a brilliant luminescent display that is triggered when the arms are touched. The luminescence and fluorescence emanate from the arm spine nerve ganglia (Emson and Herring 1985). Grober (1988a,b, 1989) has shown that luminescence, coupled with unpalatability of the species, deters predation by shrimps and crabs. In response to artificial illumination, these light-sensitive brittle stars pull their arms into sinusoidal coils before completely withdrawing into the substratum (Hendler, previously unpublished; Hendler and Byrne 1987).

The size of its egg (0.23 mm in diameter) is consistent with a lecithotrophic pattern of development (Hendler and Littman 1986), but the larva of *O. riisei* has not been reared or identified. Holmquist (1994) reported that in Florida Bay, Florida, individuals are dispersed in clumps of drift algae. The upper lethal temperature for this species is about 40°C (Singletary 1971). Hotchkiss (1982) found a specimen of *Amphiura stimpsonii* clinging to an *O. riisei*, but the association was probably incidental.

SELECTED REFERENCES: Lütken 1859b: 238, pl. 5, figs. 2a–c; Lyman 1865:150, figs. 16, 17; Koehler 1913:368, pl. 21, fig. 9.

FIGURE 50. *Ophiopsila vittata*. Photo by G. Hendler.

Ophiopsila vittata H. L. Clark
Figure 50

DESCRIPTION: A specimen with a 7 mm (0.3 in) disk diameter has arms 80 mm (3.2 in) long. *O. vittata* has the same jaw structures and specialized tentacle scales as noted for its congeners. It has up to eight arm spines, all slender, flattened, and bluntly rounded; the most ventral spines are the longest and may have a slightly flared tip. The scales covering the disk of this species are so extremely small and delicate that they are barely visible, even under a microscope.

The disk of *O. vittata* is gray or yellowish gray, with dark brown spots; the radial shields are white or gray. The arms are pale yellow and are banded with reddish brown rings every four to eight joints. A series of minute brownish spots forms a characteristic pair of discontinuous stripes on the dorsal edges of the arms. There may be a reddish brown stripe on the bottom of the arm; dark spots are lacking there and on the oral frame.

HABITAT: Usually near coral reefs; in sandy areas with seagrass, scattered colonies of coral, rubble, or sponges.

DISTRIBUTION: Previously known from the Dry Tortugas, here reported from the Florida Keys, the Bahama Islands, Panama, and Belize (Hendler, previously unpublished). DEPTH: 11–15 m (36–48 ft).

BIOLOGY: The species burrows in sand, often wedging its disk beneath a small piece of shell or rubble; it is not found in exposed coral heads or rubble. It has been collected in close proximity with other burrowing brittle star species including *Amphipholis januarii* and *Ophiophragmus pulcher*. After dusk, *O. vittata* extends several extremely long and slender arms from the sediment. The arms luminesce brilliantly when they are disturbed (Hendler, previously unpublished).

REMARKS: Thomas (1967:127, figs. 7–9) suggested that *O. vittata* may be identical to *Ophiocnida caribea* (Ljungman), but further study is required to verify this synonymy. The only known specimen of *O. caribea* was collected in deep water off Anguilla. *Ophiopsila polysticta* H. L. Clark, previously known only from Barbados, also occurs off Panama. There, it lives in soft sediments at similar depths as *O. vittata*, but at different localities (Hendler, previously unpublished).

SELECTED REFERENCE: H. L. Clark 1918:330, pl. 8, fig. 2.

FAMILY OPHIONEREIDIDAE

Ophionereis olivacea H. L. Clark
Figure 51

DESCRIPTION: This is the smallest *Ophionereis* in this region, with a disk diameter of only 6 mm (0.2 in), and arms approximately 33 mm (1.3 in) long; on the basis of size alone it could be mistaken for the young of another *Ophionereis* species. Like other members of the genus, it has a finely scaled disk with small radial shields. The disk is usually pentagonal, instead of round as in *O. reticulata* and *O. squamulosa*. The dorsal arm plates are longer than wide, roughly hexagonal, and widest near the center. The accessory dorsal arm plates adjoin the distal half of the dorsal arm plate; minute overlapping scales may be associated with the acces-

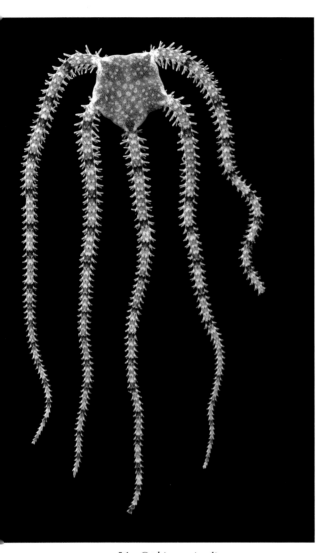

FIGURE 51. *Ophionereis olivacea.*

sory dorsal arm plates of large individuals. There are three erect, smooth arm spines. Near the edge of the disk of large individuals, at the widest part of the arm, there are several joints with elongate middle arm spines that may be nearly twice the length of the dorsal and ventral spines. They are shaped like a wooden match, nearly cylindrical with a thick tip. The tentacle scales are large; each one completely covers a tentacle pore.

The disk is gray, with gray-green blotches and an irregular dense or netlike pattern of the same color. The arms are banded; dark brownish joints and gray joints are separated at irregular intervals by dark and white mottled joints. The dorsal arm plates sometimes have dark edges and may have a series of markings forming a discontinuous, dark arm stripe.

HABITAT: Nonturbulent areas such as mangrove channels, lagoons, and the deep fore reef; generally living in clumps of algae.

DISTRIBUTION: The Florida Keys, Puerto Rico, Colombia, and Belize; questionable records (Thomas 1973) include St. John, Curaçao, and Panama. DEPTH: Intertidal to 24 m (79 ft).

BIOLOGY: *O. olivacea* is a protandric hermaphrodite (Byrne 1991) that broods its young (Hendler and Littman 1986). The smallest individuals, which are males, probably reverse sex when they attain a disk diameter of 2.2–4.0 mm. The females have six to 10 large (0.4 mm in diameter), pink, yolky eggs in each ovary. A female may brood as many as 165 embryos, all at the same stage of development. The ciliated embryo has been described as a modified vitellaria (Byrne 1991), but it lacks ophiopluteus and vitellaria features. In Florida, adults brood from January to April and release young of 0.5 mm disk diameter with two to three arm joints (Byrne 1991). Although direct development may restrict the dispersal of the species, adult individuals usually live in algae and are disseminated in clumps of drift algae (Hendler and Littman 1986; Holmquist 1994). Unfortunately, previous publications citing *O. olivacea* (e.g., Pearson 1937; Thomas 1973) are not reliable, because of the likely misidentification of this species.

SELECTED REFERENCE: H. L. Clark 1901a:248, pl. 14, figs. 10–13.

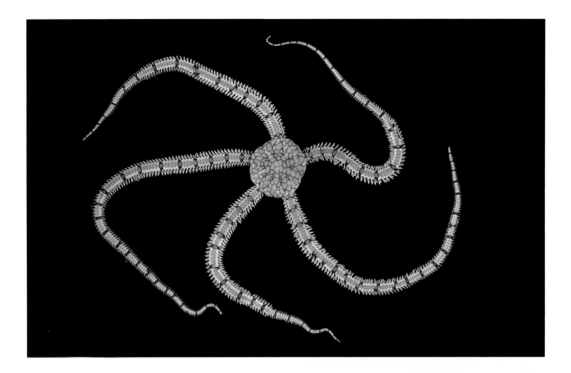

FIGURE 52. *Ophionereis reticulata.*

FIGURE 53. *Ophionereis reticulata:* detail of the disk and a commensal polychaete, *Malmgreniella variegata.*

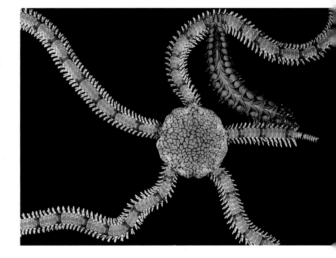

Ophionereis reticulata (Say)
Figures 30C-3, 52, 53

DESCRIPTION: Individuals can grow to 15 mm (0.6 in) disk diameter with arms over 120 mm (4.7 in) long. The disk is finely scaled and the radial shields are small; the primary plates are rarely discernable in adults, unlike the situation in some smaller species of *Ophionereis*. The proximal dorsal arm plates are approximately twice as wide as long; the accessory dorsal arm plates are as long as the adjacent dorsal arm plate. The three arm spines are smooth, compressed, blunt-tipped, and generally longer than an arm joint. The middle arm spine is the largest and is longest on joints at a distance from the disk edge equivalent to about one disk diameter. A single, large tentacle scale completely covers each tentacle pore, and the tube feet are white.

O. reticulata, as its name implies, has a well-defined, brown or reddish brown network pattern on the pale gray disk, but it is sometimes less distinct than the pattern on the disk of *O. olivacea* and *O. squamulosa*. On the white disk of small specimens there is an open, purplish brown network and prominent, paired dark markings near the base of each arm. Typically, arms are banded with blackish or purplish brown about every fourth joint, and there is a dark thin band between the intervening pale arm joints. In darkly pigmented individuals, the lighter joints are pale brown rather than white, and the arms have a brown, median arm stripe.

HABITAT: All reef zones, mangroves and seagrass beds. Usually on sand under rocks, coral rubble, or corals.

DISTRIBUTION: Bermuda, the Bahama Islands, South Carolina offshore reefs, Florida (Fort Pierce to the Florida Keys), the Dry Tortugas, Texas offshore reefs, Cuba, Jamaica, Pedro Bank, Haiti, Puerto Rico, the Virgin Islands, the Leeward Islands, Barbados, Tobago, the Netherlands Antilles, Belize, Panama, Colombia, Venezuela, and Brazil. Possibly occurs on both sides of the Atlantic (Madsen 1970). DEPTH: Less than 1–221 m (less than 1–726 ft).

BIOLOGY: *O. reticulata* is one of the most frequently encountered brittle stars in Bermuda and the Florida Keys (H. L. Clark 1942; Kissling and Taylor 1977). Individuals commonly extend two or three arms from beneath chunks of coral rubble. They are negatively phototaxic and may prefer dark surfaces (May 1925; Millott 1953).

Grave (1898b) reported that *O. reticulata* and *Ophiocoma pumila* occur together, and Nutting (1919) noted the resemblance of the two species, suggesting that they have similar "camouflage." However, Sides (1985) found that *O. reticulata* usually occurs singly and is negatively associated with *O. pumila* and *Ophiolepis impressa*. In fact, it actively displaces *O. impressa*. It uses the tube feet to excavate a burrow, maintain a respiratory current, and gather food particles, including algal filaments and diatoms, from the sediment surface (May 1925; Sides, personal communication). It also uses them for locomotion, advancing at a rate of about 50 cm/minute (May 1925). The species is most active at night. Individuals extend several of their arms, usually through separate holes, before sunset and withdraw them after sunrise (Sides, personal communication).

Hendler and Littman (1986) suggested that the species may have lecithotrophic development, based on the moderately large size of its eggs (0.25 mm in diameter). The reaction of *O. reticulata* to a battery of chemicals, its rate of oxygen uptake, and its tolerance of elevated temperatures and desiccation all have been documented (May 1925; Singletary 1971; Petersen 1976). The behavior of an individual with a severed neural ring was discussed by Fox and Gilbert (1991).

For this species, the percentage of regenerating arms (74.4%) and estimated rate of arm breakage are high (Sides 1982, 1987), possibly reflecting damage by predators. Off Texas, many individuals had regenerating arm tips, but remains of brittle stars were not identified in fish stomachs (Shirley 1982). Elsewhere, *Ophionereis* species are consumed by a squirrelfish, a triggerfish, and the sand tilefish, composing a significant prey item of the last species (Randall 1967).

The scaleworm *Malmgreniella variegata* (Treadwell) (formerly called *Harmothoe lunulata* [Delle Chiaje]) is associated with *O. reticulata* (Millott 1953; Devaney 1974b; Pettibone 1993) (Figure 53). *M. variegata* also lives on *Ophionereis annulata* (LeConte) in the Gulf of Panama. Thus, the worm–brittle star symbiosis may predate the Pliocene separation of the Atlantic and Pacific oceans (Hendler, previously unpublished). Millott (1953, 1966) found that *Malmgreniella* luminesces bluish green and reported that *Ophionereis* too is bioluminescent, an observation that has not been verified.

SELECTED REFERENCES: Say 1825:148 [as *Ophiura reticulata*]; Lütken 1859b:212, pl. 3, figs. 6a–c; A. M. Clark 1953: pl. 1, figs. 1, 2, text fig. 3a; Thomas 1973:586, figs. 1A–D, 5A.

Ophionereis squamulosa Koehler
Figure 54

DESCRIPTION: This species is decidedly smaller than similar-looking *O. reticulata*; a large individual is 6 mm (0.2 in) in disk diameter with arms 45 mm (1.8 in) long. It has conspicuous primary plates, unlike some large *Ophionereis* species. The accessory dorsal arm plates of *O. squamulosa* are almost as long as its dorsal arm plates; the latter are somewhat wider than long. The three arm spines exceed the length of an arm joint; they are less markedly compressed than the spines of *O. reticulata*.

The gray background color of the disk has brown or brownish green spots and blotches, scattered about an irregular network of reddish or greenish brown pigmentation. Most of the arm joints are darkly pigmented (almost the reverse of the color pattern of *O. reticulata*); a dark reddish or greenish brown joint and several mottled gray-brown joints alternate with single white arm joints.

HABITAT: In quiet waters behind the reef crest and also in the turbulent spur-and-groove zone; among coral heads, coral rubble, and seagrass.

DISTRIBUTION: The Bahama Islands, the Florida Keys, the Dry Tortugas, Haiti, Puerto Rico, St. Thomas, Tobago, Belize, and Brazil. DEPTH: 1–40 m (3–131 ft).

BIOLOGY: In algae where other small or juvenile brittle stars congregate, densities of this species can exceed 100 individuals per liter (Hendler and Littman 1986). At Looe Key, Florida, it occurred with *Amphipholis squamata, Amphiodia pulchella, Ophiostigma isocanthum,* and small *Ophiocoma echinata* (Hendler, previously unpublished), and it has been found with *O. reticulata* (Thomas 1973).

The white testes of ripe males and pink ovaries of ripe females are visible through the thin ventral body wall. Mortensen (1921) described the spawning of this species during April, at Tobago. All the oocytes were not released at once; a spawning female shed about 90 negatively buoyant, red eggs, of 0.2 mm diameter. They developed into yolky nonfeeding vitellaria larvae that completed metamorphosis within 6 days.

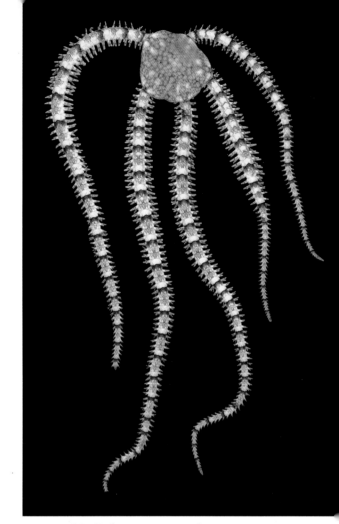

FIGURE 54. *Ophionereis squamulosa*.

SELECTED REFERENCES: Koehler 1913: 360, pl. 21, figs. 4–6 [as *Ophionereis squamata*]; Koehler 1914a:44; A. M. Clark 1953: 71, pl. 3, figs. 3, 4, text fig. 2; Thomas 1973: 588, figs. 2A–D, 5B; Hotchkiss 1982:395, pl. 175.

Ophionereis vittata Hendler
Figure 55

AUTHOR'S NOTE: The following account is a condensed version of the formal description of this species (Hendler, in press).

DESCRIPTION: This species is smaller and more delicate than *O. reticulata*, but grows to a larger size than *O. squamulosa* and *O. olivacea*. A specimen of 2.8 mm (0.1 in) disk diameter has arms 25 mm (0.9 in) long. One that is 6.7 mm (0.3 in) in disk diameter has 85 mm (3.3 in) arms, much longer than the arms of other Caribbean *Ophionereis* of a comparable disk size. The disk scales are exceedingly thin and delicate, the radial shields are very small, and the primary plates are not evident. Accessory dorsal arm plates border the posterior half of the dorsal arm plate and appear to be composed of overlapping scales. There are three smooth arm spines; the middle arm spine is elongate, and the tip sometimes broadened, just as it is in *O. olivacea*. However, in *O. vittata*, the longest spines are on arm joints at a distance from the disk equal to several disk diameters, rather than near the base of the arm.

O. vittata conforms to Thomas's (1973: 590, fig. 3A–D) description of *O. olivacea*: "Disc gray, often with gold reticulations. Arms pale tan, almost white, with dark green bands occupying single joints every 3 to 5 joints. Often a fine line of green along dorsal side of arm." However, the arm stripe is a brilliant red in the living brittle star and changes color in preservative.

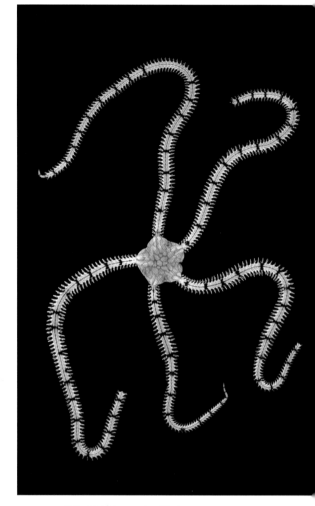

FIGURE 55. *Ophionereis vittata*.

HABITAT: A "deep-reef" species (Hendler and Peck 1988), found in clumps of algae on the fore reef slope.

DISTRIBUTION: Florida (Miami and the Florida Keys), the Bahama Islands, the Greater and Lesser Antilles, Belize, Panama, and Colombia. DEPTH: 20–126 m (65–413 ft).

SELECTED REFERENCES: Thomas 1973: 590, fig. 3A–D [as *O. olivacea* in part]; Hendler, in press.

FIGURE 56. *Ophioderma appressum*. Photo by G. Hendler.

FAMILY OPHIODERMATIDAE

Ophioderma appressum (Say)
Figure 56

DESCRIPTION: Individuals can attain 25 mm (0.98 in) disk diameter with arms 125 mm (4.9 in) long, but seldom exceed 18 mm (0.7 in) in disk diameter. The species has tiny adpressed arm spines like other members of the genus. Its disk is covered by small, rounded granules, and it has four bursal slits beside each arm. The color *pattern* of some individuals resembles that of O. *rubicundum*. Unlike the latter species, however, it is rarely reddish, and its radial shields are covered with granules. Its arms differ in overall shape from those of O. *rubicundum*; they are widest beyond the edge of the disk and remain a constant width for some distance, instead of gradually tapering from the disk edge to the arm tip. The arm tips taper gradually; their terminal joints are not longer than wide. The most ventral arm spine is tapered and is markedly larger than the dorsal spines. The posterior edge of the oral shield is straight or slightly concave and touches the bursal slit.

The coloration of O. *appressum* is extremely variable, but falls into two patterns: "uniform" and "harlequin" (Lütken 1859b; Koehler 1913; Devaney 1974b), one or another of which predominates at a given locality (Hendler, previously unpublished). The disk of "uniform" individuals is gray, green, or brown, usually with tiny clusters of salt-

and-pepper flecks. The disk of "harlequin" individuals is almost entirely white or has irregularly shaped patches of contrasting color (green, white, black, yellow). The arms are banded with the ground color and with a lighter shade or white.

HABITAT: Reef and seagrass habitats from the intertidal to the fore reef slope; under coral rubble and in branching and foliose corals.

DISTRIBUTION: Bermuda, the Bahama Islands, South Carolina, Florida, the Florida Keys, the Dry Tortugas, Texas offshore reefs, Cuba, Jamaica, Haiti, Puerto Rico, the Virgin Islands, the Leeward Islands, Barbados, Tobago, Trinidad, Curaçao, Aruba, Belize, Swan Island, Isla de Providencia, Panama, Colombia, Venezuela, and Brazil. Records from Africa are based on misidentified specimens of O. *longicaudum* (Madsen 1970; Bartsch 1974). Reports from Ascension Island and Tenerife (Canary Islands) are also questionable (Pawson 1978; Tortonese 1983). DEPTH: Intertidal to 50 m (164 ft).

BIOLOGY: *O. appressum* is one of the most abundant shallow-water brittle stars of Florida and the Caribbean (Kissling and Taylor 1977; Lewis and Bray 1983; Hendler and Peck 1988). It frequently occurs with other gregarious reef flat species such as *Ophioderma cinereum*, *O. guttatum*, *O. phoenium*, *O. rubicundum*, *Ophiocoma echinata*, and *O. wendtii* (Burke 1974; Hendler and Littman 1986; Hendler and Peck 1988; Aronson 1993). Individuals in shallow waters grow larger than those on the reef slope (Hendler and Peck 1988).

Reimer and Reimer (1975) investigated the feeding behavior of *O. appressum* in response to chemical activators. Its diet includes plant material; the stomach of an individual from Looe Key, Florida, contained calcareous and filamentous algae. Sides (personal communication) has found that the species is a nocturnal scavenger that feeds on fish feces and material collected from within its home crevice. Microscopic crescentic ossicles invest the stomach walls (Irimura 1988, 1991). Adhesive tube feet enable it to manipulate food and even to climb glass aquarium walls (Cowles 1910).

The species is exceedingly quick to escape when pursued; when restrained it is more prone to autotomize its arms than other shallow-water *Ophioderma* species (Hendler, previously unpublished). The proportion of individuals with regenerating arms may be high, and the rate of arm regeneration is moderate, but the incidence of sublethal disk damage is low (Sides 1982, 1987; Aronson 1991). Shirley (1982) reported that, off Texas, many individuals had regenerating arm tips; however, *O. appressum* was not identified in the stomach contents of 31 fish species he examined, indicating that autotomy is not necessarily elicited by predacious fish.

In Panama, *O. appressum* spawns from September through November, and some intertidal populations may release gametes in June and July (Hendler 1979a). Spawning occurs during the summer in Jamaica (Grave 1898b). Because of the size (about 0.3 mm) of yolky eggs, and the prevalence of nonfeeding larvae in the genus, it is presumed that *O. appressum* has a vitellaria larva (Hendler 1979a; Hendler and Littman 1986). The upper lethal temperature of the species is 37.7°C (Mayer 1914; Glynn 1968).

REMARKS: In addition to the color variation discussed above, this species exhibits morphological variability. For example, individuals from temperate waters are generally larger and more robust than their tropical counterparts.

SELECTED REFERENCES: Say 1825:151

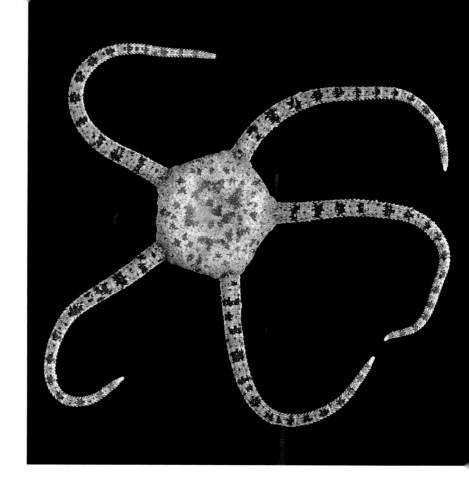

FIGURE 57. *Ophioderma brevicaudum*.

[as *Ophiura appressa*]; Lütken 1859b:194, pl. 1, figs. 4a–d [as *Ophioderma virescens*]; Devaney 1974b:141, fig. 9.

Ophioderma brevicaudum Lütken
Figure 57

DESCRIPTION: As its name implies, the arms of this species are short in relation to the disk. It grows to 22 mm (0.9 in) in disk diameter with arms 77 mm (3.0 in) long. *O. brevicaudum*, like its congeners, has four bursal slits beside each arm, one pair near the oral shield and the other near the disk margin. The granules at the center of the top of the disk are small and rounded, those at the edge of the disk are larger and often polygonal, and on the ventral surface the granules are appreciably separated from one another. The adoral shields are sparsely granulated. The small, adpressed arm spines are flat and almost triangular in outline and nearly touch each other at the base. The most ventral spine is not appreciably larger than adjacent spines; it is in contact, or nearly so, with the tentacle scale of the next distal joint. The dorsal arm plates are sometimes fragmented. Juveniles have granules on the dorsal and ventral surface of the disk and the arms, which are lost when they reach adult size.

The disk is mottled, and the arms are banded with various combinations of green, blue-green, gray, and white. Often, patches of rust red or reddish brown decorate the ventral interradii or the top of the disk. Rarely, individuals are reddish rather than green.

HABITAT: High-energy areas near breaking waves, on shallow reef crest, beach-rock platforms, and patch reef, and stunted seagrass beds.

DISTRIBUTION: Bermuda(?), South Carolina, the Florida Keys, the Dry Tortugas, the Bahama Islands, Cuba, Jamaica, Haiti, Puerto Rico, the Virgin Islands, the Leeward Islands, the Windward Islands, Barbados, Tobago, Isla La Tortuga, the Netherlands Antilles, Belize, Panama, Colombia, Venezuela, French Guiana; also Ascension Island (Koehler 1914a). DEPTH: Usually less than 1–18 m (less than 1–60 ft); reportedly as deep as 64 m (210 ft) (Koehler 1914a).

BIOLOGY: The feeding response of O. brevicaudum is elicited by a broader range of chemical activators than for many other Ophioderma species (Reimer and Reimer 1975). The lack of specificity may be related to the presence of the species in areas with heavy wave action, where the duration of chemical signals could be fleeting. The stomach wall of this brittle star is invested with microscopic crescentic ossicles (Irimura 1988, 1991).

The species spawns only between September and November in Panama, and it has a low gonadal index (Hendler 1979a). Perhaps its low fecundity is responsible for the rarity of O. brevicaudum in comparison with its commonplace congeners. Mayer (1914) found that individuals cease movement at 7.5–8.0°C and are killed by a temperature of −1.7°C.

SELECTED REFERENCES: Lütken 1856:8; 1859b:196, pl. 1, figs. 3a–c [as *Ophioderma brevicauda*]; A. H. Clark 1939a:452, pl. 54, figs. 1, 2 ["*Ophioderma brevicaudum*, *Ophiocryptus* stage"].

FIGURE 58. *Ophioderma brevispinum*. Photo by G. Hendler.

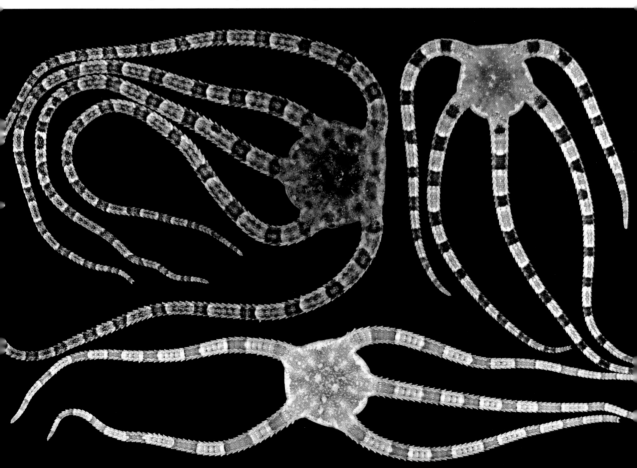

Ophioderma brevispinum (Say)
Figure 58

DESCRIPTION: Individuals grow to 15 mm (0.6 in) in disk diameter with arms 70 mm (2.8 in) long. Its coloration is variable, resembling that of several congeners that also have small, rounded disk granules and four bursal slits beside each arm, *O. appressum, O. brevicaudum,* and *O. rubicundum.* Its granule-covered radial shields distinguish it from *O. rubicundum.* In contrast with most specimens of *O. appressum,* its disk is pentagonal and its arms gradually taper from the edge of the disk to the arm tip. Like *O. brevicaudum,* its ventral arm spine is similar in length to the adjacent spine; however, it does not touch the tentacle scale of the next distal joint, and the arm spines are thin, straight-sided (peglike), and blunt-tipped. Unlike *O. brevicaudum,* it has relatively long arms and close-set disk granules of uniform dimensions.

The disk is uniformly colored or variegated with combinations of gray, green, brown, yellow, orange, pink, red, white, and black; the arms usually are banded.

HABITAT: Usually in seagrass beds (e.g., *Thalassia* in tropical waters and *Zostera* in northern waters), but also on shell, sand, or mud bottoms, and in mangroves and oyster reefs. On coral reefs it occurs in lagoon and back reef zones in clumps of branching coral, sponge, algae, and coral rubble.

DISTRIBUTION: Bermuda(?), the Bahama Islands, Massachusetts to Florida and the Gulf coast, Cuba, Jamaica, Haiti, Puerto Rico, Swan Island, the Virgin Islands, the Leeward Islands, the Windward Islands, Barbados, Aruba, Belize, Colombia, Venezuela, French Guiana, and Brazil. Also Ascension Island (Koehler 1914a). DEPTH: 1–223 m (3–732 ft) reported; however, deep records may be based on misidentified *O. januarii* Lütken.

BIOLOGY: This species is frequently abundant (Tabb and Manning 1961; Aronson and Harms 1985; Corvea et al. 1985). It often occurs together with *Ophioderma appressum* and *Ophiothrix angulata.* Perhaps the success of the species stems from its tolerance of physical stress. Reportedly (though questionably) it can survive 16 hours in fresh water and temperatures of −0.6 to 37.7°C (Mayer 1914). It has been suggested, but never proved, that the highly variable coloration of *O. brevispinum* serves as defensive camouflage and conforms with the substratum (Grave 1900; Larrauri 1981). Glaser (1907: 206; see also Grave 1900), impressed with its agility, stated that this species "moves in practically all the ways in which it is possible for a pentaradiate animal of its construction to move."

O. brevispinum has been characterized as a sluggish scavenger (Grave 1900), but it can capture small, rapidly moving crustaceans in the delicate coils of the arm tip (Hendler 1982b). Aronson and Harms (1985) observed it pursuing *Ophiothrix orstedii.* In Massachusetts, during the winter, *O. brevispinum* is torpid and does not respond to food, but in the summer its feeding activity is intense. In its stomach contents are found small crustaceans, worms, sponges, debris, and algae (Hendler 1982b). The stomach wall is invested with microscopic ossicles, including cresentic and triradiate forms (Irimura 1988, 1991).

O. brevispinum is the only brittle star whose rate of digestion, approximately 1 mg/hr, has been measured (Hendler 1982b). Transport of nutrients by the hemal system of *O. brevispinum* and the importance of the uptake of dissolved nutrients by the integument have been traced with radioactive labels (Ferguson 1985). Its oxygen uptake is relatively independent of oxygen pressure (Mangum and Van Winkle 1973; compare Allee 1927). Its rate of nitrogen excretion, as ammonia and urea, has been determined (Stickle et al. 1982; Stickle 1988). Other studies bearing on the species describe ciliated muscle cells within the tube feet, which may act as sensory structures (Gardiner and Rieger 1980). In addition, Rockstein (1971) analyzed high-energy phosphagens in this brittle star and discussed their significance for phylogenetic analysis.

The pattern of reproduction of the species seems in part related to temperature, based on comparison of populations from temperate and warm waters (Stancyk 1974; Hendler and Tyler 1986). Ripe females may have as many as 1,385 eggs of 0.30–0.35 mm diameter (Stancyk 1973, 1974; Hendler and Tyler 1986). Both brown and green pigmented oocytes may occur in an individual (Hendler and Tyler 1986; compare Grave 1899, 1900). Grave (1916:441) observed that spawning *O. brevispinum* stand on arched arms, their buoyant eggs rising to the surface like "ascending streams of minute bubbles of air." Presumably then, the initial developmental stages of *O. brevispinum* float near the surface. Its development, including behavior of the yolky vitellaria larva and its metamorphosis and settlement at 8 days, has been described perfunctorily (Grave 1899, 1900, 1914, 1916; Webb 1989). Certain characteristics of the larva were thought to be adaptations for estuarine habitats (Stancyk 1973, 1974), but the similarity of *O. brevispinum* vitellariae to the larvae of reef-dwelling *Ophioderma* argues against that hypothesis.

As Ayres (1851b:135) noted, the species "manifests very little disposition to dismember itself upon being handled." Therefore, the prevalence of *O. brevispinum* with regenerating arms (Aronson 1987, 1991) seems to attest to the effect of predators. It is consumed by the buffalo trunkfish, *Lactophrys trigonus* (Linnaeus) (Randall 1967). The process of arm regeneration can be disrupted by insecticides and antifouling paints (Walsh et al. 1986).

Symbionts associated with *O. brevispinum* include four genera of amphipods (Hendler 1982b) and a parasitic gastropod (*Vitreolina* sp.) (Warén, personal communication).

REMARKS: Populations of *O. brevispinum* from the northeastern United States have been called *Ophiura* [=*Ophioderma*] *olivacea* Ayres (Ayres 1851b), and some from Florida and the Caribbean Region have been referred to *Ophioderma serpens* Lütken (Lütken 1856). In addition, *Ophioderma holmesii* (Lyman), collected off South Carolina, was thought to "stand between *O. brevispina* and *O. olivacea*" (Lyman 1865:22). A comprehensive systematic analysis of *Ophioderma* populations between Bermuda and Brazil is required to clarify the taxonomy of the putatively wide-ranging brittle star currently known as *O. brevispinum*. Species-level differences among northern and southern populations would not be unexpected. *Ophioderma januarii* Lütken, a species previously known from Brazil, but also occurring in deep waters of the Gulf of Mexico (Hendler, previously unpublished), closely resembles *O. brevispinum* (Ziesenhenne 1955; Monteiro 1992). It is distinguished by the sharply gabled shape of the tops of arm joints near the disk.

SELECTED REFERENCES: Say 1825:149 [as *Ophiura brevispina*]; Lütken 1859b:198, pl. 1, fig. 6a,b [as *Ophioderma serpens*]; Lyman 1865:21, pl. 1, fig. 7 [as *Ophiura holmesii*].

Ophioderma cinereum Müller & Troschel
Figures 30C-6, 59

DESCRIPTION: This robust species often grows to 29 mm (1.1 in) in disk diameter, with arms 148 mm (5.8 in) long; an excep-

BRITTLE STARS AND BASKET STARS

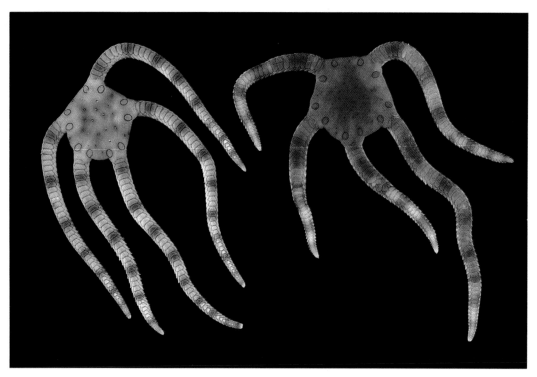

FIGURE 59. *Ophioderma cinereum.*

tionally large individual of 37 mm (1.5 in) disk diameter with arms 210 mm (8.3 in) long has been reported (Rathbun 1879). The top of the disk, except for the conspicuous radial shields, is covered with microscopic, rounded granules. Granules surround the bursal slits and oral shield, with often just a single row of granules along the distal edge of the shield. The oral shield is rounded-triangular; the distal edge is not strongly convex. There are four bursal slits beside each arm, as in other *Ophioderma* species. A useful diagnostic characteristic of individuals larger than about 10 mm (0.4 in) in disk diameter is the fragmentation of numerous dorsal arm plates on joints near the disk; the plates are irregularly dissected by fissures running from the proximal to the distal edge. The arm spines are small, adpressed, bluntly pointed, and gradually increase in size toward the ventral surface.

The ground color in *O. cinereum* ranges from light to dark gray or brown; juveniles may be pale pink, rose, or lavender. Most individuals, except the darkest, have a conspicuous, dark border that demarcates the edge of the radial shields. Usually, tiny clusters of black and white (or yellow) specks are scattered on the disk. The arms are conspicuously banded, and the dorsal arm plates have an irregular reticulate pattern.

HABITAT: All reef zones, as well as mangrove and seagrass habitats.

DISTRIBUTION: Bermuda(?), the Bahama Islands, the Florida Keys, the Dry Tortugas, Cuba, Jamaica, Haiti, Puerto Rico, the Virgin Islands, the Leeward Islands, the Windward Islands, Barbados, Tobago, Curaçao, Aruba, and the coast of Central and South America to Brazil. DEPTH: Typically intertidal to 24 m (78 ft), but reported at 358 m (1,176 ft) and even 1,719 m (5,640 ft) (Koehler 1914a).

BIOLOGY: H. L. Clark (1942:379) stated that O. *cinereum* "occurs only under slabs of broken coral and rocks, and makes no attempt to get into interstices and crevices." It often lives under the same shelters with other large brittle stars such as *Ophiocoma wendtii*, *O. echinata*, *Ophioderma appressum*, *O. brevicaudum*, and *O. brevispinum* (Rathbun 1879; H. L. Clark 1901a; Hendler, previously unpublished). Juveniles occasionally are found in seagrass debris and rubble (Hendler and Littman 1986; Emson et al. 1985a).

Individuals usually remain concealed, even at night, but a Belizean population lives in the open on mangrove channel banks (Hendler, previously unpublished). At night, its arms are extended from the entrance of the crevice it occupies (Sides, personal communication). During periods of mass mortalities caused by severe low tides (Hendler 1977b), individuals emerge during the day to scavenge on an intertidal reef flat. Sides (personal communication) found that individuals "rush" out of their crevices to feed, often on "faeces just released by passing fish." Stomach contents include the carapaces and limbs (possibly molts) of crabs and other crustaceans. Individuals in Jamaica and Panama sometimes occupy octopus dens where middens of mollusc and crustacean parts accumulate (Hendler, previously unpublished, Sides, personal communication). The feeding habits and diet of *O. cinereum* were reported by Tewes (1984), and Reimer and Reimer (1975) described the feeding response elicited by chemical activators. The stomach wall of this species is invested with microscopic rod-shaped spicules (Irimura 1991). Evidently, arm damage caused by predation is relatively unimportant in the ecology of this stout-bodied brittle star. Up to 34% of the arms of individuals in a population may be regenerating at a given time, but this reflects the slow rate of regeneration as well as the frequency of autotomy (Sides 1982, 1987). Aronson (1993) found that population densities of *O. cinereum* and the more delicate *O. appressum* were not reduced by hurricanes. Sides (personal communication) has found that *O. cinereum* is a vagile species and described its competitive interactions with other crevice-dwelling brittle stars and with damselfish.

In Panama, different populations of *O. cinereum* breed asynchronously; spawning peaks between October and May, but continues all year long. The yolky eggs, approximately 0.3 mm in diameter, give rise to rapidly developing, yolky, vitellaria larvae (Hendler 1979a). The respiration of this brittle star (Petersen 1976), and the proportions of magnesium carbonate and strontium in its calcite skeleton, have been examined (Clarke and Wheeler 1915; Pagett 1985). Ostracods, probably *Pontocypria hendleri* Maddocks (R. F. Maddocks, personal communication), were associated with an individual from Looe Key, Florida (Hendler, previously unpublished), and a copepod symbiont has been reported, *Pseudanthessius deficiens* Stock, Humes & Gooding (Stock et al. 1963).

REMARKS: The juveniles of *O. cinereum*, which have their disk and arms covered by granules, were originally described as *Ophiocryptus hexacanthus* H. L. Clark. *Ophioderma saxatilis* Duchaissang is probably the young of *O. cinereum* as well (Lyman 1871; H. L. Clark 1933).

SELECTED REFERENCES: Müller and Troschel 1842:87; Lütken 1859b:190, pl. 1, figs. 1a–c [as *Ophioderma antillarum*]; Koehler 1914a:8, pl. 2, figs. 1, 2 [as *Ophioderma* sp., young].

Ophioderma guttatum Lütken
Figures 60, 61

DESCRIPTION: The disk of this brittle star can reach 45 mm (1.8 in) in diameter, and the relatively short, thick arms, which grow to 143 mm (5.6 in), are flattened and blunt at the tip. The species has four bursal slits beside each arm, as is typical of *Ophioderma* species. However, its distinctive, flattened disk granules resemble only those of *O. squamosissimum*. They completely cover the adoral shields and closely surround the oral shield and bursal slits, separating the oral shields from the bursal slits. Each dorsal arm plate is composed of a mosaic of larger and smaller nonoverlapping scales; the smallest scales are usually at the distal edge of the composite plate. The arm spines are small, and the dorsal spines are more slender than typical for *Ophioderma* species; they are particularly delicate and sharp-tipped in small individuals.

The dorsal surface of the disk and arms is greenish gray, embellished with numerous, small, uniformly spaced black spots that are bordered by pale granules. The ventral surface is a contrasting yellow, orange, or brownish yellow, with orange or brown spots.

HABITAT: Near the crest and the spur-and-groove zone of coral reefs; beneath slabs of rubble and under large brain corals.

DISTRIBUTION: The Florida Keys, the Bahama Islands (Hendler, previously unpublished), Jamaica, St. Thomas, Barbados, Tobago, Belize, and Colombia. DEPTH: Less than 1–18 m (less than 1–60 ft).

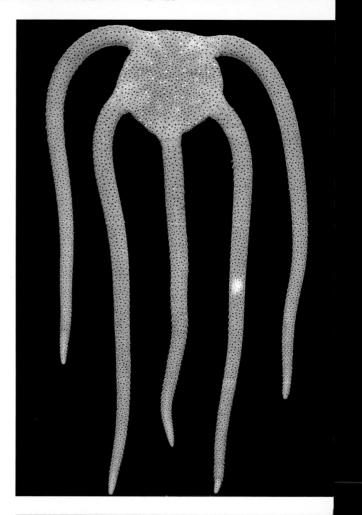

FIGURE 60. *Ophioderma guttatum:* dorsal surface.

FIGURE 61. *Ophioderma guttatum:* ventral surface.

BIOLOGY: Individuals of *O. guttatum* are rare, with the exception of a dense population reported from Enmedio and Lobos reefs in the southwestern Gulf of Mexico (Henkel 1982; Britton and Morton 1989). This brittle star may occur in the same shelters as *Ophioderma anitae* Hotchkiss, *O. appressum*, *O. phoenium*, and *Astrophyton muricatum* (Hendler, previously unpublished). It secretes mucus when injured (Hendler and Miller 1984b). Microscopic ossicles found in the stomach wall include cresentic shapes, rods, and perforated disks (Irimura 1991).

SELECTED REFERENCES: Lütken 1859b: 197, pl. 1, fig. 8a,b [as *Ophioderma guttata*]; Thomas 1962a:65, fig. 1A.

Ophioderma phoenium H. L. Clark
Figure 62

DESCRIPTION: Individuals grow to 23 mm (0.9 in) in disk diameter, with arms 90 mm (3.5 in) long. Granules usually, but do not always, cover the radial shields; they completely cover the adoral shields and surround the oral shield and the bursal slits. The oral shield does not have a strongly convex distal edge. There are four bursal slits beside each arm, as is characteristic for all *Ophioderma* species. The arms are thick, tapering near the distal end, and terminating in a blunt, flattened tip. The arm spines are adpressed, the dorsal arm spines short and pointed; ventrally, they become gradually longer, larger, and more blunt-tipped.

The dissimilar coloration of the disk and arms, although sometimes subtle, is a valuable guide for field identification. The disk is usually red or brown (the specific name derives from the Greek for "blood red"); the arms are contrasting brown, pinkish, gray, or green and generally have dusky to dark bands. Small individuals have relatively brighter and more contrasting coloration. However, pure red and purely green individuals have been reported. *O. rubicundum* is also red, but it lacks the tiny scattered clusters of white or white and black flecks found on the disk of *O. phoenium*. Furthermore, in *O. phoenium* the entire dorsal and ventral surfaces of the disk are the same color, but the ventral interradii are "two-toned" in *O. rubicundum*.

FIGURE 62. *Ophioderma phoenium*.

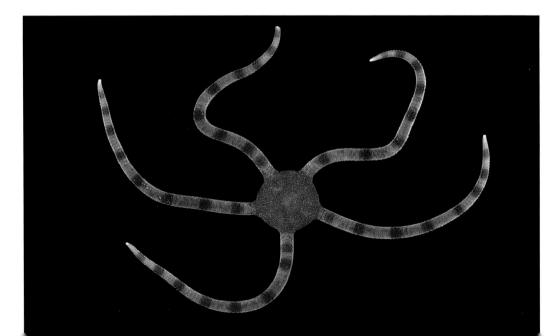

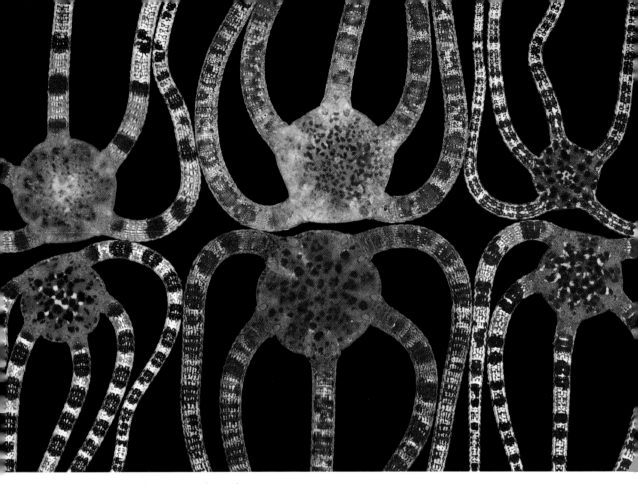

FIGURE 63. *Ophioderma rubicundum*.

HABITAT: Turbulent sites, generally seaward of the reef crest, especially in the spur-and-groove zone; under large slabs of coral rubble on clean, coarse sand.

DISTRIBUTION: The Florida Keys, the Bahama Islands (Hendler, previously unpublished), Cuba, Tobago, Barbados, Belize, and Panama. DEPTH: 1–14 m (3–45 ft), the greatest depth recorded at Looe Key, Florida (Hendler, previously unpublished).

BIOLOGY: Groups of more than a few individuals of this species rarely are found together. They sometimes live under rubble slabs with relatively uncommon congeners including *Ophioderma guttatum* and *O. anitae*, and also with other *Ophioderma* and *Ophiocoma* species, *Ophiomyxa flaccida*, and *Ophiothrix orstedii*. *O. phoenium*, like other Caribbean *Ophioderma* species, probably has lecithotrophic development (Hendler and Littman 1986). The microscopic ossicles in the stomach wall resemble those of *O. brevicaudum* (Irimura 1991).

SELECTED REFERENCES: H. L. Clark 1918:333, pl. 6, figs. 1, 2; 1919:69, pl. 3, fig. 1.

Ophioderma rubicundum Lütken
Figure 63

DESCRIPTION: Adults attain a disk diameter of 23 mm (0.9 in), with arms 135 mm (5.3 in) long. The radial shields are usually bare of the rounded granules covering the rest of the disk. Flanking each arm there are four bursal slits, as is typical for *Ophioderma* species. The oral shield has a convex distal edge that touches the edge of the proximal bursal slits. The dorsal arm plates of *O. rubicundum* are not fragmented and are usually arched near the disk. The arm tip is very thin and tapered, nearly as high as broad. The arm spines are small and adpressed. The lowest arm spine, especially near the disk, is considerably broader and longer than the dorsal spines; in large individuals it nearly touches the tentacle scale of the adjacent arm joint. Juvenile *O. rubicundum* have granules covering more of the body than the adults.

As its scientific name implies, *O. rubicundum* is typically reddish, but an exceptional bright yellow individual has been reported (Fontaine 1953). The brownish to purplish red ground color, retained for a considerable time in alcohol-preserved specimens, is usually variegated with black, gray, brown, and sometimes with irregular white patches on the disk and mottled bands on the arms. The "two-toned" ventral interradii consist of a whitish proximal region and a reddish periphery.

HABITAT: All reef zones; in clumps of living coral and under rubble.

DISTRIBUTION: The Bahama Islands, the Florida Keys, the Dry Tortugas, Texas offshore reefs, Cuba, Jamaica, Puerto Rico, St. Thomas, St. John, Guadeloupe, Barbados, Tobago, Curaçao, Panama, Colombia, and Venezuela. DEPTH: 1–31 m (3–100 ft).

BIOLOGY: Individuals remain hidden during the day, but are seen in the open at night. H. L. Clark (1933:71) surmised that this species is "far from common, apparently living in water a little too deep for the shore collector and not deep enough for the dredge." Indeed, *O. rubicundum* may comprise only 5% of brittle stars collected on the shallow reef, but represents 45% of those on the fore reef slope (Hendler and Peck 1988). However, its abundance varies markedly among localities; Kissling and Taylor (1977) reported that it is scarce in the the Florida Keys, although it was common in collections from Looe Key, Florida (Hendler, previously unpublished). In shallow water it resides under rubble slabs with *Ophioderma* and *Ophiocoma* species, *Ophiolepis impressa, Ophiothrix angulata,* and *O. orstedii,* and on the fore reef tends to be found in corals with *Ophiomyxa flaccida* and *Ophioderma appressum.*

Its body size is related to the type of substratum, decreasing from rubble, to coral, to algae. Juveniles sometimes exceed 100 individuals per liter in algae (Hendler and Littman 1986). Reimer and Reimer (1975) reported that feeding behavior of *O. rubicundum* is elicited by relatively few chemical activators. Embedded in the stomach wall of this species are hooked triradiate ossicles, rods, and perforated disks (Irimura 1988, 1991). Randall (1967) found *O. rubicundum* in the stomach contents of a wrasse, a pufferfish, and a trunkfish.

In Panama, this species spawns between September and November, the timing varying among populations (Hendler 1979a). At Curaçao, breeding individuals in densities exceeding 50 per square meter spawned at night in late September. Van Veghel (1993), who mistakenly identified them as *Ophiomyxa flaccida,* observed groups of 10–15 spawning brittle stars atop corals just before spawning of the corals. It is presumed to have a yolky nonfeeding larva, similar to those of *O. cinereum* and *O. brevispinum* (Hendler and Littman 1986).

SELECTED REFERENCES: Lütken 1856:8; 1859b:192, pl. 1, fig. 2a–c [as *Ophioderma rubicunda*]; Hotchkiss 1982:404, fig. 180a,b [as *Ophioderma* species, juvenile].

Ophioderma squamosissimum Lütken
Figures 30C-5, 30D-13, 64

DESCRIPTION: This, the most rarely seen *Ophioderma* species in the region, is remarkable for its large size, with a disk diameter of 42 mm (1.2 in) and arms 200 mm (7.9 in) long. Its disk granules are flattened and polygonal, creating a mosaic surface that is conspicuous under the microscope. Each dorsal arm plate appears to be divided into a transverse series of plates, such that each arm joint is covered by a small, central (roughly diamond-shaped) scale and several symmetrically placed pairs of lateral scales. The scales on adjacent joints overlap, giving the arm the appearance of snakeskin. Lateral arm plates near the arm tip have a granular texture because of the presence of expanded peripheral trabeculae (EPT). Like its congeners, *O. squamosissimum* has smooth arms with small adpressed arm spines and four bursal slits at the base of each arm.

Large living individuals are entirely scarlet red or orange-hued vermilion; the brilliant pigment quickly and completely bleaches in alcohol-preserved specimens. Small individuals have pale orange patches on the disk.

HABITAT: Reef slope and patch reefs; within or beneath large structures such as brain corals.

DISTRIBUTION: The Bahama Islands, the Florida Keys, Texas offshore reefs, St. Barthélemy, and Belize. DEPTH: 3–85 m (10–279 ft).

BIOLOGY: Thomas (1962a:67) described *O. squamosissimum* as a "giant among the ophiodermatids." Some immunity from predation might accrue from its large size, but many individuals have regenerating arms

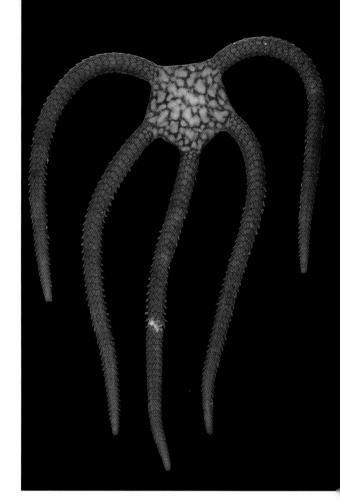

FIGURE 64. *Ophioderma squamosissimum*, small individual.

(Hendler, previously unpublished), and the species is consumed by spanish wrasse (Randall 1967). Large *O. squamosissimum* ought to be capable of ingesting food of a wide size range, but the stomach contents of the only specimen examined consisted entirely of sponge (Hendler, previously unpublished). In the Bahama Islands, active individuals were observed at night deep within cavities in eroded brain corals (Hendler and Miller, previously unpublished), but Burke (1974) reported this brittle star on sand bottom and coral.

Hendler and Miller (1984b) found that it secretes mucus when injured, possibly a defensive behavior; the "gelatinous cuticle" observed on *O. squamosissimum* by Hotchkiss (1982) may be the dried mucous secretion. Zeiller (1974) claimed that a bisected individual of this species can regenerate two complete brittle stars.

SELECTED REFERENCES: Lütken 1856:8; 1859b:194, pl. 1, figs. 7a,b [as *Ophioderma squamosissima*]; H. L. Clark 1918:335, pl. 4, fig. 1, pl. 6, figs. 3, 4; 1919:69, pl. 3, fig. 2; Thomas 1962a:67, fig. 1B.

FAMILY HEMIEURYALIDAE

Sigsbeia conifera Koehler
Figures 30C-2, 30D-15, 65

DESCRIPTION: This small brittle star seldom exceeds 5 mm (0.2 in) in disk diameter. The thick disk is covered by large radially situated plates with conical protuberances and a stellate array of small, irregular scales. The arms are short and bear minute spines, thick accessory dorsal arm plates, and widely separated ventral arm plates. Surfaces of the major plates are covered by glassy bumps, visible under the microscope. The disk is a combination of pink to purple and cream color, and the arms are irregularly banded with pink, brown, and purple.

FIGURE 65. *Sigsbeia conifera*.

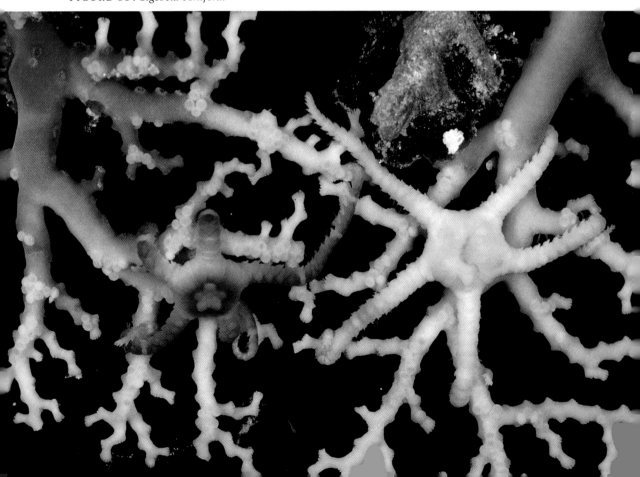

HABITAT: Small recesses in the massive coral structures of the spur-and-groove and fore reef zones; associated with hydrocorals.

DISTRIBUTION: The Bahama Islands, Cuba, Belize, and Panama. Neither this brittle star not its stylasterine host (Cairns 1987) are reported from Florida waters. However, the recent record of *Stylaster roseus* (Pallas) in the Gulf of Mexico (Horta-Puga and Carricart-Ganivet 1990) suggests that the brittle star and hydrocoral may both be more widely distributed than previously thought. DEPTH: 4–24 m (13–99 ft), and there is a reliable record of its occurrence at 221–368 m (726–1,206 ft) (Koehler 1914a).

BIOLOGY: *S. conifera* is the only shallow-water hemieuryalid species in this region. Typical of the family, it has coiling, prehensile arms with which it clings to a "coral" host. This species generally lives on the pink or purple hydrocoral *Stylaster roseus* and is reported from deep-water *Stylaster filogranus* Pourtalès (Koehler 1914a). It is seldom seen because of its small size, and because *S. roseus* grows in small, dimly lit reef caves and crevices. Moreover, it closely matches the color of its host. During the day, individuals remain tightly wrapped around hydrocoral branches, and at night their arms are lifted, possibly to suspension feed (Hendler, previously unpublished). *S. conifera* has large yolky eggs, 0.8 mm in diameter (Byrne 1991). It broods up to 11 embryos, as large as 0.8 mm in disk diameter with eight arm segments (Hendler and Littman 1986; Byrne 1991), and because of the host specificity of adults, the crawl-away babies are automatically provided with an appropriate hydrocoral host.

REMARKS: This species has sometimes been mistaken for a large, deep-water congener, *Sigsbeia murrhina* Lyman.

SELECTED REFERENCES: Koehler 1914a: 133, pl. 14, fig. 7, pl. 17, fig. 6; Verrill 1899a: 72, pl. 2, figs. 1, 1a [as *Sigsbeia murrhina* in part]; Hotchkiss 1982:391, fig. 172 [as *Sigsbeia murrhina*].

FAMILY OPHIACTIDAE

Hemipholis elongata (Say)
Figure 66

DISTRIBUTION: A large specimen of 10 mm disk diameter has arms approximately 90 mm long, but individuals seldom grow to even half that size. The species has five arms and a single oral papilla. It lacks evident scales on the ventral surface of the disk, and the dark brown color of the stomach shows through the thin ventral body wall. The dorsal surface of the disk is covered by large scales. The conspicuous radial shields have been described aptly as "pear-seed shaped" (Lyman 1865), and a few papillae at the outer edge of the radial shields are the only spines on the disk. There are three arm spines and a single tentacle scale. Individuals are usually brown, tan, or gray, sometimes with touches of blue or green, and the arms are banded light and dark. The papillose tube feet are red.

HABITAT: Protected embayments, oyster beds, often in soft, poorly oxygenated, unvegetated sediments; also the sand plains fringing coral reefs.

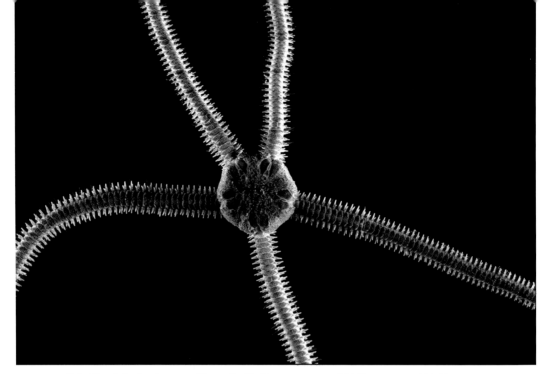

FIGURE 66. *Hemipholis elongata*. Photo by R. Fox, courtesy of E. Ruppert and R. Fox.

DISTRIBUTION: South Carolina, Georgia; Florida (Brevard County) and the Florida Keys (Biscayne Bay) (based on HBOI museum collections); the Florida Gulf coast, Louisiana, Mississippi, Texas, Cuba, Puerto Rico, Trinidad, Panama, and Brazil.
DEPTH: 2–35 m (8–115 ft).

BIOLOGY: Stimpson (1852:226) noted that *H. elongata* is "gregarious, living in companies of twenty to thirty. The existence of these groups is indicated at low water by spaces of about a foot in diameter covered with small holes, looking very much as if a charge or shot had been fired into them. If these spots are watched as the tide rises, from each hole an arm of one of the star-fishes will be seen to protrude and wave about in the water, with red tentacular filaments [tube feet]."

The species often lives in benthic communities with other burrowing brittle stars such as *Amphiodia atra* (Stimpson), *A. planispina*, *Amphioplus albidus* (Ljungman), *Amphipholis gracillima*, and *Ophiophragmus wurdemanii* (Lyman) (Heatwole 1981; Absalao 1990). Populations may persist for 15 or more years, with densities reaching 2,438 individuals per square meter (Valentine 1991a,b). Juveniles may number 100 individuals per liter in shallow-water algae (Boffi 1972) and are sometimes much more numerous than adults (Valentine 1991a,b). However, 3% or fewer recruits survive for more than 1 year (Valentine 1991a).

Young *H. elongata* look so different from adults that they were described as a distinct species, *Ophiolepis uncinata*, by Ayres (1852) and were placed in the genus *Ophionyx* by Stimpson (1852). The young have hooked arm spines with ventrally directed teeth, and their arms tend to flex downward (Turner and Miller 1988). They often cling to algae or to adults until they are large enough to burrow, and for a time the species was thought to brood its young (Lyman 1882; Ludwig 1904). However, the larva is free-living, and the adults, lacking bursae, definitely do not brood (Mortensen 1920).

Populations in Mississippi Sound can spawn more than once during the summer (Valentine 1991b). The spawning individuals raise the disk above the sediment with their arms and shed streams of gametes through a series of fleshy "genital papillae" on the ventral interradii of the disk. The size and number of the papillae vary with the gender of the individual, a rare example of sexual dimorphism in echinoderms (Heatwole 1981; Heatwole and Stancyk 1982). The ovaries contain lavender-colored eggs, about 0.11 mm in diameter. They develop, within 48 hours after fertilization, into a four-armed ophiopluteus; the larval skeleton, similar to that of an *Ophiactis* species, has a double recurrent rod (Hendler, previously unpublished).

The role of the tube feet of this species in locomotion, respiration, and as secretory/sensory structures has been considered in an investigation of ultrastructure (Hajduk 1992). The tube feet bear numerous multicellular complexes consisting of paired secretory cells flanking a central, sensory, ciliated cell; the complexes are surrounded by microvilli and capped with an attenuated layer of cuticle. The water vascular system of *H. elongata* gives rise to elaborate bundles of tubules called Simroth's appendages, which are visible through the ventral disk wall. Fluid within Simroth's appendages and the water vascular system contains hemoglobin-carrying cells that are circulated by contractions of the tube feet (Hajduk and Cosgrove 1975; Heatwole 1981). The oxygen transported by the hemoglobin may compensate for the absence of respiratory bursae, enabling *Hemipholis* to live in poorly oxygenated sediments. The association of individuals with polychaete worm tubes (Ruppert and Fox 1988) might also bring the brittle star in contact with oxygenated water. *H. elongata* tolerates reduced salinities (Ferguson 1948; Binyon 1966; Sheridan and Badger 1981) and often lives in eutrophic or polluted environments where it may accumulate toxic mercury and methylmercury compounds (Windom and Kendall 1979; Camargo 1982).

SELECTED REFERENCES: Say 1825:146 [as *Ophiura elongata*]; Lyman 1882:158, pl. 40, figs. 8–12, pl. 44, figs. 13–16 [as *Hemipholis cordifera*]; Thomas 1962b:686, fig. 22.

Ophiactis algicola H. L. Clark
Figure 67

DESCRIPTION: A typical specimen of this small, six-armed species is 1.7 mm (0.07 in) in disk diameter with arms 9.4 mm (0.4 in) long; individuals greater than 3 mm (0.1 in) in disk diameter are rare. The radial shields of *O. algicola*, distinctly less than half the radius of the disk, help distinguish it from *O. savignyi*. Another diagnostic feature, visible under a microscope, is the uniform granular texture of the disk scales and of the dorsal and lateral arm plates. In contrast, the scales of *O. savignyi* have minute, irregularly arranged punctae. There is a single oral papilla (seldom two papillae) on each adoral shield, and there may be a tiny proximal papilla high on each side of the jaw. The dorsal arm plates are separated or barely touch, and their proximal edges diverge. The species has four slender arm spines; the middle one is largest, broadest, and has a thorny tip.

O. algicola is lightly pigmented; the disk is orange, brown, and/or gray, and the whitish arms have bands (sometimes also a stripe) of orange, tan, or brown. The arm spines usually have an internal dusky spot and sometimes are encircled by a brown or orange band. The less dramatically pigmented of the two individuals depicted here is probably the more typical of the species.

HABITAT: On pilings, rocks, and on the back reef and fore reef; in clumps of coralline algae and dense algal turf and also among encrusting animals such as bryozoans and in sponges.

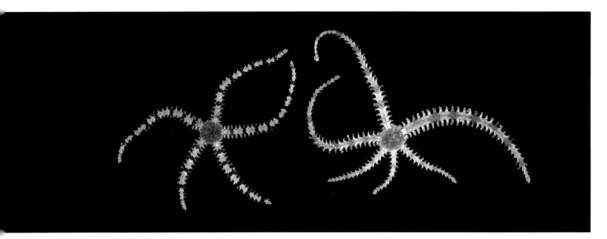

FIGURE 67. *Ophiactis algicola*.

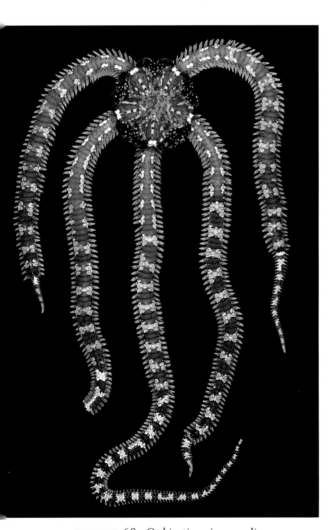

FIGURE 68. *Ophiactis quinqueradia*.

DISTRIBUTION: Bermuda, St. Lucie County, Florida (based on HBOI museum specimens), the Dry Tortugas, Puerto Rico(?), Jamaica, Tobago, Belize, Panama, and Colombia. DEPTH: Shallow water to 24 m (80 ft).

BIOLOGY: This species is fissiparous, reproducing asexually by splitting in two and regenerating the lost body parts.

REMARKS: Available keys to the species of *Ophiactis* are flawed (H. L. Clark 1918, 1933), and *O. algicola* has been confused with *O. lymani* Ljungman, *O. plana* Lyman, and *O. loricata* Lyman, which are all deep-water species.

SELECTED REFERENCES: H. L. Clark 1933:38, 56; Hotchkiss 1982:393, fig. 174a,b; H. L. Clark 1918:303, pl. 4, figs. 5, 6 [as *Ophiactis lymani*].

Ophiactis quinqueradia Ljungman
Figure 68

DESCRIPTION: This is the largest Caribbean *Ophiactis* species and is exclusively five-armed. It attains 13 mm (0.5 in) disk diameter with arms 100 mm (3.9 in) long. The disk spines are small and usually restricted to the ventral and lateral sides of the disk; those spines on the dorsal surface are usually granulelike; often there is a relatively large spine below the outer end of the radial shield. *O. quinqueradia* has two (sometimes three) oral papillae; the most proximal one on the side of the jaw is long and slender; there are one or two rounded scales on the jaw or adoral shield. The dorsal arm plates are subellipsoidal, wider than long. There are usually six arm spines; the middle spines are longest, flattened, with a broad, spinulose tip; the dorsal spines are shorter, rounded, and more pointed; and the most ventral spine is smallest and nearly circular in outline.

The disk is dark reddish brown, purplish brown, or black, variegated with gray and white; the radial shields usually have pale outer tips. The arms are banded with gray, black, tan, or brown; bands often are more evident in small or pale specimens than in darkly pigmented large ones.

HABITAT: Pier pilings, rock reef, and coral reef habitats where appropriate sponges are present.

DISTRIBUTION: The Bahama Islands, the Florida Keys and the Dry Tortugas, Texas offshore reefs, Puerto Rico, the Virgin Islands, the Windward Islands, the Leeward Islands, Barbados, Curaçao, Belize, Panama, and Brazil. DEPTH: 2–618 m (8–2,028 ft), but most records are for depths less than 73 m (less than 240 ft).

BIOLOGY: *O. quinqueradia* is an endocommensal of sponges such as *Agelas* spp., *Verongia lacunosa* (Lamarck), and *Neofibularia nolitangere* (Duchassaing & Michelotti) (Hopkins et al. 1977; Kissling and Taylor 1977; Hendler 1984a). Reports that *O. quinqueradia* usually occurs in algae or seagrass are mistaken (e.g., Voss 1976), but individuals occasionally are found outside sponges, such as one seen inside the test of a dead sea urchin at Looe Key, Florida (Hendler, previously unpublished). *Agelas clathrodes* (Schmidt) and *N. nolitangere* are toxic and are avoided by sponge-eating fishes (Hoppe 1988), possibly to the advantage of the resident brittle stars. Other brittle stars occur on and in the same sponges with *O. quinqueradia,* including *Ophiactis savignyi, Ophiothrix angulata, O. suensonii, O. lineata,* and *O. orstedii* (Hendler 1984a).

At night, *O. quinqueradia* extends pairs of arms from sponge oscula, apparently to suspension feed; when illuminated by dive lights, the arms swiftly press against the surface of the sponge and withdraw (Hendler and Miller, previously unpublished). The species does not reproduce by fission, and the small size of its violet-colored oocytes (0.07 mm in diameter) is indicative that it has planktotrophic larval development (Hendler and Littman 1986). Some large individuals have sessile, ciliate protozoa attached to their arm spines (Hendler, previously unpublished).

SELECTED REFERENCES: Ljungman 1871:628 [as *Ophiactis muelleri* Lütken var. *quinqueradia*]; H. L. Clark 1915:263, pl. 11, figs. 5, 6 [as *Ophiactis muelleri*].

Ophiactis rubropoda Singletary
Figure 69

DESCRIPTION: The disk diameter of this six-armed species reportly reaches 5 mm (0.2 in). It has disk spines only on the interbrachial and ventral surfaces, one pair of oral papillae, and four to five short, pointed arm spines. The body is brown, greenish brown, and gray, and the arms have alternating pale and dark bands of the same color. A notable feature, and a helpful diagnostic characteristic of living individuals, is their red tube feet.

FIGURE 69. *Ophiactis rubropoda*. Photo by G. Hendler.

HABITAT: Among encrusting sponges, cnidarians, and barnacles.

DISTRIBUTION: Originally reported from pier pilings at Virginia Key in Biscayne Bay, Florida (Singletary 1973), and recently collected again at the same locality (Hendler and Pawson, previously unpublished). DEPTH: Shallow subtidal.

BIOLOGY: *O. rubropoda* reproduces asexually by fission. The distinctive color of its tube feet is imparted by red fluid in the water vascular system, presumably a hemoglobin compound, but the site (probably intracellular) and characteristics of the red pigment have not been determined. Individuals retract groups of the tube feet in waves, along the length of the arm; the contracted feet turn white momentarily as pigmented water vascular fluid is expelled.

SELECTED REFERENCES: Singletary 1973:175, figs. 1–4; Ruppert and Fox 1988: 71, fig. 112 left side.

Ophiactis savignyi (Müller & Troschel)
Figures 32-4, 70

DESCRIPTION: A typical specimen of this six-armed species is 3.8 mm (0.1 in) in disk diameter with arms 16.3 mm (0.6 in) long; individuals as large as 5 mm (0.2 in) in disk diameter are relatively rare. The length of the radial shield typically exceeds half the radius of the disk. Small rough-tipped spines, scattered over the disk, are evident on larger specimens. The two (sometimes one or three) oral papillae of *O. savignyi* are flattened and scalelike. The five to six arm spines are of roughly similar size, except for the relatively small ventral spine, and are almost as wide at the tip as at the base. The spine tip is distinctly spinulose, frequently with minute "teeth" at opposite corners.

Individuals are various combinations of greenish, greenish brown, brown, and cream; there is usually a white patch at the outer tip of the darkly pigmented radial shield. The dorsal arm plates typically have pairs of dark marks on the distal edge, which sometimes demarcate a small median distal lobe on the plate (Mortensen 1933a). H. L. Clark (1938) described an Australian "variety" of *O. savignyi*, based on an anomalous specimen with a brown disk and bright yellow arms.

HABITAT: All coral reef zones, seagrass and mangrove areas, and in fouling communities on marine structures and ships' bottoms. It lives in algae, corals, sponges, and rubble.

DISTRIBUTION: In warm waters through-

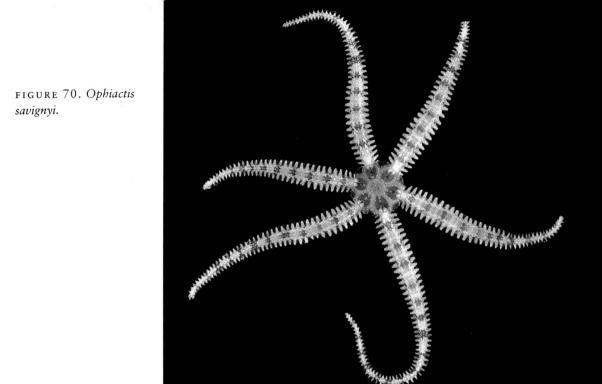

FIGURE 70. *Ophiactis savignyi*.

out the Indo-Pacific, eastern Pacific, and on both sides of the Atlantic. Off the east coast of the Americas it is reported from South Carolina and Bermuda, throughout the Caribbean and Gulf of Mexico, southward to Brazil. DEPTH: Commonly shallow water, but reportedly to 518 m (1,698 ft) (H. L. Clark 1946).

BIOLOGY: *O. savignyi* occurs, sometimes with other *Ophiactis* species, in sponges and algae. It lives in densities approaching 3,000 individuals per liter and up to 855 individuals per sponge, sometimes filling the excurrent canals of the host (A. M. Clark 1967a; Henkel 1982; Hendler 1984a; Emson et al. 1985a; Hendler and Littman 1986; Mladenov and Emson 1988). At least 10 genera of Caribbean sponges harbor *O. savignyi*, and regarding its absence from some sponges, Mortensen (1933b:442) could only suggest that they lack "the right flavour." The population density of *O. savignyi* in coral formations depends on the size, growth form, and other properties of the coral (Sloan 1982; Tsuchiya et al. 1986). Individuals are usually ensconced only in the dead coral skeleton, because they can be killed by living coral polyps (Hendler and Littman 1986). Boffi (1972) described *O. savignyi* as "detritophagous" based on its stomach contents, which largely consist of detritus and sand grains, some bryozoans, foraminiferans, and the remains of small gastropods. Emson and Mladenov (1992) characterized it as a suspension feeder that uses its tube feet to catch current-borne particles and selectively transfer them to the mouth. Mortality of the species caused by high temperature has been reported (Glynn 1968).

Emson and Wilkie (1980) reviewed asexual fission in *O. savignyi*. There is not a consistent plane of fission through the disk. Following fission, two arms grow from the healed edge of the disk and a third arm bud arises in the space between them (H. L. Clark 1914). Therefore, individuals generally have several large arms and several smaller regenerating arms.

O. savignyi of 4 mm disk diameter and

larger may develop gonads; the largest females can produce 10,000 oocytes (0.10 mm in diameter) (Emson et al. 1985a). Mortensen (1931) described the ophiopluteus of this species, but the larva has not been reared through metamorphosis, and it is not known with certainty if newly settled brittle stars have five or six arms. Emson and Wilkie (1984) speculated that large numbers of extremely small six-armed individuals in some sponges were newly recruited and suggested that individuals in single sponges are clonal. As expected, there may be greater genetic similarity among the individuals in a given sponge than among individuals in algal turf, a microhabitat more accessible to immigration (Mladenov and Emson 1990).

Its dual capacity for sexual and asexual reproduction could account for the abundance and widespread distribution of this species (Hyman 1955; Emson and Wilkie 1980). H. L. Clark (1933, 1946:210) ventured that "this is the most common brittlestar in the world." He also pointed out that its pattern of distribution is probably "artificial," because the species can be rafted, or carried on ship bottoms (H. L. Clark 1919).

REMARKS: Several fissiparous *Ophiactis* species recorded from the Caribbean region are superficially similar to, and could be easily confused with, O. *savignyi*.

Ophiactis muelleri Lütken (1856:12) is a six-armed species with one oral papilla and four arm spines. Individuals attain 3–4 mm (0.1–0.2 in) disk diameter with arms 10–15 mm (0.3–0.6 in) long. Its distinctive features, as illustrated and described by Mortensen (1933b, fig. 16), include the almost hexagonal, and somewhat concave, shape of the ventral arm plates, and disk spines that can have hyaline tips. The disk is a dull bluish green with white and dark blue markings; the arms may be striped and banded. It is recorded from the Virgin Islands, Antigua, Barbados, Colombia, Venezuela, Brazil, and other Caribbean localities. But, "in view of the uncertainty as to the nature and limits of the species, it is futile to discuss the distribution" (H. L. Clark 1933:59). Apparently, many citations of O. *muelleri* in the literature actually refer to O. *savignyi* and O. *quinqueradia*. In addition, *Ophiactis cyanosticta* H. L. Clark and *Ophiactis maculosa* Martens probably are synonyms of O. *muelleri* (Mortensen 1933b; A. M. Clark 1976a). The species has been illustrated by several authors (H. L. Clark 1918:307, photo pl. 4, figs. 3, 4 [as O. *cyanosticta*]; Mortensen 1933b:445, fig. 16a–c; de Roa 1967:288, fig. 19).

Ophiactis lymani Ljungman (1871:629) is another six-armed species with a single oral papilla and four arm spines and has been mistaken for O. *muelleri* and O. *savignyi* (see A. M. Clark 1955). The largest specimens have a disk diameter of 5 mm. It is said to differ from O. *muelleri* by its coloration, brownish red or greenish, variegated with white. The radial shields are distinctly less than half the disk radius. Unlike O. *savignyi*, its disk spines (if present) are peripheral and ventral, and have a sharp, hyaline tip. Evidently, the disk spine is the only character that reliably distinguishes this species, but the spines may be absent or broken (Alvà and Vadon 1989). O. *lymani* has been reported from St. Thomas, Brazil, and Bermuda (Mladenov and Emson 1988; Alvà and Vadon 1989). It also occurs off St. Helena, Ascension Island, the Gulf of Cadiz, and the Cape Verde Islands to Angola, Africa, and is reported from the intertidal to over 110 m (361 ft) depth. It has been illustrated (Koehler 1926:24, pl. 5, figs. 1, 2; Mortensen 1933b: 442, figs. 15a–d, 16d–e), and useful figures and commentary on the species are provided by A. M. Clark (1955:35, fig. 12a,b) and Madsen (1970:208, fig. 34).

O. *lymani* may be identical with *Ophiactis plana* Lyman (Lyman 1869; Mortensen 1924; A. M. Clark 1974; A. M. Clark and Courtman-Stock 1976; Guille and Ribes

1981; Alvà and Vadon 1989), which has been reported on both sides of the Atlantic. *O. plana* has been reported off the Florida Keys and the Dry Tortugas, from 24 to 183 m (78–600 ft) (Lyman 1869). It is illustrated in H. L. Clark (1915: pl. 10, figs. 1, 2) and Alvà and Vadon (1989: fig. 5a,b).

Ophiactis notabilis H. L. Clark (1939: 415, pl. 52, figs. 1, 2) was described from a single six-armed specimen collected at 6 m (21 ft) depth off Puerto Rico and has not since been reported. It has a single oral papilla and four thin, blunt arm spines. It is distinguished by very broad dorsal arm plates, the absence of disk spines, and adoral shields that are in contact interradially and conceal the first ventral arm plate. Its color in life is not known.

SELECTED REFERENCES: Müller and Troschel 1842:95 [as *Ophiolepis savignyi*]; Koehler 1922:193, pl. 64, figs. 5, 6, pl. 96, fig. 2; A. M. Clark 1967a:148, fig. 1A; 1976a:112, pl. 1, fig. 4; Irimura 1982:28, fig. 17, pl. 1, fig. 6.

FAMILY AMPHIURIDAE

Most amphiurids burrow; their long arms extend to the surface through a mucus-lined channel. Many species in the family look alike, having a relatively tiny disk, slender arms, and muted color patterns. The amphiurids voluntarily cast off and gradually regenerate the disk, further complicating species recognition. The principal features used for amphiurid identification are microscopic structures such as the oral papillae and disk spines. These features are depicted for each of the species in Figures 100–106. Specimens to be identified should be "keyed out" before consulting descriptions and photographs in the text.

Key to Genera of Floridian and Bahamian Amphiurid Brittle Stars Found at Depths Less Than 30 m (98 ft)

1a. Most of disk lacks scales and appears very dark .... ***Ophionephthys limicola* and *Amphiura (Ophionema) intricata*** (Figures 101H, 103K)
 b. Dorsal surface of disk with spines or tubercles and scales 2
 c. Dorsal surface of disk with scales, spines lacking 3
2a. Spines ("fence papillae," Figure 30D-8) around the dorsal margin of disk ....
 .... ***Ophiophragmus***
 b. Scattered spines or tubercles on dorsal surface of disk ....
 .... ***Ophiostigma* and *Ophiocnida***
3a. Five or four oral papillae on each side of jaw (Figure 32-1) .... ***Amphioplus***
 b. Three oral papillae on each side of jaw (Figure 32-2) 4
4a. Infradental and distal oral papillae separated by a gap; middle papilla set relatively deep in mouth; distal oral papilla often spine-shaped (Figure 32-3) (Note: Sometimes four oral papillae in large individuals) .... ***Amphiura***

b. Oral papillae attached at edge of buccal slit and nearly contiguous; distal papilla often the longest .... **Amphiodia** and **Amphipholis**[1,2]

[1] *Ophiophragmus* species with regenerated disks lacking fence papillae may key out here.
[2] The assignment of various *Amphipholis* and *Amphiodia* species to the nominal genera *Axiognathus*, *Diamphiodia*, and *Microphiopholis* (Fell 1962; Thomas 1966; Turner 1985) is not based on convincing taxonomic evidence (A. M. Clark 1970). Therefore, the latter three names are rejected, pending a systematic revision of both *Amphipholis* and *Amphiodia*.

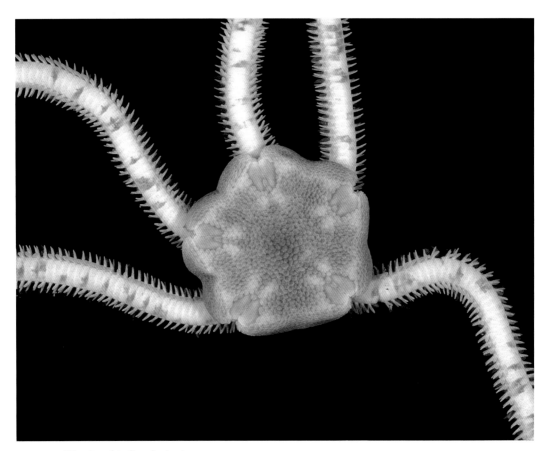

FIGURE 71. *Amphiodia planispina*.

Amphiodia planispina (Martens)
Figures 71, 100A,B,C

DESCRIPTION: This species is moderately large for an amphiurid; a specimen of 10 mm (0.4 in) disk diameter has arms over 100 mm (3.9 in) long. The disk is inflated and covered by opaque, imbricating scales. The pairs of radial shields are almost circular in outline, and they flank a notch at the edge of the disk. There is a characteristic pit at the center of each jaw of preserved specimens. *A. planispina* has three laterally compressed arm spines that are widest below the tip and very blunt. The dorsal spine is longest on joints near the disk; the ventral spine is longest near the arm tip. There are two tentacle scales. Tube feet at the tips of the arms have annuli studded with minute papillae.

The disk is bluish gray. The radial shields are brown or dark gray, with white tips; they are surrounded by whitish scales. In some individuals the arms are whitish or pale gray, irregularly banded with sparse patches of reddish brown and brown (Hendler, previously unpublished); however, the arms of the preserved specimens upon which the original description of the species was based are dusky with purplish gray bands (Thomas 1962b).

HABITAT: Mud and seagrass roots, and unvegetated sandy mud or sand-and-shell locales.

DISTRIBUTION: Florida: Biscayne Channel and Looe Key (Hendler, previously unpublished), the Dry Tortugas, Cuba, Panama, Brazil, and Argentina. There are less reliable records for Puerto Rico, Grenada, and Barbados. DEPTH: 1–49 m (3–161 ft).

BIOLOGY: Individuals readily autotomize the disk and arms, even when collected with utmost care. The species occurs with other burrowing brittle stars such as *Ophiophragmus pulcher* at Looe Key, Florida, and *Amphioplus albidus* (Ljungman) and *Hemipholis elongata* off Brazil (Absalao 1990). Individuals from Looe Key, Florida, had white testes or pale salmon-colored ovaries; the color of the ripe gonads was visible through the disk wall (Hendler, previously unpublished).

REMARKS: Specimens from the Dry Tortugas and Looe Key, Florida, have narrower radial shields and smaller, more numerous disk scales than "typical" *A. planispina*, most of which have been collected from the coast of South America (H. L. Clark 1918; Hendler, previously unpublished). The relationship between *A. planispina* and the Floridian brittle stars requires more extensive taxonomic study (Thomas 1962b; Parslow and A. M. Clark 1963).

Amphiodia atra is distinguished from *A. planispina* as noted under *Remarks* for *A. trychna*.

SELECTED REFERENCES: Martens 1867b: 347 [as *Amphiura planispina*]; H. L. Clark 1915:248, pl. 8, figs. 8, 9; Thomas 1962b: 648, fig. 8A–D; Bernasconi and D'Agostino 1977:93, pl. 2, figs. 2, 3.

Amphiodia pulchella (Lyman)
Figures 72, 100J,K

DESCRIPTION: Individuals seldom exceed 5 mm (0.2 in) in disk diameter with slender arms 40–50 mm (1.6–2.0 in) long. The disk is covered with fine scales, and the pairs of radial shields are closely joined. The primary plates are prominent on the disk of smaller individuals, but generally absent in larger ones. The presence of only a single tentacle scale makes this brittle star unique among shallow-water Caribbean *Amphiodia* and *Amphipholis* species. The middle arm spine is dorsoventrally flattened, with a truncate, echinulate tip; the other two spines are bluntly rounded. Also distinctive are the elongate, blunt-tipped shape of the two distal pairs of oral papillae. The tube feet have a bulbous tip.

The disk is gray to brownish gray; often the primary plates and other large scales are reddish or purplish gray; the radial shields are dark proximally and have a pale distal tip. Reddish brown pigmentation of the stomach may be visible through the thin body wall. The arms are pale gray, blotched or banded with dark gray, brown, or reddish brown.

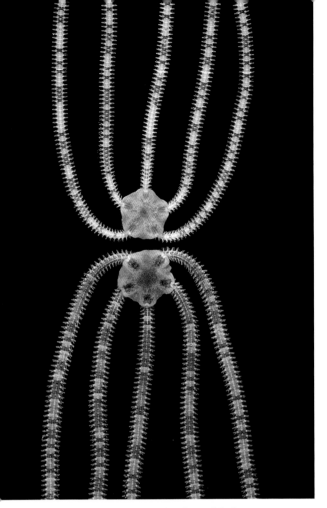

FIGURE 72. *Amphiodia pulchella.*

The arm spines usually have a dusky internal spot. The light middorsal arm stripe noted by Thomas (1962b) was probably from the vertebrae, which are visible through the thin dorsal arm plates.

HABITAT: Soft sediments, usually in back reef areas with seagrass or sponges; also, associated with clumps of algae such as *Halimeda* and the interstices of coral colonies.

DISTRIBUTION: Bermuda, the Bahama Islands, the Florida Keys and the Dry Tortugas, the Gulf coast, Cuba, Jamaica, Puerto Rico, the Windward and Leeward Islands, Tobago, Mexico, Belize, Brazil, and Argentina. DEPTH: Less than 1–71 m (less than 3–234 ft).

BIOLOGY: This small, cryptic species is easily overlooked. However, it occurs in densities of over 100 individuals per liter of algae and four individuals per 0.22 square meter of sediment (McNulty 1961; Hendler and Littman 1986). It can be particularly abundant in poorly oxygenated sediment in areas with the seagrass *Halodule* (O'Gower and Wacasey 1967). At Looe Key, Florida, it was collected from sand and seagrass with *Ophiocnida scabriuscula* and *Ophiophragmus pulcher,* and from filamentous algal turf with *Ophiostigma isocanthum, Amphipholis squamata, Ophionereis squamulosa,* and juvenile *Ophiocoma echinata* (Hendler, previously unpublished).

A. pulchella occupies shallow burrows in soft substrate and extends two to four arms across the sediment surface, with its arm tips slightly raised. When disturbed, or falling through the water, individuals fold their arms in zig-zag fashion and hold them against the top of the disk (Hendler, previously unpublished). This reaction may be similar to the "balling" response of *Amphipholis squamata* described by Emson and Wilkie (1982) and probably serves the same hydrodynamic function. The species has been found in the stomach contents of the sea star *Luidia senegalensis* and the fish *Serranus subligarius* (Cope) (Halpern 1970; Shirley 1982). Based on its very small (0.65 mm) oocytes, *A. pulchella* is presumed to have a planktonic, feeding larval stage (Hendler and Littman 1986).

REMARKS: A similar congeneric species, *A. atra,* is distinguished from *A. pulchella* as noted under Remarks for *A. trychna.*

SELECTED REFERENCES: Lyman 1869: 337 [as *Amphiura pulchella*]; Lyman 1875: pl. 5, fig. 75 [as *Amphiura pulchella*], p. 18, pl. 3, figs. 38–40 [as *Amphiura repens*]; Thomas 1962b:641, fig. 5.

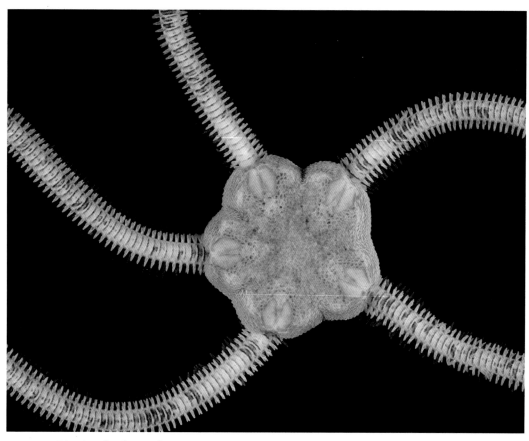

FIGURE 73. *Amphiodia trychna*.

Amphiodia trychna H. L. Clark
Figures 73, 100D,E,F,G

DESCRIPTION: This is a big amphiurid, with arms ranging in length from 13 to over 24 times the disk diameter; a large specimen of 10 mm (0.4 in) disk diameter has arms 250 mm (9.8 in) long. *A. trychna* has closely joined radial shields about twice as long as wide and thick, irregular disk scales. The dorsal arm plates are over twice as wide as long. Its tentacle scales and arm spines are similar in number and shape to those of *A. planispina*; however, the middle arm spine is compressed dorsoventrally, rather than laterally as in *A. planispina*. The adoral shields of *A. trychna* usually are broadly joined proximal to the oral shield; those of *A. planispina* are separated.

The body is mostly tan. The disk is mottled with tan and brown scales. Each dorsal arm plate has a band of dark tan abutting a white line on the plate's distal edge; series of darker and lighter joints confer a banded pattern on the arm (Thomas 1962b).

HABITAT: Unvegetated sand or mud, *Thalassia* seagrass beds, channels in mangrove cays, and reef-associated sand plains (Hendler, previously unpublished).

DISTRIBUTION: Florida (Destin), the Dry Tortugas, Cuba, Puerto Rico, Tobago, Belize, Panama, Venezuela, and Brazil(?). DEPTH: 1–160 m (2–525 ft).

BIOLOGY: H. L. Clark (1918:289) collected *A. trychna* "in company with half a dozen other species of amphiurans" in Buccoo Bay, Tobago. It burrows deep in mud and slowly moves its arm tips across the surface of the sediment to deposit feed (Hendler, previously unpublished). Some individuals have a scaleworm associate, *Malmgreniella puntotorensis* Pettibone (Pettibone 1993).

REMARKS: A similar, moderate-sized, gray-colored species, *A. atra*, is found in lagoonal or estuarine waters, generally intertidal to 30 m (98 ft) depth (see Figures 77, 100*H,I*). Although reported from Puerto Rico, Trinidad, Brazil, and from South Carolina to Texas, it appears to be absent from southern Florida (including the Florida Keys) and the Bahama Islands. A specimen of *A. atra* with 7.6 mm (0.3 in) disk diameter has arms about 100 mm (3.9 in) long (Thomas 1964; Ruppert and Fox 1988). The three arm spines are long, slender, pointed, and of nearly equal size. The oral plate is diamond-shaped in large individuals. Unlike *A. trychna*, its disk scales are very small, except for an enlarged marginal series. Contiguous adoral shields distinguish it from *A. planispina*. It differs from *A. pulchella* in having two tentacle scales, the larger of which extends almost the entire length of the ventral arm plate. *Amphiodia atra* was first described by Stimpson (1852:225 [as *Ophiolepis atra*]) and is illustrated in H. L. Clark (1915:245, pl. 8, figs. 5–7; 247, pl. 8, figs. 3, 4; 245 [as *A. limbata*], pl. 7, figs. 1–4 [as *A. gyraspis*]) and in Thomas (1964:160, figs. 2a–b, 3a–c). Growth stages of the juveniles are depicted in Turner and Miller (1988).

SELECTED REFERENCES: H. L. Clark 1918:289, pl. 3, figs. 1–3 [as *Amphiodia trychna*]; 290, pl. 2, fig. 6 [as *Amphiodia tymbara*]; Thomas 1962b:645, figs. 6A,B, 7A1–A3; Parslow and Clark 1963:30, fig. 9c–e [as *Amphiodia trychna*]; 30, fig. 9a,b [as *Amphiodia tymbara*].

Amphioplus coniortodes H. L. Clark
Figures 74, 101*A,B,C*

DESCRIPTION: The disk diameter of a large individual is 7 mm (0.3 in), and the arms, about 20 times the diameter of the disk, are long and thin compared with *A. sepultus* and *A. thrombodes*. The dorsal disk scales are extremely delicate. The ventral surface of the disk is virtually without scales, and the brown stomach is visible through the naked, transparent body wall. The three arm spines are slender and acutely pointed, and there are two tentacle scales. The disk is gray, and the arms generally are tan, blotched with dark gray.

HABITAT: Usually unvegetated mud or sandy mud, sometimes seagrass beds.

DISTRIBUTION: The Florida Keys and Biscayne Bay, Florida, and possibly Cuba and Venezuela. DEPTH: 2–5 m (7–16 ft).

BIOLOGY: *A. coniortodes* is found in densities as high as 57 individuals per square meter and has been collected at the same sites as *A. sepultus*, *A. thrombodes*, *Amphipholis gracillima*, and *Ophionephthys limicola* (H. L. Clark 1933; McNulty et al. 1962; Thomas 1962b; O'Gower and Wacasey 1967; Singletary 1970, 1980; Singletary and Moore 1974).

In a laboratory setting, Singletary (1970, 1971, 1980) found that this species is diurnally active, a selective detritus feeder, and that it can tolerate a temperature range of 0–39°C. It typically burrows to 10 cm in the sediment and maintains a respiratory current through the burrow by undulating its arms (Woodley 1975). Many individuals show the effects of predation; 87% in one population were regenerating arms, and 3% were regenerating disks (Singletary 1970, 1980).

A. coniortodes breeds seasonally, spawning between September and November; juveniles recruit between November and

BRITTLE STARS AND BASKET STARS

FIGURE 74. *Amphioplus coniortodes*. Photo by G. Hendler.

January. The surprisingly precocious recruits are thought to reach sexual maturity within just 3 months (Singletary 1970, 1980).

SELECTED REFERENCES: H. L. Clark 1918:291, pl. 7, figs. 3, 4; Thomas 1962b: 656, fig. 10B; A. M. Clark 1970:21, 49, fig. 9d,e.

Amphioplus sepultus Hendler
Figures 75, 101D,E

AUTHOR'S NOTE: The following account is a condensed version of the formal description of this species (Hendler, in press).

DESCRIPTION: A large individual is 9 mm (0.4 in) in disk diameter with arms 110 mm (4.3 in) long. The dorsal and ventral surfaces of the disk are covered with thin, overlapping scales. The dorsal and ventral arm spines are blunt-tipped and laterally flattened, and the middle spine is dorsoventrally flattened with a broad, blunt tip. There are two tentacle scales.

Individuals from the Gulf coast and from nearshore bays are brown, patterned with light gray, reddish tan, and dark gray. Their arms are mottled and sometimes have a pale middorsal stripe; scattered black scales may be present on the disk, and the radial shields may have a pale outer tip. Individuals from reef habitats are smaller and more pale.

FIGURE 75. *Amphioplus sepultus*.

HABITAT: Soft sediments ranging from mud to sand with shell fragments; most abundant in fine sediments associated with seagrass.

DISTRIBUTION: Florida (from Biscayne Bay to the Dry Tortugas on the Atlantic coast and Flamingo to Destin on the Gulf coast).
DEPTH: Intertidal to 82 m (269 ft).

BIOLOGY: Thomas (1962b:654), referring to this species as *Amphioplus abditus* (Verrill), characterized it as "the most common intertidal amphiurid of South Florida," and a density of 410 individuals per square meter has been reported (McNulty et al. 1962). It seems to be distinctly less common in coral reef habitats than in bays (Hendler, in press).

The portion of the arm in the sediment undulates to circulate water through the burrow. The pumping arms have a "respiratory fringe" of mucus-coated arm spines, tube feet, and adhering sediment particles, forming a gasket between the arm and the burrow (Woodley 1975). This species ingests sediment and plant material, algae, pollen, fecal pellets, and microscopic invertebrates (Hendler 1973).

A. sepultus spawns during the winter. Sexually ripe individuals have whitish testes or ovaries that are whitish to gray, sometimes tinged with yellow or green. Females produce up to 9,000 eggs, about 0.18 mm in diameter. The larva probably has an abbreviated pattern of development (Hendler 1973, 1977a). This species is preyed upon by the sea star *Luidia senegalensis* (Halpern 1970). It tolerates an upper lethal temperature of approximately 40°C (Singletary 1971).

A. sepultus hosts a menagerie of symbionts including the copepods *Parophiopsyllus ligatus* Humes & Hendler and *Presynaptiphilus amphiopli* Humes & Hendler (Humes and Hendler 1972), a rotifer, the scaleworm *Malmgreniella maccraryae* Pettibone (Pettibone 1993), a juvenile leptonacean clam, and internal parasites including an unidentified copepod and a flatworm metacercaria (Hendler 1973).

REMARKS: This species has mistakenly been referred to as *Amphioplus abditus* (Verrill), a congener that occurs from Maine only as far south as Georgia (Hendler 1973). Furthermore, *A. sepultus* and *A. abditus* have been confused with *Amphioplus macilentus* (Verrill), a smaller, offshore species living at depths of 97–210 m (318–689 ft) from Massachusetts to North Carolina.

SELECTED REFERENCES: Thomas 1962b: 651, fig. 9A,B [as *A. abditus*]; Hendler 1973: 40, figs. 5d–f, 8C–D, 11a–f, 12a–h; Hendler, in press.

Amphioplus thrombodes H. L. Clark
Figures 32-1, 76, 101F,G

DESCRIPTION: A specimen of 5 mm (0.2 in) disk diameter has arms 60 mm (2.4 in) long. *A. thrombodes* is deceptively similar to *A. sepultus* in overall appearance. However, bumps on the edge of the dorsal scales give its disk a characteristic papillate texture. Most arm joints have a single pair of tentacle scales; a few proximal arm joints have two pairs. Of the three arm spines, the middle one is thickest and somewhat compressed; its tip is not broadened as in *A. sepultus*.

The disk is pale gray, with scattered brownish scales and white-tipped radial shields, and the arms are pale tan, blotched with dark gray.

HABITAT: Sandy mud sediment in seagrass beds (typically of *Halodule wrightii* Ascherson).

DISTRIBUTION: Previously reported from the Gulf coast and the Florida Keys; the range extends to Fort Pierce in northern

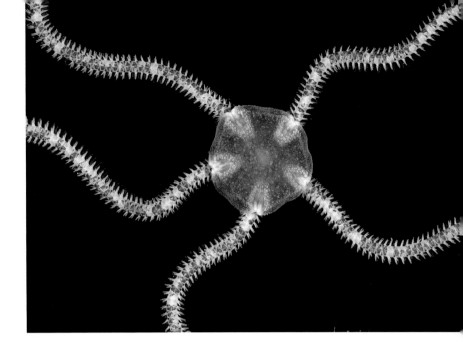

FIGURE 76. *Amphioplus thrombodes*.

Florida (Miller, previously unpublished).
DEPTH: 0.3–0.6 m (1–2 ft).

BIOLOGY: *A. thrombodes* can occur with other burrowing brittle stars such as *A. sepultus*. Small rotifers sometimes attach to its arm spines, similar to those associated with *A. sepultus*, *Ophiophragmus pulcher*, and *A. gracillima*, tentatively identified as *Zelinkiella synaptae* (Zelinka) (Hendler, previously unpublished; Ruppert, personal communication).

SELECTED REFERENCES: H. L. Clark 1918:292, pl. 7, figs. 1, 2; Thomas 1962b: 654, fig. 10A.

Amphipholis gracillima (Stimpson)
Figures 77, 102A,B

DESCRIPTION: An individual of 4.6 mm (0.2 in) disk diameter has arms 52 mm (2 in) long, but the disk may grow to 8 mm (0.3 in) with the arm length reaching an estimated 20 times the disk diameter. This delicate amphiurid has a round, inflated disk and long, almost threadlike arms. The disk is covered with extremely fine scales of uniform size. The thin radial shields, four to six times longer than wide, are set in notches at the disk edge. The distal oral papilla is shaped like a long isosceles triangle, with the narrow base proximal and the long sides tapering to

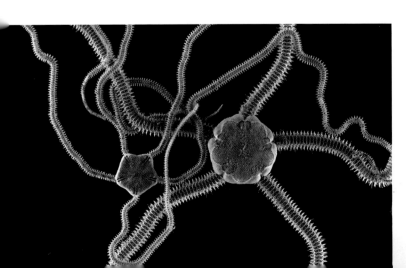

FIGURE 77. *Amphipholis gracillima* (left) and *Amphiodia atra* (right). Photo by R. Fox, courtesy of E. Ruppert and R. Fox.

a point at the distal end. *A. gracillima* has as many as five arm spines near the base of the arm, but there are three slender, pointed spines on most arm joints. The middle spine is slightly flattened and longer than the others. There are two narrow tentacle scales. The tube feet are smooth, with a rounded terminal bulb.

The disk is tan or gray, and the radial shields are often a dark contrasting hue and have a white distal tip. The brown stomach may show through the thin disk wall. The arms are irregularly banded with tan and gray or black and sometimes have a thin, whitish, middorsal stripe.

HABITAT: Usually reported from unvegetated sand or mud, but collected at Looe Key, Florida, from sediment around *Syringodium* seagrass (Hendler, previously unpublished). Prefers sediments with intermediate grain size and high organic content (Zimmerman et al. 1988).

DISTRIBUTION: Bermuda, the Bahama Islands, South Carolina, Virginia, the Florida Keys, Puerto Rico, the Virgin Islands, Tobago, Curaçao, and Belize. DEPTH: Shallow water to 26 m (84 ft).

BIOLOGY: *A. gracillima* occurs in densities up to 56 individuals per square meter in Biscayne Bay, Florida, where it is often the most numerous macroscopic benthic invertebrate. It may occur with *Amphioplus coniortodes* and *Ophionephthys limicola* (Thomas 1962b; Singletary and Moore 1974). Its tolerance of temperature stress (Singletary 1971) may partially explain the local abundance and wide geographic distribution of this species. Some individuals carry symbionts including scaleworms, *Malmgreniella maccraryae* and *M. taylori* Pettibone on the disk, the bivalve *Mysella* sp. C on the arm spines, and a rotifer, *Zelinkiella synaptae,* on the tube feet (Ruppert and Fox 1988; Pettibone 1993).

Stimpson (1852) was the first to discover that this "singular species" burrows 20 cm deep in the mud with one or two arms extending to the surface of the sediment, and that it "throws" off its disk when disturbed (compare Zimmerman et al. 1988). Normally, *A. gracillima* extends its arms to feed on suspended particles or bottom material (Singletary 1970, 1980); it preferentially selects items less than 0.15 mm in diameter with an organic or bacterial coating (Clements and Stancyk 1984).

It is likely that predators, such as goatfish, induce autotomy of the exposed arms more often than of the buried disk; individuals with regenerating disks (1.5–70%) are somewhat less common than those with regenerating arms (77–85%) (Singletary 1970, 1980; Stancyk et al. 1994). A disk with a functioning stomach can be regenerated within 14 days after autotomy (Dobson 1986), a growth process that may entail the uptake of dissolved nutrients from seawater (Clements 1985, 1986; Clements et al. 1988; Dobson et al. 1991). In the absence of external nutrient, autotomized structures are regenerated at different rates, depending on the severity of tissue loss (Fielman et al. 1991). The rate of regeneration is also temperature dependent. The high frequency and rapidity of regeneration in warm waters suggest that the species may provide a significant source of food for pelagic or benthic predators (Stancyk et al. 1994).

A. gracillima spawns throughout the year in Florida; the ovaries contain few large, pink eggs at any one time, because the ripe gametes are continuously released (Singletary 1970, 1980). The larva of this species is an eight-armed ophiopluteus (Webb, personal communication). Singletary (1970, 1980) estimated that the recently metamorphosed juveniles have a remarkably rapid growth rate, reaching sexual maturity in only 2 months.

SELECTED REFERENCES: Stimpson 1852: 224 [as *Ophiolepis gracillima*]; H. L. Clark 1915:241, pl. 6, figs. 5, 6; John and Clark 1954:155, fig. 8; Thomas 1962b:660, fig. 12A,B.

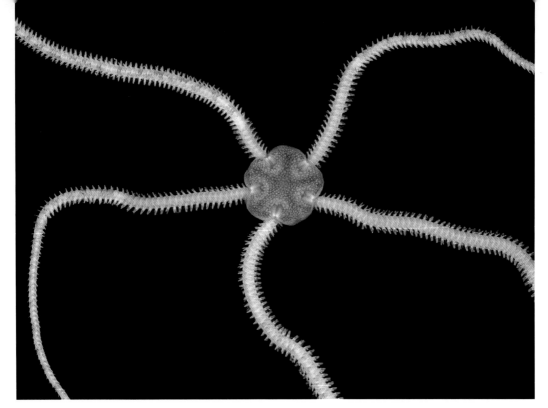

FIGURE 78. *Amphipholis januarii*.

Amphipholis januarii Ljungman
Figures 78, 102C,D,E

DESCRIPTION: *A. januarii* is a moderate-sized amphiurid; a specimen of 4 mm (0.2 in) disk diameter has arms 40–50 mm (1.6–2.0 in) long. Its toothed arm spines set the species apart from other *Amphipholis* and *Amphiodia* species described in this book. On joints with three arm spines, the middle spine is longest, flattened, and has a prominent, flesh-covered, subterminal tooth (sometimes several teeth) on each edge. On joints with four arm spines, the second from the bottom is similarly modified. The other spines are rounded, terminating in a blunt point. The tentacle scale on the ventral arm plate is distinctly broader than the associated scale on the lateral arm plate. The proximal tube feet have a bulbous tip, and those at the outer end of the arm bear several small, ventrally directed papillae.

The disk and arms are light brown, tan, light greenish yellow, or gray, and there are usually incomplete bands of dusky gray, gray-green, or brown on the arms (H. L. Clark 1933; Thomas 1962b). The center and edge of the disk scales may be different hues; the radial shields, with whitish distal tips, are usually darker than the rest of the disk. Sometimes a pale, thin, middorsal stripe on the arm is produced by the vertebrae showing through the thin dorsal plates. Within the arm spines there is often a spot of brown pigment. The oral frame and proximal arm plates may be darker brown than the remainder of the ventral surface of the arm.

HABITAT: In algae, under stones, in bryozoan colonies, within small crevices in rock, rubble, and coral, and in shelly mud or sand. It tends to be more abundant in seagrass beds of *Thalassia* than of *Halodule* (O'Gower and Wacasey 1967).

DISTRIBUTION: South Carolina, Florida (based on HBOI museum collections) and the Florida Keys, Texas, Cuba, Puerto Rico, the Virgin Islands, Tobago, Brazil, and an uncertain record for Barbados. DEPTH: 1–55 m (2–180 ft).

CLASS OPHIUROIDEA

BIOLOGY: *A. januarii* has a penchant for protected crevices in buried objects such as rubble or shell. At Looe Key, Florida, an individual was found beneath a buried shell fragment together with *Ophiopsila vittata*, and another in a small piece of buried rubble with *Ophiophragmus pulcher* (Hendler, previously unpublished). The species also occurs with *Amphiura* (*Ophionema*) *intricata* and *Ophiocnida scabriuscula* (H. L. Clark 1933; Carrera 1974). Small individuals, particularly abundant in algae, can attain densities of 7.5 individuals per liter (Boffi 1972).

Woodley (1975) found that the "hatchet spines" produce a mucus, which traps a fringe of sediment on the arms, helping to pump water through the burrow. The eggs of the species are gray, and ripe and "nearly ripe" individuals have been reported from different localities in May, July, and November (Tommasi 1970; Carrera 1974; Hendler, previously unpublished).

SELECTED REFERENCES: Ljungman 1867:165; Thomas 1966:827; and as *Amphipholis pachybactra* in: H. L. Clark 1918: 284, pl. 1, figs. 3–5; Thomas 1962b:657, fig. 11A,B; Tommasi 1967:1, fig. 1; A. M. Clark 1970:28, fig. 6b–d.

Amphipholis squamata (Delle Chiaje)
Figures 79, 102F,G

DESCRIPTION: The largest individuals from this region are only about 3 mm (0.1 in) disk diameter with arms 11 mm (0.4 in) long, but the species grows to at least 5 mm (0.2 in) in disk diameter in temperate waters. Because of its small size, it is sometimes mistaken for the juvenile of larger species. However, it has coarse scales on top of the disk, primary plates usually lacking, and pairs of radial shields that touch one another and are nearly circular in outline. Its middle oral papilla is rounded and markedly smaller than

FIGURE 79. *Amphipholis squamata*. Photo by G. Hendler.

the infradental papilla and the elongate, opercular distal oral papilla. The dorsal and ventral arm plates of successive joints are separated by the lateral arm plates, and the projecting ridges of the lateral arm plates give the arm a "beaded" appearance. Each ridge bears three erect arm spines (sometimes four near the disk) that have a bulbous base. There are two tentacle scales. The adoral shields are broadly in contact proximal to the oral shield.

The disk is white, yellow, orange, pale brown to reddish, or gray (dark pigmentation of the stomach sometimes shows through the body wall), and the radial shields commonly have a white distal tip; the arms are pale brown, yellowish, or white, sometimes with thin dusky, green, or brown bands.

HABITAT: Rock or coral reef, seagrass and algae beds, mangroves, estuaries, and numerous other habitats, even brackish and hypersaline waters (Tortonese and Demir 1960; Emson and Foote 1980; Pina Albuquerque and Pérez Ruzafa 1984; Pina Albuquerque 1985). It lives beneath stones on sand bottoms, under rubble, and in colonies and clumps of sponges, bryozoans, mussels, and marine plants.

DISTRIBUTION: "Cosmopolitan," occurring in all except the extreme polar regions, though not reported from every Caribbean island. DEPTH: Intertidal to 1,330 m (4,363 ft); those found at the greatest depths are not sexually mature (Gage et al. 1983).

BIOLOGY: A. squamata is a luminescent, viviparous, self-fertilizing hermaphrodite. It may occur in aggregations exceeding 500 individuals per square meter (Rumrill 1982; Pérez Ruzafa 1989) or even 500 per liter of algae (Boffi 1972; Zavodnik 1972; Rumrill 1982; Witman 1982). The dense assemblages are maintained, in part, by negative phototaxic behavior (Martin 1968) and by the propensity of individuals to roll into a ball and sink to the bottom (Corry 1974; Austin and Hadfield 1980; Emson and Wilkie 1982). Movements of the tube feet and of the arms are both used for locomotion (Fell 1946; Martin 1968; Austin and Hadfield 1980). Studies have revealed specialized muscular, sensory, and secretory structures of its spines and tube feet (Buchanan 1963; Pentreath 1970; Corry 1974; Whitfield and Emson 1983; De Voss 1985) and have localized the source and neurological control of the brittle star's green luminescence and fluorescence (Brehm and Morin 1977; Mallefet et al. 1989, 1992).

This diminutive brittle star feeds on suspended material, bottom detritus, and tiny animals and plants (e.g., Martin 1968; Pentreath 1970; Austin and Hadfield 1980; Jones and Smaldon 1989) and is one of the brittle star species involved in the reef's fish feces food web (Rothans and Miller 1991). Its mode of nitrogen excretion was documented by Stickle (1988).

In temperate regions, A. squamata lives for 1–2.5 years and reproduces seasonally, adults brooding up to 25 embryos. In the Tropics individuals have fewer embryos and breed year-round (Johnson 1972; Hendler 1975; Rumrill 1982; Buckland-Nicks et al. 1984; Emson and Whitfield 1989; Emson et al. 1989; Jones and Smaldon 1989; Alvà and Jangoux 1992). Embryos removed from bursae and reared to maturity in isolation produced viable offsping, demonstrating that the species can self-fertilize (or is parthenogenetic) (Strathmann and Rumrill 1987; cf. Binaux and Bocquet 1971).

The eggs of A. squamata are tiny (0.1 mm in diameter). Its embryonic development takes from 3 to 7 months (Johnson 1972; Hendler 1975; Rumrill 1982). It has been suggested that nutrient is transferred directly from the mother to the embryo (Fell 1940, 1946), taken up by the embryo from dissolved substances (Fontaine and Chia 1968; Lesser and Walker 1992), and supplied by endosymbiotic bacteria in the tissues of the bursa and the embryo (Walker and Smith 1985; Lesser 1986; Walker and Lesser 1989; Lesser and Walker 1992). Advanced embryos with an active embryonic mouth and esophagus feed while in the bursa (Bernasconi 1926; Johnson 1972; Oguro et al. 1982; Walker and Fineblit 1982; Walker and Lesser 1989).

The dispersal of A. squamata is not hampered by its lack of a planktonic larval dispersal stage. The species is thought to have spread along shallow land bridges (Fell 1946). However, it can raft even to isolated oceanic islands and has been found en route in floating kelp (Mortensen 1933b, 1941).

Fish such as gobies and wrasses prey upon A. squamata and may account for some instances of arm autotomy (Martin 1968; Emson and Wilkie 1980, 1982), but arm regeneration is also attributable to physical disturbance (Alvà and Jangoux 1990). Several related species of copepod parasites (genus Cancerilla) cling to the brittle star's arms and may induce a growth anomaly in the host (Mortensen 1933a). The species is also attacked by gall-forming copepods (genus *Parachordeumium* [Goudey-Perrière 1979; Boxshall 1988]). *Parachordeumium* and *Rhopalura ophiocomae* Giard, a microscopic mesozoan parasite, disrupt reproduction and growth (Rader 1982; Jangoux 1984; Emson et al. 1988; Whitfield and Emson 1988). Ciliate protozoans, copepods, and a polychaete also are associated with this brittle star (Johnson 1972; Zavodnik 1972; Barel and Kramers 1977; Martin and Alvà 1988).

A. *squamata* sometimes clings to large brittle stars such as *Ophiocoma dentata* Müller & Troschel and *O. erinaceus* Müller & Troschel (H. L. Clark 1921; Ely 1942; Devaney 1974a). It has also been detected on *O. echinata,* on the biscuit urchin *Clypeaster rosaceus,* and the sea cucumber *Isostichopus badionotus* (Hendler, previously unpublished). Perhaps the association is "more than coincidental" (Devaney 1974a), but without more information that cannot be determined.

REMARKS: Since its discovery, this species has been assigned at least 25 different scientific names; some modern treatments cite it as *Axiognathus squamatus*. Historically, *Amphipholis* populations from varied geographic regions have been regarded as separate species for a time and then referred to *A. squamata*. Further investigation is required to determine the genetic relatedness of different populations and whether *A. squamata* is truly a single biological species (Mortensen 1924).

SELECTED REFERENCES: Delle Chiaje 1828:74, 77, 79, pl. 34, figs. 1–4 [as *Asterias squamata*]; Coe 1912:81, pl. 17, figs. 1–5; Bernasconi 1926:146, pl. 2, figs. 1–3, pl. 3, figs. 1–5, pl. 4, figs. 1–9; Thomas 1962b:662, fig. 13A,B; Irimura 1982:41, fig. 26A,B, pl. 2, fig. 1.

Amphiura fibulata Koehler
Figures 80, 103A,B,C,D

DESCRIPTION: Typical specimens grow to 8 mm (0.3 in) disk diameter, and the arms exceed 75 mm (3 in) in length. The disk is completely scaled, but scales in the ventral interradii are minute. The radial shields are three times longer than wide and bluntly rounded at both ends. The proximal dorsal arm plates are nearly circular in outline. There is only one tentacle scale per tentacle pore. Individuals have five to eight broad, blunt arm spines, which generally have an internal brown spot. The ventral spine is longest, tapered, and somewhat flattened. The next two spines are somewhat pickax shaped, with terminal, flesh-covered teeth that are directed to opposite sides. The following two may have distally directed, terminal thorns; other spines are distinctly flattened, with prickly tips, and diminish in length toward the top of the arm. A yellowish or orange stripe may extend the length of the arm of preserved specimens (H. L. Clark 1933; Thomas 1962b). The arm of a living individual from Looe Key, Florida, had a brown dorsal stripe and was tan on the ventral surface.

HABITAT: Unconsolidated substrates (sand, sand and shell hash), in sand-filled shells, and possibly among seagrass roots.

DISTRIBUTION: The Florida Keys and the Dry Tortugas, Barbados, and Brazil.
DEPTH: Generally 2–10 m (7–33 ft). Based on a synonymy suggested by Thomas in an unpublished manuscript, the species may occur off Cuba and at depths up to 239 m (784 ft).

BIOLOGY: The well-developed teeth on the arm spines, as in *A. palmeri,* probably contain glands that secrete a burrow-stabilizing mucus. Randall (1967) found *A. fibulata* in the stomach contents of the margate, a bottom-feeding fish.

REMARKS: The distinctive double-toothed arm spine serves to distinguish *A. fibulata* from the somewhat similar, thorny-spined *Amphiura palmeri*. Both species have rounded dorsal arm plates. However, the latter species has two tentacle scales, and *A. fibulata* has single scales (rarely two on some of the proximal arm joints). *Amphiura semiermis* Lyman and *Amphiura kinbergi* Ljungman are two congeneric Caribbean species that have toothed arm spines (Thomas 1965a). The former occurs at bathyal depths off the Dry Tortugas; the latter lives at shallow depths but is not known from Florida (Koehler 1914a; Thomas 1965a).

SELECTED REFERENCES: Koehler 1913: 359; 1914a:56, pl. 7, figs. 3–5; Thomas 1962b:638, fig. 4A–C.

FIGURE 80. *Amphiura fibulata*. Photo by G. Hendler.

Amphiura (Ophionema) intricata
(Lütken)
Figures 81, 103*K,L*

DESCRIPTION: This is a large amphiurid; specimens of 6–8 mm (0.2–0.3 in) disk diameter have slender arms about 140–150 mm (5.5–5.9 in) long. It has a dark-colored disk and contrasting, pale, bar-shaped radial shields. The disk is naked, like that of *Ophionephthys limicola*, but lacks the peripheral chains of scales that characterize that species. *A. intricata* has a deep cavity in the middle of each jaw and three oral papillae; the infradental is larger than the middle papilla, and the distal papilla is very small. There are four or five arm spines on arm joints near the disk; all are slender and acutely pointed, with finely serrate edges; the two most ventral arm spines are the longest. The ventral arm plates near the disk are rectangular, longer than wide. Tentacle scales are lacking.

FIGURE 81. *Amphiura (Ophionema) intricata*. Photo by G. Hendler.

There are dark brown dashes along the dorsal midline of the arm, brown marks on the lateral arm plates at the base of the arm spines, and yellow lines radiating from the oral frame to the proximal ventral arm plates (Thomas 1964). There may be a discontinuous purple stripe on each edge of the arm and irregular purple bands on the yellowish white ventral surface of the arm (Lütken 1869; H. L. Clark 1918).

HABITAT: Mud, sandy mud, and sand sediments.

DISTRIBUTION: The Florida Keys, the Virgin Islands, and Tobago. DEPTH: 1–8 m (2–20 ft) and a single record for 328 m (1,076 ft) off St. Croix. Presumably, this species most often occurs "beyond the range of the shore and shallow water collector" (Thomas 1964:160).

BIOLOGY: It is possible that *A. intricata* is rarely reported because it is so unlikely to be recovered intact; its delicate disk and arms are prone to damage and autotomy. Individuals probably burrow deep in the sediment. H. L. Clark (1918:283) noted that an animal in a basin of water thrust a single arm outward and contracted it in an "S" to pull the disk forward, "the process being rapidly repeated led to a speed not often attained by an echinoderm. . . ." At St. Thomas, *A. intricata* was reported by Lütken (1869) to occur with two other infaunal species, *Ophionephthys limicola* and *Amphipholis gracillima*.

SELECTED REFERENCES: Lütken 1869: 27, unnumbered fig. [as *Amphiura intricata*]; H. L. Clark 1918:282, pl. 2, figs. 1–3; Nielsen 1932:265, figs. 6a, 7a; Thomas 1964: 159, fig. 1a,b [as *Ophionema intricata*]; A. M. Clark 1970:19, fig. 3h [as *Amphiura (Ophionema) intricata*].

FIGURE 82. *Amphiura palmeri*.

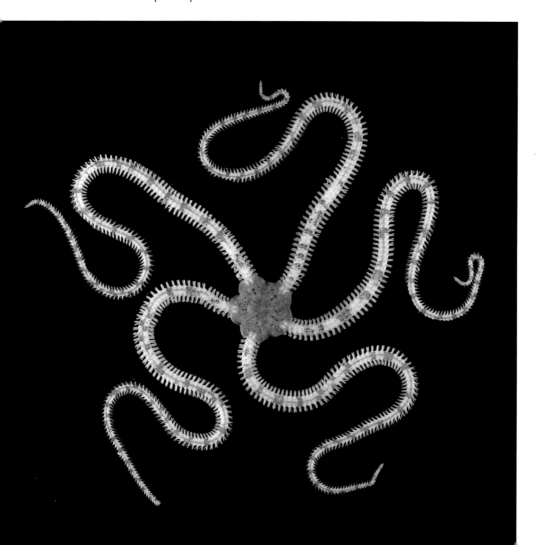

Amphiura palmeri Lyman
Figures 82, 103E,F,G

DESCRIPTION: *A. palmeri* is a moderate-sized amphiurid, reaching 8 mm (0.3 in) disk diameter; a specimen of 5.3 mm (0.2 in) disk diameter has arms 60 mm (2.4 in) long. Similarly to *A. fibulata,* it has proximal dorsal arm plates that are nearly circular and a completely scale-covered disk. The radial shields are at least three times longer than wide, with an acute proximal angle. Unlike *A. fibulata,* it has two tentacle scales at each tentacle pore; the scale on the lateral arm plate is larger than, and overlaps, the one on the ventral arm plate. *A. palmeri* is further distinguished by the shape of its six to seven flattened arm spines, which gradually, but markedly, increase in length from the dorsal to the ventral side of the arm. All spines except the very upper and lowermost have a spinulose tip and a distally directed subterminal tooth. The beaklike apex of these spines is covered with transparent flesh.

The species has been described as "light tan when alive" (Thomas 1962b:635). Individuals from Looe Key, Florida, have the disk gray, light tan, or yellowish, accented with some larger brown scales and white-tipped radial shields. The arms are tan to orange, banded with light brown and gray, often with a dark middorsal stripe near the tip of the arm; there may be a dark brown pigment spot within the arm spines.

HABITAT: Coarse calcareous sand and among the roots of *Thalassia* seagrass.

DISTRIBUTION: Georgia, the Florida Keys and the Dry Tortugas, Puerto Rico, the Virgin Islands, Barbados, and possibly Venezuela. DEPTH: Less than 6 m (less than 20 ft) near Florida and the Dry Tortugas, but elsewhere reported from 183 to 479 m (600–1,572 ft).

BIOLOGY: Individuals burrow in the sediment and may wedge in fissures in buried rubble (Hendler, previously unpublished). It has been collected with other burrowing and crevice-dwelling brittle stars such as *Amphioplus thrombodes, A. sepultus, Amphiura stimpsonii, Ophiostigma isocanthum, Ophiocnida scabriuscula,* and *Ophionereis reticulata* (Koehler 1914a; Thomas 1962b). Woodley (1975) found that mucus produced by its "hatchet arm spines" retains a fringe of sediment on the arms that acts as a gasket, enabling the arm to pump water through the burrow. The proximal tube feet of *A. palmeri* have bulbous tips, and the distal tube feet have ventrally directed, microscopic papillae, which are probably involved in suspension feeding (Hendler, previously unpublished).

SELECTED REFERENCES: Lyman 1882: 143; Koehler 1914a:55, pl. 18, figs. 1, 3; Thomas 1962b:633, fig. 2A,B; Koehler 1913:356, pl. 20, figs. 1–4 [as *A. kuekenthali*].

Amphiura stimpsonii Lütken
Figures 83, 103H,I,J

DESCRIPTION: This is the smallest shallow-water *Amphiura* species from the Caribbean region, rarely larger than 4 mm (0.2 in) disk diameter, with arms 15–20 mm (0.6–0.8 in) long. Because it has single tentacle scales, it might be mistaken for juvenile *A. fibulata* (photographs of the two species are shown side by side in Koehler [1913]). However, it is distinguished by having three to five arm spines; its thorny-tipped middle spines often have diminutive distally and proximally directed thorns. It lacks the exaggerated subterminal teeth characteristic of *A. fibulata* and *A. palmeri* arm spines (compare Figure 103J,G, and C). The ventral arm plates of *A. stimpsonii* are markedly longer than wide, and the dorsal arm plates have

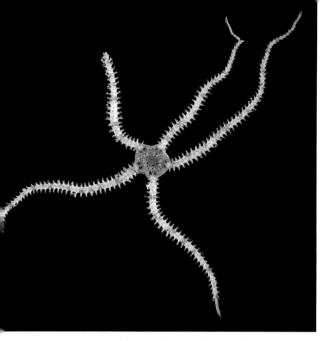

FIGURE 83. *Amphiura stimpsonii*.

straight lateral edges that converge proximally and a convex distal edge.

The disk is grayish or reddish, and the large scales, primary plates, and radial shields are darkly pigmented; outer tips of the radial shields usually are pale. The arms usually are more pale than the disk, light yellow or white, sometimes irregularly banded with brown or gray. Some individuals, described as *Amphiura vivipara annulata* by H. L. Clark (1918), have arms with red bands.

HABITAT: From the back reef to the fore reef slope; in sheltered substrate such as dense mats of algae, thickets of branching coral, coral fragments, and shell rubble.

DISTRIBUTION: The Florida Keys and the Bahama Islands (Hendler, previously unpublished), the Dry Tortugas, Texas offshore reefs, Jamaica, Haiti, Puerto Rico, the Virgin Islands, the Leeward Islands, Barbados, Tobago, the Netherlands Antilles, Belize, and Brazil. DEPTH: 1–126 m (2–414 ft).

BIOLOGY: *A. stimpsonii* commonly occurs in clumps of coralline algae with other species of small, live-bearing, or asexually reproducing brittle stars (H. L. Clark 1918; Hendler and Littman 1986). It has large yolky eggs (0.7 mm in diameter), broods embryos in its bursae, and releases "crawl-away" young that have a disk diameter of 0.8 mm and up to 10 arm joints (Hendler 1988a; Byrne 1991). It is hermaphroditic; an ovary and a testis are connected to each bursa (Mortensen 1920, 1921; Hendler 1975).

SELECTED REFERENCES: Lütken 1859b: 218; Koehler 1914a:64, pl. 7, figs. 1, 2 [as *A. stimpsoni*]; H. L. Clark 1918:268, pl. 1, figs. 1, 2 [as *A. vivipara*]; Thomas 1962b:636, fig. 3A,B [as *A. stimpsoni*]; Hendler 1988a:20, figs. 1A–C, 2A–E.

Ophiocnida scabriuscula (Lütken)
Figures 84, 104A,B

DESCRIPTION: A moderate-sized amphiurid; the disk can attain 9 mm (0.4 in) in diameter, but the arms are relatively short. An individual of 6.6 mm (0.3 in) disk diameter has arms 62 mm (2.4 in) long. This species is unique among shallow-water Caribbean amphiurids because most of its disk is covered with scattered, short, pointed spinules. The proximally diverging pairs of radial shields are bare of spinules and flanked by overlapping scales that are distinctly larger than those at the center of the disk. Three thick arm spines with truncate tips give the arm a robust appearance. The dorsal and ventral spines are laterally compressed. The middle spine is dorsoventrally flattened, and near the middle of the arm it is longer than the other spines. The dorsal and ventral arm plates are wider than long. The dorsal arm plates of adjacent arm joints overlap, but ventral arm plates do not. There are two flattened tentacle scales.

The disk is light yellow to yellowish brown, with contrasting areas of gray, yellow,

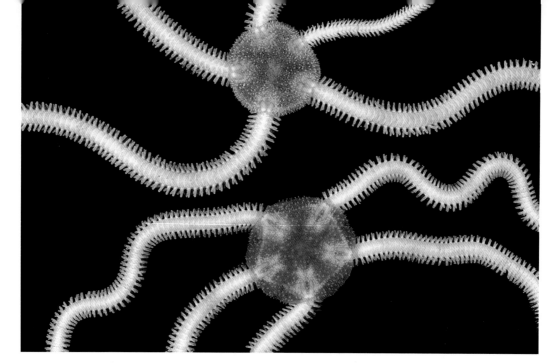

FIGURE 84. *Ophiocnida scabriuscula*.

or pink; the outer tips of the radial shields are white. The arms are yellowish brown, the distal portion sometimes with thin purplish brown bands between joints, and purple spots and yellow patches on the dorsal arm plates. The arm spines may be brownish. The dorsal arm plates of individuals from Puerto Rico have a pale or white distal border (Carrera 1974).

HABITAT: Shallow seagrass habitats, where sandy sediments are mixed with coral rubble or stones.

DISTRIBUTION: Bermuda, Florida (based on HBOI museum specimens), the Florida Keys, the Dry Tortugas, Jamaica, Puerto Rico, the Virgin Islands, the Leeward Islands, Tobago, Venezuela, and Brazil. DEPTH: Less than 2 m (less than 7 ft).

BIOLOGY: *O. scabriuscula* reportedly covers itself with a layer of sediment and does not burrow (Kissling and Taylor 1977), but it has been found associated with buried seagrass rhizomes. Sediment clings to the disk and arm spines of living individuals, which suggests the presence of mucus-secreting structures on those surfaces (Hendler, previously unpublished). Individuals can autotomize and regenerate both the arms and the disk, and the regenerating disks possess typical spines and radial shields (Koehler 1907; Hendler, previously unpublished). Sexually ripe specimens have been reported from Puerto Rico in April (Carrera 1974) and from Key Biscayne, Florida, in January (Hendler, previously unpublished). The moderate size of the eggs, 0.21 mm in diameter, is indicative that the species has an abbreviated larval development (Hendler 1975).

SELECTED REFERENCES: Lütken 1859b: 220, pl. 3, fig. 4a–c [as *Amphiura scabriuscula*]; H. L. Clark 1915:251, pl. 9, figs. 3, 4; Thomas 1962b:684, fig. 21A,B.

Ophionephthys limicola Lütken
Figures 85, 101H,I

DESCRIPTION: This brittle star attains an impressive size for an amphiurid, 13 mm (0.5 in) disk diameter with arms over 250 mm

FIGURE 85. *Ophionephthys limicola.*

(9.8 in) long. The disk appears dark green-brown or yellow-green, because the color of the stomach shows through the diaphanous body wall. The few scales on the disk that are visible to the naked eye are clustered at the proximal ends of the radial shields and in a series linking the distal ends of the shields. *O. limicola* has four to five oral papillae, arranged as in *Amphioplus* species. The ventral arm plates are slightly longer than wide, usually with an indented distal end. There are up to five acute, slender, slightly flattened arm spines and a single, minute tentacle scale.

Individuals from Belize, Panama, and Florida have pale tan arms with a thin, black, middorsal stripe that is crossed at irregular intervals by dark hatch-marks (Hendler, previously unpublished); those from St. John lack the dorsal stripe (Thomas 1962b).

HABITAT: Bays, mangrove channels, and reef-associated sand plains, in soft sediment with little or no fixed vegetation; rarely in beds of seagrass or coarse sandy bottoms. Most abundant in low-energy muddy environments, but also found in seasonally turbulent localities (McNulty 1961; Hendler, previously unpublished).

DISTRIBUTION: The Florida Keys, the Dry Tortugas, the Florida Gulf coast, the Virgin Islands, Belize, and Panama. DEPTH: 1–12 m (3–40 ft) (Hendler, previously unpublished).

BIOLOGY: *O. limicola* often occurs with other burrowing amphiurids such as *Amphioplus coniortodes*, in densities of 60 individuals per square meter, composing up to 20–30% of total benthic biomass. A population studied in Biscayne Bay, Florida, persisted for at least 13 years (McNulty 1961; McNulty et al. 1962; Thomas 1962b; Singletary and Moore 1974).

Adults burrow 14 cm below the surface, but leave their arm tips exposed. Thus, subject to partial predation, most individuals have regenerating arms; population mortality has been estimated at about 15% per month (Singletary 1970, 1980). Individuals readily autotomize the disk in reaction to relatively mild physical disturbance (Thomas 1962b) and can regenerate it in a month (Singletary 1970, 1980).

This species is a selective detritus feeder (Singletary 1970, 1980) that uses its long arms to collect particles from the surface of the sediment (Hendler, previously unpublished). Its tube feet are employed for feeding, burrowing, and for creating a fringe of mucus-bound sediment that helps propel a respiratory current through the burrow (Woodley 1975). Two scaleworm species live on this brittle star, *Malmgreniella hendleri* Pettibone and *M. puntotorensis* (Pettibone 1993).

In Biscayne Bay, Florida, *O. limicola* has one or two major spawning episodes between August and November (Singletary 1970, 1980), and most juveniles recruit during the winter and spring (Singletary 1970, 1980). The species is capable of surviving temperatures of 0–40°C (Singletary 1971).

SELECTED REFERENCES: Lütken 1869: 25, text fig. p. 25; Nielsen 1932:265, figs. 8a,b; Thomas 1962b:681, fig. 20; A. M. Clark 1970:19, fig. 3i.

Ophiophragmus cubanus
(A. H. Clark)
Figures 86, 87, 105A,B

DESCRIPTION: This is a large amphiurid; a specimen of 13.9 mm (0.5 in) disk diameter may have arms 223 mm (8.9 in) long. It is one of the few *Ophiophragmus* species that have spinelike papillae on the ventral surface of the disk, as well as on the disk margin. Small individuals also have some papillae on

FIGURE 86. *Ophiophragmus cubanus.*

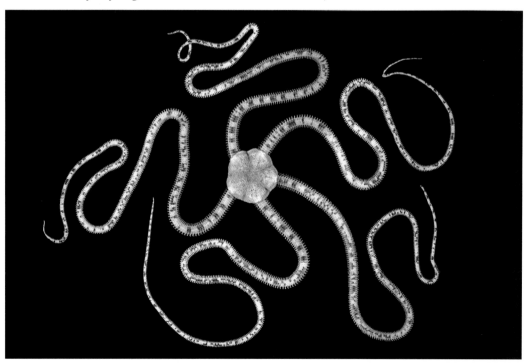

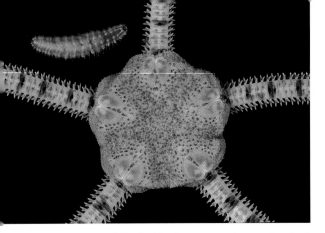

FIGURE 87. *Ophiophragmus cubanus:* detail of the disk and a commensal polychaete, *Malmgreniella puntotorensis*.

SELECTED REFERENCES: A. H. Clark 1917:69 [as *Ophiocnida cubana*]; H. L. Clark 1933:55, pl. 7, figs. a, b [as *Ophiocnida cubana*]; Thomas 1963:218, figs. 1–3 [as *Ophiophragmus cubanus*].

Ophiophragmus filograneus (Lyman)
Figures 88, 105C,D,E

DESCRIPTION: The disk diameter can reach 9 mm (0.3 in); an individual with 6 mm (0.2 in) disk diameter has arms 80 mm (3.1 in) long. Tiny, conical papillae are closely packed in triangular patches on the lateral and ventral surface of the disk. The only other Caribbean *Ophiophragmus* species with ventral disk papillae is *O. cubanus*. At the edge of the disk, 20 or more small, pointed fence papillae crowd between each pair of arms. The three blunt arm spines are considerably longer near the disk than at the middle of the arm. The ventral spine is longest, and the middle spines of arm joints under the disk may have very small terminal teeth.

the top of the disk. The lateral edges of the ventral arm plates are characteristically indented. The three arm spines are pointed and flattened; the middle spine is strongly compressed, with a broad base and an abruptly narrowed tip.

The disk is brown or gray, with dark spots on the scales. The arms have a pinkish orange, beige, or cream ground color, handsomely marked with narrow brown or greenish brown bands and broad greenish brown bands with black crescentic streaks (Thomas 1963; Hendler, previously unpublished).

HABITAT: Among seagrass roots and in mangrove channels with unvegetated mud (Thomas 1963; Hendler, previously unpublished).

DISTRIBUTION: The Dry Tortugas, Cuba, the Virgin Islands, and Belize. DEPTH: 1–18 m (3–59 ft).

BIOLOGY: This species burrows deep in soft mud and slowly moves its long arm tips across the surface of the sediment to deposit feed (Hendler, previously unpublished). It harbors a scaleworm, *Malmgreniella puntotorensis*, which also is symbiotic on other amphiurid species (Pettibone 1993) (Figure 87).

The disk is a dark gray or brown; the edges of many disk scales are often lighter than the base; the outer tips of the radial shields are usually white. The arms are irregularly blotched with dark and light gray.

HABITAT: Soft mud, sometimes in sediment rich in wood fragments. This species is restricted to the brackish-water shoal grass (*Halodule wrightii*) community in the protected bays and inlets of Florida (Turner and Meyer 1980).

DISTRIBUTION: From the southern tip of the Florida peninsula to Pensacola Bay on the northwest coast and to Mosquito Lagoon (near Cape Canaveral) in the northeast. Not known from the Florida Keys, but present in Biscayne Bay and Whitewater Bay (Pearson

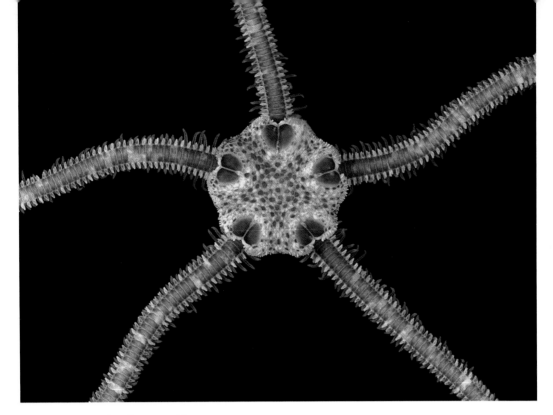

FIGURE 88. *Ophiophragmus filograneus.*

1937; Thomas 1961, 1962b). DEPTH: Less than 3 m (less than 8 ft).

BIOLOGY: Its abundance and endemic distribution make *O. filograneus* a worthy candidate for the "Florida State Brittle Star." Because few brittle star species survive in brackish or hypersaline waters, *O. filograneus* has gained notoriety for its ability to withstand salinities from 7.7 to 42 ppt (Thomas 1961, 1962b; Turner and Meyer 1980).

Its egg size (0.22 mm in diameter) and moderate fecundity (about 6,000 eggs per individual) suggest that *O. filograneus* has a modified mode of development, but its larval form is not known (Stancyk 1973, 1974). Individuals grow to maturity within a year and live for at least 2 years, spawning annually in the late fall (Turner 1985). Turner (1985) described the asymmetrical growth pattern of the young brittle stars less than 8 months old, which have two long arms and three shorter arms.

O. filograneus can autotomize the disk by voluntarily weakening specialized connective tissue attachments between the disk and arms (Dobson 1985; Dobson and Turner 1989). About 25% of individuals in a population are estimated to autotomize their disks each year. Depredation by the Atlantic stingray is one possible cause of disk loss, and the sea star *Luidia clathrata* and the southern pufferfish also prey on *O. filograneus* (Turner et al. 1982). A normally functioning stomach forms within 2 months after disk loss, with regeneration apparently utilizing stored nutrient (Turner and Murdoch 1976). However, nearly 20% of an individual's net metabolism may be supported by the uptake of amino acids from seawater (Ferguson 1982a,b). In fact, Ferguson (1982a,b) suggested that the uptake of dissolved organics, rather than particulate matter, is the primary nutritional resource of *O. filograneus.*

REMARKS: Two congeners similar in appearance to *O. filograneus*, *O. wurdemanii*

(Lyman) and O. *moorei* Thomas, occur in oceanic waters off Florida. O. *wurdemanii* can grow to 10 mm (0.4 in) disk diameter with arms 125 mm (5 in) long (see Figure 105H,I). It has widely separated tentacle scales like those of O. *filograneus* and rarely has ventral disk papillae. The uppermost fence papillae, slender and blunt, number 17–30 between adjacent arms. It has smooth, acutely pointed middle arm spines beneath the disk and shows a characteristic decrease in length of the arm spines beyond the seventh arm joint. Its disk is pale gray or tan, peppered with gray, and the arms are white, with irregularly spaced black bands. The ventral surface is pale, except for the almost black oral shields, jaws, and proximal disk scales. O. *wurdemanii* was originally described by Lyman (1860:196 [as *Amphiura wurdemanii*]) and was discussed and illustrated by H. L. Clark (1933:47, fig. 1, pl. 6, figs. a, b), Thomas (1962b:675, fig. 18A,B [as O. *wurdemani*]), and A. M. Clark (1970: 26, text fig. 5h [as O. *wurdemanni*]). The species ranges from North Carolina to the Gulf coast of Florida, Mexico, and possibly Venezuela, from near shore to 47 m (154 ft). It may occur together with O. *filograneus* on the Gulf coast of Florida (Alexander and Haburay 1977).

Preserved specimens of O. *moorei* are light tan or gray, with closely spaced dark gray or tan bands on the dorsal surface of the arms and with dark ventral arm plates (see Figure 105F,G). The species was described, illustrated, and contrasted with O. *wurdemanii* by Thomas (1965b:850, fig. 1a–d, e2). O. *moorei* is found on the Gulf coast be-

FIGURE 89. *Ophiophragmus pulcher.*

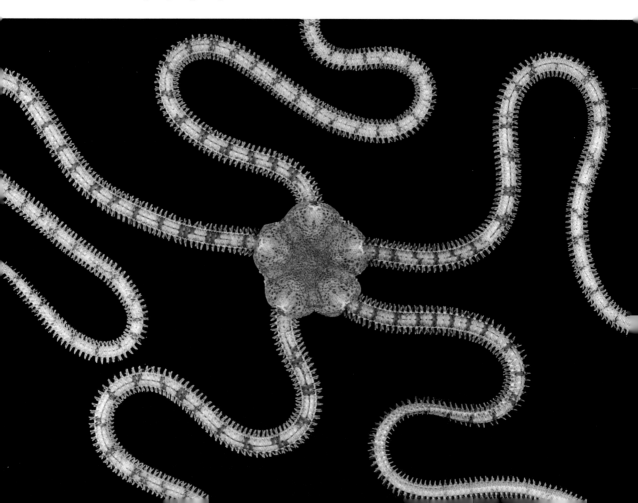

tween the Mississippi River and western Florida, at depths from 1 to 31 m (3–100 ft).

SELECTED REFERENCES: Lyman 1875: 20, text figs. 88, 89 [as *Ophiocnida filogranea*]; H. L. Clark 1918:274, pl. 2, figs. 4, 5; Thomas 1962b:677, fig. 19A,B.

Ophiophragmus pulcher H. L. Clark
Figures 30*D*-8, 32-2, 89, 106*A*,*B*

DESCRIPTION: An individual of 7 mm (0.3 in) disk diameter has arms about 80 mm (3.1 in) long. The blunt fence papillae at the edge of the disk usually number 6–12 between each pair of arms, but sometimes fewer. The dorsal and ventral arm spines are vertically compressed and shorter than the horizontally flattened middle spine. All the spines have blunt tips; the tip of the middle spine is usually broadened. Tube feet at the arm tip have papillose annuli.

H. L. Clark (1918) named the species "pulcher" (beautiful, in Latin) for its "notable coloration." The pale gray to reddish disk may have areas of lavender or pinkish hue and greenish or red-brown spots; the fence papillae are white, and the radial shields usually tipped with white. A thin middorsal stripe of red, blue, or green and narrow greenish brown bands adorn the drab tan to yellowish arms. There may be a similarly colored, discontinuous stripe on the ventral surface of the arm.

HABITAT: Reef, seagrass, and mangrove settings; in a variety of sediments including *Halimeda* sand, shell, or rocks and sand, and "finger-coral" rubble.

DISTRIBUTION: The Bahama Islands, the Florida Keys and the Dry Tortugas, Puerto Rico, Aruba; also in Belize (Hendler, previously unpublished) and Brazil. DEPTH: 1–13 m (2–42 ft).

BIOLOGY: Individuals burrowed in loose sediment are often wedged beneath buried shells, rubble, or twigs. Their arms extend upward several centimeters, and the tips stretch across the surface of the sediment; the tube feet transfer particles of food to the mouth (Hendler, previously unpublished). Freshly collected specimens have sediment adhering to the arm spines, probably remnants of a "respiratory fringe" (sensu Woodley 1975). The species sometimes occurs with other burrowing amphiurids including *Amphioplus thrombodes* and *Amphipholis januarii* (Thomas 1962b). The gonads of the females are pink, and the males have white testes (Hendler, previously unpublished).

O. pulcher from Looe Key, Florida, had commensal rotifers on the arms and disk (Hendler, previously unpublished), probably *Zelinkiella synaptae* (Ruppert, personal communication). A eulimid gastropod, *Oceanida* sp. (Warén, personal communication), was found on *O. pulcher* at the same locality, its proboscis embedded in the brittle star's arm. In Belize, the scaleworm *Malmgreniella puntotorensis* lives on some individuals (Pettibone 1993), and a minute commensal bivalve occurs on others (Hendler, previously unpublished).

SELECTED REFERENCES: H. L. Clark 1918:274, pl. 8, fig. 1 [as *Ophiophragmus pulcher*], 288, pl. 8, fig. 4 [as *Amphiodia rhabdota*]; Thomas 1962b:675, figs. 16a,b, 17a,b; Parslow and A. M. Clark 1963:35, fig. 10e–g [as *Ophiophragmus pulcher*], p. 34, fig 10a–d [as *Amphiodia rhabdota*].

Ophiophragmus riisei (Lütken)
Figures 90, 106*C*,*D*,*E*

DESCRIPTION: An individual of 8.8 mm (0.3 in) disk diameter has arms approximately 50 mm (2 in) long. The central disk scales are large, thick, and not overlapping.

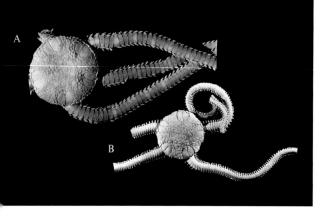

FIGURE 90. *Ophiophragmus riisei*: *A*, USNM 7583; *B*, UZM, Copenhagen (type, *Amphiura riisei* Lütken). Photo by G. Hendler.

The fence papillae of the disk (if present) are short, blunt, and number 1–25 between each pair of arms. There are three arm spines, and the middle spines on joints beneath the disk usually have one to two and up to five small, glassy teeth at the tip. This is probably the most helpful diagnostic feature for this species. As described by Thomas (1962b), the disk is white and the arms are blotched with light reddish brown. This pattern is evident in dried specimens from Florida (MCZ 6683) (Hendler, previously unpublished). However, Tommasi (1970) noted that Brazilian individuals are chestnut brown, with pale irregular patches on the arms and the distal portion of the radial shields.

HABITAT: Medium-grain sand.

DISTRIBUTION: Off Florida (Sombrero Key and Miami), St. Thomas, between Haiti and Jamaica, and possibly Panama and Puerto Rico. DEPTH: 12–87 m (39–285 ft).

REMARKS: Thomas (1962b, 1965c) suggested that *Ophiophragmus brachyactis* H. L. Clark is synonymous with *Amphiura riisei* Lütken (as per Lyman 1860:258) and proposed the new combination *Ophiophragmus riisei* (Lütken) for the species in an unpublished manuscript. Thomas's suggestion was verified with a side-by-side comparison of the holotypes of *A. riisei* and *O. brachyactis* (Hendler, previously unpublished). Therefore, the name *Ophiophragmus riisei* is adopted here, and information on its habitat and distribution reflects records for *O. brachyactis* and *A. riisei*. However, because of the scarcity of specimens, the taxonomic status of this brittle star remains unclear. New material, which better illustrates the manner in which color and structure vary geographically, could prompt a reconsideration of Thomas's synonymy.

SELECTED REFERENCES: Lütken 1859b: 222, pl. 3, fig. 2a,b [as *Amphiura cordifera*]; H. L. Clark 1915a:238, pl. 10, figs. 13, 14; Thomas 1962b:666, fig. 14A,B.

Ophiophragmus septus (Lütken)
Figures 91, 92, 106F,G

DESCRIPTION: A large individual is 9.3 mm (0.4 in) in disk diameter with arms 190 mm (7.5 in) long. The species has slender, sharply pointed fence papillae on the disk, about 20 between adjacent arms. The shape of the middle arm spine sets this species apart; its curved distal edge and slightly concave proximal edge make the spine appear to bend toward the disk.

The disk is gray and brown, with some scales and the radial shields usually much darker than the rest. The arms are mottled with black, brown, or yellow-brown, darker near the disk than at the tip. A dark middorsal arm stripe is typical in populations from Fort Pierce, Florida, and Puerto Rico (Thomas 1962b; Hendler, previously unpublished). However, a white median arm stripe enclosed by two black stripes is characteristic for *O. septus* from Belize and Panama, as shown in the individual depicted here (Hendler, previously unpublished).

HABITAT: Sandy peat in Belizean mangrove

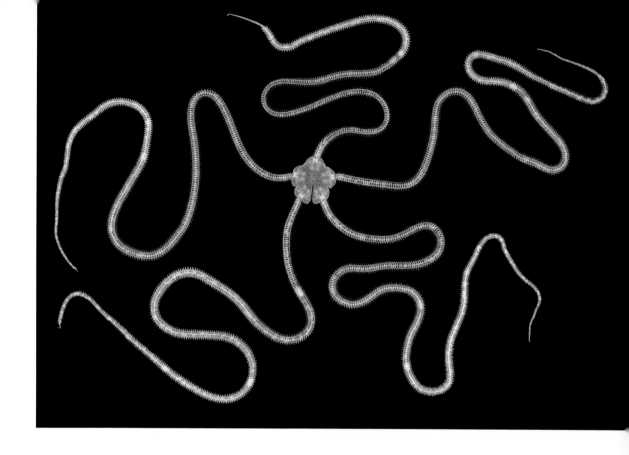

FIGURE 91. *Ophiophragmus septus.*

FIGURE 92. *Ophiophragmus septus:* detail of the disk and a small commensal bivalve.

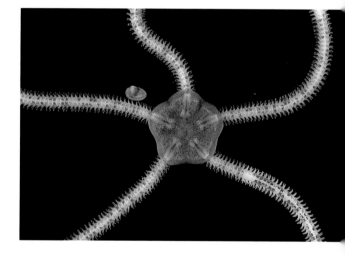

channels, near-reef sand plains in Panama, and shelly sand and mud in central Florida.

DISTRIBUTION: North Carolina and the east coast of Florida (Fort Pierce to Miami Beach), but not the Florida Keys. Also reported from Puerto Rico, St. Thomas, Tortola, Tobago, Colombia, as far south as Brazil. DEPTH: Less than 1–100 m (less than 1–328 ft).

BIOLOGY: Individuals generally extend one or two arms above the burrow, and they can simultaneously suspension and deposit feed, with one arm in the water and another on the surface of the sediment (Hendler, previously unpublished). Two scaleworm species, *Malmgreniella puntotorensis* and *M. galetaensis* Pettibone, are associated with *O. septus* (Pettibone 1993), and a commensal bivalve (Figure 92) lives on some individuals in Belize (Hendler, previously unpublished).

REMARKS: In an unpublished manuscript, Thomas synonymized *Ophiophragmus luetkeni* (Ljungman) with *O. septus,* based on Koehler's (1914a) discussion of the two species. According to Thomas, *Amphiodia erecta* Koehler is also identical with *O. sep-*

tus. A systematic reexamination of *O. septus* might clarify this unresolved synonymy and explain the basis for the geographic variation in pigmentation that was noted above.

SELECTED REFERENCES: Lütken 1859b: 222 [as *Amphiura septa*]; 1872:85, pl. 1, 2, figs. 3a–c [as *Amphipholis septa*]; Koehler 1914a:67, pl. 6, figs. 4–7 [as *Amphiodia erecta*]; Thomas 1962b:669, fig. 15A,B.

Ophiostigma isocanthum (Say)
Figures 93, 104C,D

DESCRIPTION: A typical specimen of 4.2 mm (0.2 in) disk diameter has five arms that reach 14.2 mm (0.6 in) in length, and individuals with disk diameters up to 7 mm (0.3 in) are known. The disk is covered by numerous short, blunt tubercles that obscure the scales and may cover the radial shields. Usually several of the tubercles near the radial shields are markedly larger than the rest. It is sometimes mistakenly identified as *Ophiocnida scabriuscula* because of its disk spination. However, the distalmost of oral papillae, which set it apart from the latter species, are long, opercular, and close the gaps between the jaws. The three arm spines are blunt and somewhat flattened. There are two small, slender tentacle scales. Dorsal arm plates near the disk are subovoidal; they touch or overlap one another. The adoral shields overgrow the first ventral arm plates, both series thereby forming a nearly continuous circle of plates around the mouth. The grainy texture of the lateral arm plates contrasts with the smooth dorsal and ventral arm plates.

Particles of fine sediment and debris, caught in mucus, usually adhere to the dorsal

FIGURE 93. *Ophiostigma isocanthum*.

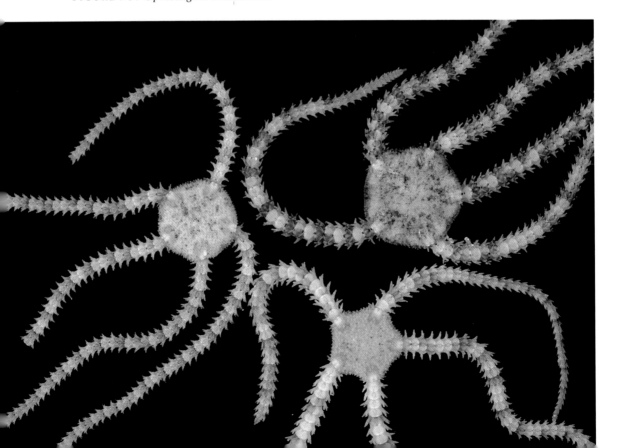

surface of the disk and arms. The animal itself is "sandy colored," gray to brown with reddish brown, gray-brown, orange, or black markings. The outer ends of the radial shields are usually white. Tips of the arm spines are reddish or orange. The tops of the arms often have a complex, dusky, chainlike pattern, and appear banded.

HABITAT: All reef zones, and seagrass beds; under stones and rubble, and in branching coral and algae.

DISTRIBUTION: Bermuda, the Bahama Islands, North Carolina to Florida, the Florida Keys, the Dry Tortugas, Texas offshore reefs, Cuba, Jamaica, Puerto Rico, the Virgin Islands, the Leeward Islands, Barbados, Tobago, Curaçao, Aruba, Costa Rica, Panama, Colombia, Venezuela, and Brazil. DEPTH: Less than 1–223 m (less than 3–732 ft). However, because this species has been confused with a six-armed congener, *O. siva*, there is uncertainty regarding these records (Hendler, in press).

BIOLOGY: *O. isocanthum* and the fissiparous six-armed *O. siva* were found in the same clumps of algae at Looe Key, Florida, and other localities (Hendler, in press). Individuals can autotomize and regenerate the disk, and the regenerated disks have spinules and radial shields of typical shape (Hendler, previously unpublished). Information regarding its reproductive biology is contradictory. A five-armed specimen from Belize had eggs 0.20 mm in diameter (Hendler, previously unpublished), but Emson et al. (1985a) estimated 5,000 eggs, 0.12 mm in diameter, in another individual (number of arms unspecified). Singletary (1971) found the upper lethal temperature for *O. isocanthum* to be 39.8°C, but he may have been examining the related fissiparous species.

SELECTED REFERENCES: Say 1825:150 [as *Ophiura isocantha*]; Koehler 1913:363, pl. 20, figs. 6, 7; Thomas 1962b:689, fig. 23A,B [as *Ophiostigma isacanthum*].

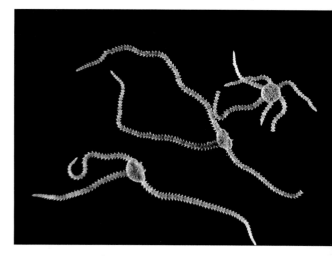

FIGURE 94. *Ophiostigma siva*. Photo by G. Hendler.

Ophiostigma siva Hendler
Figures 94, 104E,F

AUTHOR'S NOTE: The following account is a condensed version of the formal description of this species (Hendler, in press).

DESCRIPTION: Individuals have six arms, generally three longer arms and three shorter regenerating arms. The arms of *O. siva* are relatively longer in relation to the disk diameter than those of *O. isocanthum*. A specimen of 2.5 mm (0.1 in) disk diameter has arms up to 16.4 mm (0.6 in) long. The disk tubercules are longer than wide, especially at the edge of the disk. There are three erect, bluntly pointed arm spines. They appear to curve toward the disk, because of their concave proximal and convex distal edges. Dorsal arm plates near the disk are subtriangular and are separated by the lateral arm plates. The distal edge of the adoral shield touches the first ventral arm plate, but does not overgrow it.

This six-armed brittle star is similar in coloration to *O. isocanthum,* but generally a lighter hue. The arms are less darkly pigmented than the disk, with a faint, dusky, chainlike pattern, and they may be faintly banded.

HABITAT: The same as for *O. isocanthum.*

DISTRIBUTION: Bermuda, the Florida Keys and the Dry Tortugas, Jamaica, Puerto Rico, St. Thomas, and Belize. DEPTH: Less than 1–42 m (less than 3–138 ft).

BIOLOGY: H. L. Clark (1933) noted that *O. isocanthum* exhibits "autotomous reproduction," and Hotchkiss (1982) suggested that the five- and six-armed individuals are different species. *O. isocanthum* is exclusively sexually reproductive, but *O. siva* reproduces asexually through fission; nothing is known regarding its sexual reproduction.

From Florida southward, this species seems more numerous than its five-armed congener (Carrera 1974; Emson et al. 1985a; Hendler, in press). Population densities of *O. siva* in algae may exceed 100 individuals per liter (Hendler and Littman 1986; Mladenov and Emson 1988). It is frequently collected with other small brittle stars including *Amphiodia pulchella, Amphipholis squamata,* and *Ophionereis squamulosa* and with small individuals of *Ophiothrix angulata, Ophiocoma pumila,* and *O. echinata.*

SELECTED REFERENCES: Hotchkiss 1982:392, fig. 173a,b [as *Ophiostigma* sp.]; Hendler, in press.

FAMILY OPHIOTRICHIDAE

Ophiothrix angulata (Say)
Figures 30D-11, 95

DESCRIPTION: This moderately small species grows to 10 mm (0.4 in) in disk diameter with arms 80 mm (3.1 in) long. The spine arrangements and colors on the disk seem as intricate and varied among different individuals as the designs of Persian carpets. It differs from other *Ophiothrix* in this region by having numerous short, delicate, bifid and trifid spines, sometimes interspersed among long thin spines on the disk and sometimes on the radial shields. The longest arm spines near the disk are the flattened, toothed, middle spines; the longest spines near the tip of the arm are the smooth, slender topmost spines. As in other *Ophiothrix* species, the jaws bear a terminal clump of dental papillae and lack oral papillae. The tube feet are papillose.

The ground color on the dorsal surface may be pink, rose, violet, blue, purplish, orange-red, crimson, brown, gray, or green. The arms generally have a median stripe or a lacy pattern of black, white, or a contrasting color; the ventral surface of the arms is usually white. The overall color pattern frequently resembles one typical of other *Ophiothrix* species. The uniformly orange-red individual depicted here is a relatively uncommon color form.

HABITAT: Pier pilings, oyster beds, man-

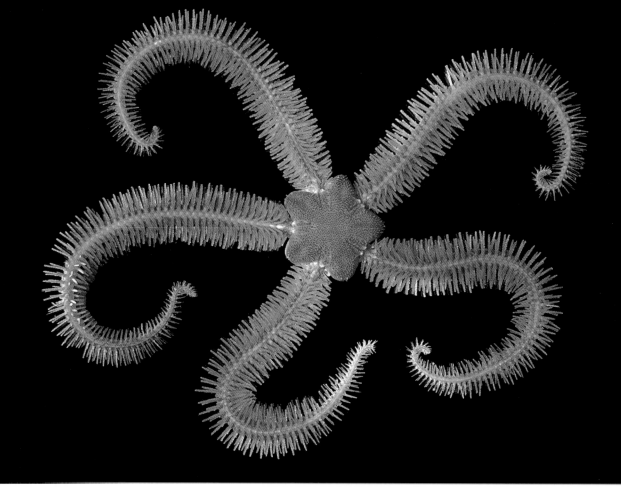

FIGURE 95. *Ophiothrix angulata*.

grove habitats, seagrass beds, and all coral reef zones; on substrates including rubble, corals, *Millepora* sp., gorgonians, algae, sponges, and other sessile biota.

DISTRIBUTION: Bermuda, the Bahama Islands, North Carolina to the Dry Tortugas and the Florida Gulf coast, the Texas coast and offshore reefs, the Greater and Lesser Antilles, and Central and South America to Uruguay. A record of the species from West Africa is probably incorrect (Madsen 1970). DEPTH: 1–540 m (3–1,772 ft).

BIOLOGY: This is one of the most common brittle stars of Florida, the Bahama Islands, and Caribbean shallow waters (H. L. Clark 1914; Avent et al. 1977; Dugan and Livingston 1982; Hendler and Peck 1988). Ubiquitous and abundant, it may have been recognized and named, as early as the eighteenth century, "Stella marina minor echinata purpurea" (Lyman 1880). Its population densities can exceed 100 individuals per liter of algae (Hendler and Littman 1986) and 16 individuals per gram of dry sponge (Boffi 1972). On sponges it often occurs with other species of *Ophiothrix* and with *Ophiactis* (H. L. Clark 1914; Devaney 1974b; Hendler 1984a; Wendt et al. 1985).

O. angulata has been characterized as a short-lived, "fugitive species" capable of rapid dispersal and colonization. Perhaps an opportunistic life-style accounts for its success in estuarine environments despite its low tolerance of temperature and salinity stress (Glynn 1968; Singletary 1971; Stancyk 1975; Stancyk and Shaffer 1977; Donachy and Watanabe 1986).

Each individual has 10 gonads (Brito

1960); the ovaries contain whitish eggs 0.10 mm in diameter. The species probably spawns year-round in the Tropics and seasonally in temperate environments (Wilson 1900; Stancyk 1975; Stancyk and Shaffer 1977). The differences noted in the breeding periodicity among color "varieties" (see *Remarks*) at the same locality (H. L. Clark 1918, 1933) may reflect the reproductive asynchrony that prevails among populations of tropical brittle stars (Hendler 1979a). Mortensen (1921:127) observed that during spawning "the eggs were shed all at the same time, almost as if by an explosion." Its larval development is rapid (Mortensen 1921; Stancyk 1973, 1974), but the ophiopluteus has not yet been reared through metamorphosis.

Species of *Ophiothrix*, eaten by over 20 species of Caribbean fish, compose more than 10% of the stomach contents in half of those species (Randall 1967). The seeming preponderance of the genus *Ophiothrix* in gut samples may reflect the ease with which their spines can be identified. *O. angulata* has been found in the stomach contents of seven fish species and proved an important component in the diet of leopard toadfish and belted sand bass (Shirley 1982). This brittle star is parasitized by *Vitreolina arcuata* (C. B. Adams), a small eulimid snail (Warén 1984).

REMARKS: H. L. Clark (1918) described "varieties" of *O. angulata*, based primarily on their coloration, named *violacea*, *atrolineata*, *megalaspis*, *phoinissa*, *phlogina*, and *poecila*. Tommasi (1970) listed 21 different color forms of this species! Because the relationship between the coloration and morphology of *O. angulata* is not consistent, the subspecific names are seemingly irrelevent. The nominal species *Ophiothrix pallida* Ljungman was reported to differ from *O. angulata* only by having a purplish or reddish arm stripe (H. L. Clark 1918) and therefore appears to be a color "variety," rather than a discrete species.

SELECTED REFERENCES: Say 1825:145 [as *Ophiura angulata*]; Lyman 1865:162, pl. 2, figs. 1–3; Koehler 1913:377, pl. 20, fig. 5, pl. 21, figs. 10, 11 [as *O. pallida*]; H. L. Clark 1918:312, pl. 5, fig. 3, pl. 8, fig. 3.

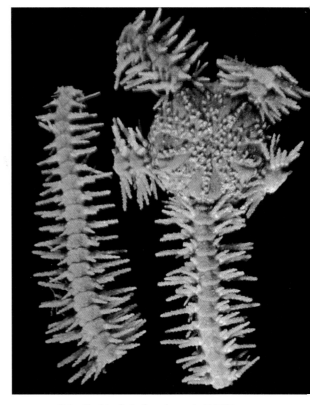

FIGURE 96. *Ophiothrix brachyactis* (MCZ 2480, holotype). Photo by G. Hendler.

Ophiothrix brachyactis H. L. Clark
Figure 96

DESCRIPTION: This is a small and distinctly flattened brittle star, with a disk diameter less than 5 mm (0.2 in) and relatively short arms less than 18 mm (0.7 in) in length. Covering the disk are large, convex radial shields and conspicuous coarse scales that each bear a short stump crowned with sharp points. This species lacks the trifid stumps that characterize O. *angulata*. The oral shield is much wider than long, with rounded lateral edges; the adoral shields are broadly in contact proximal to the oral shield. Successive arm plates are barely in contact. The ventral arm plates are squarish, with angular corners and a slightly concave distal margin. The dorsal arm plates are triangular with a broad, thickened, and slightly concave outer edge. The arm spines are echinulate; the longest spines are borne on joints well beyond the edge of the disk. Near the disk, the upper arm spines are flattened, and broader at the base than near the tip; the lower arm spines are slender and rounded. The jaws, typical for *Ophiothrix* species, have an apical cluster of dental papillae and lack oral papillae.

A dried, preserved specimen is bluish gray; the arms are a somewhat lighter shade than the disk, with dark and light bands. The dorsal arm plates have inconspicuous whitish margins; the radial shields have whitish outer tips, and the ventral surface of the body is nearly white.

HABITAT: On or associated with coral reefs; in rubble and coralline algae.

DISTRIBUTION: Florida: Biscayne Bay, the Dry Tortugas, Puerto Rico, St. Barthélemy, Grenada, Barbados, and Tobago. DEPTH: 1–6 m (3–21 ft).

BIOLOGY: This species is so rarely seen that virtually nothing is known about its natural history. It may be found together with O. *angulata* (H. L. Clark 1915).

REMARKS: *Ophiothrix hartfordi* A. H. Clark is a similar species that has a flat disk with coarse scales and short, almost granulelike, prickly-tipped stumps that are set between the large radial shields. A specimen of 4.3 mm (0.2 in) disk diameter has arms 13 mm (0.5 in) long. Unlike O. *brachyactis*, it has numerous stumps and thick integument that obscure the scales at the center of the disk. It is also distinguished by the shape of its arms. The long spines on the arm segments closest to the disk nearly touch the spines on adjacent arms. Beyond the arm base the size of the arm joints and spines decreases abruptly. The proximal arm spines are slender and echinulate, only slightly broader at the base than at the tip. The dorsal arm plates are triangular to fan-shaped; the ventral arm plates have a deeply concave distal edge. The color of the only existing specimen, which is dried, is blue-gray; the arms have faint bands of light and dark pigment. It was found off Puerto Rico, and is illustrated in A. H. Clark (1939a: pl. 53, figs. 1, 2).

SELECTED REFERENCES: H. L. Clark 1915:269, pl. 12, figs. 1, 2; 1918:320; 1933: 63; de Roa 1967:290, fig. 20.

Ophiothrix lineata Lyman
Figure 97

DESCRIPTION: This moderately large species reaches at least 12 mm (0.5 in) in disk diameter with arms 120 mm (4.7 in) long. The disk is divided into contrasting sectors; areas of large, flat scales each bearing one to several granulelike stumps, alternate with the nearly bare, triangular radial shields. The arms are broad near the disk, narrowing abruptly and terminating as a filament. The flattened appearance of this brittle star is accentuated by its horizontally directed proximal arm spines. The finely echinulate spines are conspicuously longer than twice the width of the arm. Those near the disk are dorsoventrally flattened, almost as broad at the rounded tip as at the base; the distal spines are considerably more slender, and the short ventral spines have double-hooked tips. The proximal ventral arm plates are distinctly wider than long. O. *lineata*, like its

FIGURE 97. *Ophiothrix lineata*.

congeners, has jaws bearing dental papillae and lacking oral papillae. The tube feet are papillose.

The disk is red-brown or violet-brown, and gray. There is a continuous blackish or dark brown stripe at the center of the arm, accompanied on both sides by a pale gray stripe, the basis for the species name. Arm spines near the disk are violet or rose-hued.

HABITAT: Back reef to the fore reef slope, wherever suitable sponges occur.

DISTRIBUTION: The Florida Keys and the Dry Tortugas, Barbados, Cuba, Belize, Colombia, and a questionable record from Curaçao. DEPTH: Shallow water to 49 m (162 ft).

BIOLOGY: This brittle star is an obligate commensal of coral reef sponges such as *Callyspongia vaginalis* (Lamarck) and *Verongia lacunosa*, and sometimes occurs together with other sponge-dwelling *Ophiothrix* and *Ophiactis* species (Devaney 1974b; Kissling and Taylor 1977; Hendler 1984a). Although its association with sponges helps to protect the brittle star from fish predators (Hendler 1984a), it has been identified in the stomach contents of pufferfish (Randall 1967).

By day, large individuals occupy the spacious central cavity of tube sponges; juveniles occur on the outside of the sponge. At night, the disk of adults remains hidden; the arms move slowly across the outer surface of the sponge to gather particles of sediment, detritus, and tiny organisms. Its feeding activity cleans the surface of the sponge, and may thereby enhance the sponge's ability to filter feed (Hendler 1984a).

Details of its reproductive biology are unknown, but ripe females with beige-white, and yellowish eggs were found at Looe Key, Florida, in May (Hendler, previously unpublished). A scale worm, *Lepidonopsis humilis* (Augener), was associated with *O. lineata* at Looe Key, Florida; the same polychaete species also lives on sand dollars and other substrates (Pettibone 1993).

REMARKS: *Ophiothrix platyactis* H. L. Clark resembles *O. lineata* in having a flat disk with coarse disk scales and rounded granules, and a pale stripe on the dorsal surface of the arm. However, a specimen 6 mm (0.2 in) in disk diameter has arms only 30 mm (1.2 in) long, half the arm length of *O. lineata* of that disk diameter. The dorsal arm plates of *O. platyactis* are ovoid to fan-shaped, not hexagonal, and the length of dorsal arm spines at the base of the arm is less than twice the arm width. The upper spines have a broad base and markedly taper toward the tip; they have a thickened median ridge and flattened, bladelike, echinulate edges. The ventral arm spines are slender and rounded. *O. platyactis* is reported from Barbados, at 7 m (24 ft) depth, and the only known specimen was illustrated by H. L. Clark (1939: pl. 52, figs. 3, 4).

SELECTED REFERENCES: Lyman 1860: 201; H. L. Clark 1915:273, pl. 12, fig. 4; Hendler 1984a:9, figs. 1–3.

Ophiothrix orstedii Lütken
Figure 98

DESCRIPTION: A brittle star of modest size; a large specimen is 12 mm (0.5 in) in disk diameter with arms 65 mm (2.6 in) long. The numerous long, thin spines on dorsal interradial sectors of the disk have sharp tips terminating in two or three microscopic spinelets. There are similar but many fewer spines on the radial shields. The squat spines on the edge and ventral surface of the disk may have bifid or trifid tips. The topmost arm spine is thin, rounded, nearly smooth, and sharply pointed; lower spines are distinctly flattened and have a jagged, toothed edge and a broadened, spinulose tip. Characteristically for an *Ophiothrix* species, the jaws lack oral papillae but bear a proximal cluster of dental papillae. The tube feet are papillose.

The ground color of *O. orstedii* is green, brown, red-brown, blue, purple, or gray. Bands of the ground color on the arms are interrupted by thin yellow (or whitish) lines that are bordered on each side by a thin black line. The distinctive and striking black and yellow pattern may continue over the radial shields and onto the disk. Preservation in alcohol turns specimens blue and white, but the characteristic banded pattern remains evident. H. L. Clark (1918) named a variety, *lutea*, for an exceptional orange-colored specimen of *O. orstedii* collected off Tobago.

HABITAT: All reef zones and in seagrass beds; under rocks and rubble slabs, in the interstices of corals and reef rock; sometimes epizoic on sponges and fire corals.

DISTRIBUTION: The Bahama Islands, the Florida Keys and the Dry Tortugas, Texas, Cuba, Jamaica, Haiti, Puerto Rico, the Leeward and Windward Islands, Barbados, Tobago, Isla la Tortuga, the Netherlands Antil-

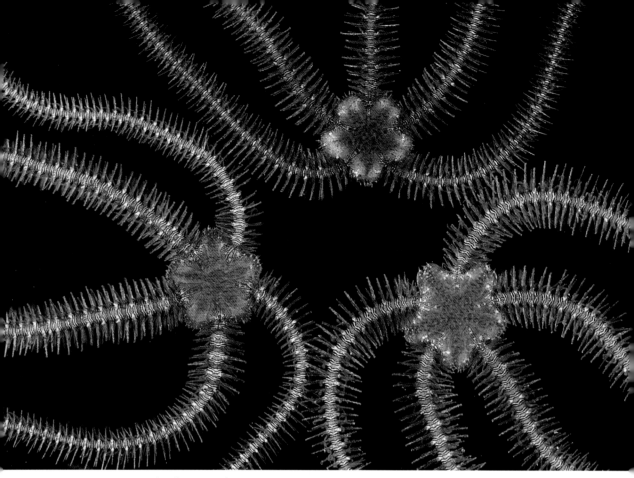

FIGURE 98. *Ophiothrix orstedii*.

les, Belize, Isla de Providencia, Panama, Colombia, and Venezuela; notably absent from Brazil (Tommasi 1970). DEPTH: Shallow water to 31 m (100 ft).

BIOLOGY: This species is an extremely abundant and conspicuous inhabitant of coral reef rubble habitats (Grave 1898b; Pearson 1937; Kissling and Taylor 1977; Lewis and Bray 1983; Emson et al. 1985a; Hendler and Littman 1986; Hendler and Peck 1988). Individuals in algae and branching corals tend to be smaller than those in the lettuce coral, *Agaricia* sp. (Hendler and Littman 1986), and large animals are more abundant in shallow water than on the reef slope (Hendler and Peck 1988). Densities as high as 72–400 individuals per square meter and 100 per liter of algae have been reported (Bray, cited in Mladenov 1979; Aronson and Harms 1985).

This species is relatively unpalatable to fish in comparison with some reef brittle stars, and it has long arm spines and defensive, cryptic behavior (Aronson 1988). It is rare among Caribbean coral reef ophiuroids for its habit of "freezing" in place when a shadow moves above it (Cobb, personal communication). *O. orstedii* is attacked by fish (wrasses, parrotfish, and pufferfish), the polychaete worm *Eunice rubra*, and the brittle star *Ophioderma brevispinum* (Aronson and Harms 1985; Aronson 1987). Its frequent injuries are attributable to predators, rather than to damage from physical disturbances such as storms (Aronson 1992).

The reproductive biology of *O. orstedii* was studied by Mladenov (1979, 1983, 1985b). The spawning period persists from August to mid-December in Barbados and may begin earlier at Jamaica; gonadal development is synchronous among members of a population. Individuals have 10 whitish testes or brownish ovaries. A ripe female may have over 9,000 cream-colored, yolky eggs that are 0.25 mm in diameter, only some of which are released during a spawning event. Spawning individuals raise the disk off the bottom and sometimes wave an arm, dispersing the gametes. The larva is a short-lived, two-armed, yolky ophiopluteus. After about 4 days of development, metamorphosed juveniles detach from the larval arms, which may continue to float for several more days.

O. orstedii can suspension feed (Aronson and Harms 1985), but its constant presence beneath rubble suggests that it can also deposit feed. Individuals require 4 months to completely regenerate an autotomized arm (Aronson 1987). They tolerate water temperatures up to 38°C (Singletary 1971).

REMARKS: Accurate identification of this species, particularly of small specimens, requires microscopic examination to distinguish it from O. angulata.

SELECTED REFERENCES: Lütken 1856: 15; 1859b:251, pl. 4, figs. 3a–e [as Ophiothrix Örstedii and O. Ørstedii]; H. L. Clark 1918:314; 1919:69, pl. 2 [as Ophiothrix oerstedii var. lutea].

Ophiothrix suensonii Lütken
Figures 30D-10, 32-7, 99

DESCRIPTION: This fairly large brittle star has a small disk, strikingly long, needlelike arm spines, and slender, boldly striped arms. A specimen of 10 mm (0.4 in) disk diameter has arms 101 mm (4.0 in) long. Only a few long, slender spines project from the smooth integument of the disk. The radial shields are nearly bare of spines; paired shields are in contact. A row of scales, contrasting in color with the surrounding surface of the disk, is arrayed at the outer edge of each radial

FIGURE 99. *Ophiothrix suensonii.*

shield; parallel rows of smaller scales and minute scattered spines cover the edge of the disk. The dorsal arm spines near the disk are several times longer than the width of the arm, very slender, glassy, finely toothed, and sharply pointed. Integument obscures the dorsal arm plates. The lateral sides of the ventral arm plates are clearly defined; those nearest the disk are square, but succeeding plates are distinctly longer than wide. Dental papillae are present at the apex of each jaw, and oral papillae are lacking, as in other *Ophiothrix* species. The tube feet are papillose.

A dark, dorsal stripe, particularly dramatic on pale-colored individuals, runs the length of the arm. It is usually black, purple, or crimson. There is often a dark stripe on the bottom of the arm. The ground color of the disk and arms is gray, tawny, lavender, pink, orange, red, or dark purple (appearing almost black).

HABITAT: From shallow to deep-reef zones, in mangroves, and even in silty lagoons, where there are suitable hosts such as gorgonians, sponges, and fire corals (*Millepora* sp.); rarely in coral rubble.

DISTRIBUTION: Bermuda, the Bahama Islands, the Florida Keys and Gulf coast, Texas and Mexican offshore reefs, the Greater and Lesser Antilles, and the coast of Central and South America to Brazil. DEPTH: Probably to several hundred feet; reported depths to 479 m (1,572 ft) are probably based on a yet-undescribed deep-water *Ophiothrix* species (Hendler, previously unpublished).

BIOLOGY: According to H. L. Clark (1933: 62), "This is undoubtedly one of the most beautiful of West Indian brittle stars and has stirred the enthusiasm of every collector who has been so fortunate as to find an adult specimen." It also is the most frequently photographed Caribbean brittle star, as much because of its conspicuousness as for its attractive appearance. Individuals may adopt a somewhat more elevated feeding posture at night, but they are quite evident on their hosts during the daytime (Hendler, previously unpublished; contra Colin [1978]). Some accounts state that *O. suensonii* is an active species, but when provoked it tightens its hold on the substrate and stiffly erects its long dorsal arm spines (Hendler, previously unpublished). The spines on the basal arm joints, directed proximally, help shield the delicate disk. The spines are sharp enough to penetrate the skin of an overly enthusiastic collector (Hendler, personal observation), but they do not always protect the brittle star from predaceous fish (Randall 1967).

Its hosts include at least five sponge genera, and *O. suensonii* often cohabits sponges with other species of *Ophiothrix* and *Ophiactis* (Devaney 1974b; Kissling and Taylor 1977; Henkel 1982; Mladenov 1983; Hendler 1984a). It is interesting that individuals on "fire sponge," *Neofibularia* sp., are usually dark purple. During the day, they cling to the exposed outer surface of the brown sponge, their behavior contrasting with the crypsis of individuals living with tube sponges. Though possibly protected by the toxic properties of *Neofibularia*, they have a high incidence of arm regeneration, indicative of significant damage from predation (Hendler 1984a).

Mladenov (1983, 1985b) described the reproductive biology of *O. suensonii* at Barbados. It breeds year-round with seasonal spawning peaks; within populations, individuals have asynchronous breeding cycles. The 10 gonads in ripe adults are multilobed; the testes are pink, and ovaries are orange. Females have up to 47,500 small eggs, 0.13 mm in diameter. The larva develops into a four-armed ophiopluteus in 10 days; it has been reared for up to 49 days without metamorphosing.

This species appears to be a passive suspension feeder. Fine particles, adhering to mucus on the spines and tube feet, are compacted and relayed to the mouth by the tube feet (Hendler, previously unpublished). The skeleton of this brittle star has been analyzed to find the relative concentrations of calcium and strontium (Pagett 1985).

SELECTED REFERENCES: Lütken 1856: 16; 1859b:250, pl. 4, fig. 2a–e; A. M. Clark 1967b:648, text fig. 1j, pl. 10, fig. 5 [as *Ophiothrix* (*Acanthophiothrix*) *suensoni*].

FIGURE 100. Details of amphiurid brittle stars photographed at low magnification, showing the dorsal surface of the disk and arms and the ventral surface of the oral frame and base of the arms of dry preserved specimens. *Amphiodia planispina:* A, dorsal, large individual; B, dorsal, small individual; C, ventral; *Amphiodia trychna:* D, dorsal; E, dorsal view of large individual; F, dorsal view of juvenile individual; G, ventral; *Amphiodia atra:* H, dorsal; I, ventral; *Amphiodia pulchella:* J, dorsal; K, ventral. Photos by G. Hendler.

FIGURE 101. Details of amphiurid brittle stars photographed at low magnification, showing the dorsal surface of the disk and arms and the ventral surface of the oral frame and base of the arms of dry preserved specimens. *Amphioplus coniortodes:* A, dorsal; B, ventral; C, ventral view of the disk showing the sparse scale covering on the ventral interradii; *Amphioplus sepultus:* D, dorsal; E, ventral; *Amphioplus thrombodes:* F, dorsal; G, ventral; *Ophionephthys limicola:* H, dorsal; I, ventral. Photos by G. Hendler.

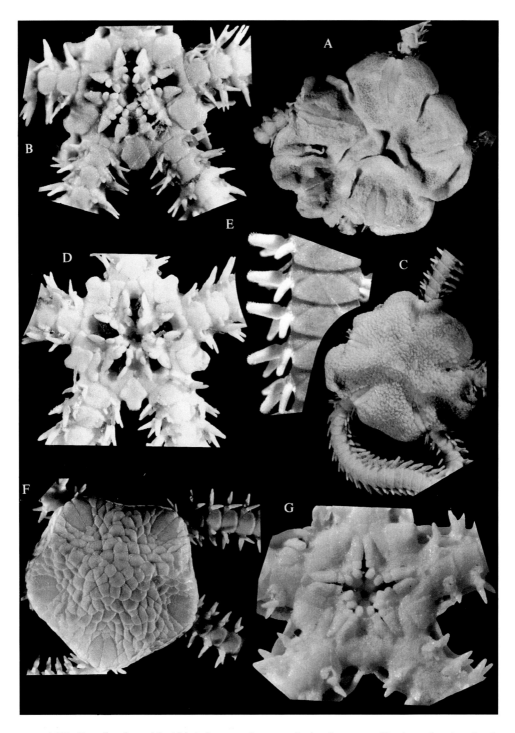

FIGURE 102. Details of amphiurid brittle stars photographed at low magnification, showing the dorsal surface of the disk and arms and the ventral surface of the oral frame and base of the arms of dry preserved specimens. *Amphipholis gracillima:* A, dorsal; B, ventral; *Amphipholis januarii:* C, dorsal; D, ventral; E, detail of arm spines in dorsal view; *Amphipholis squamata:* F, dorsal; G, ventral. Photos by G. Hendler.

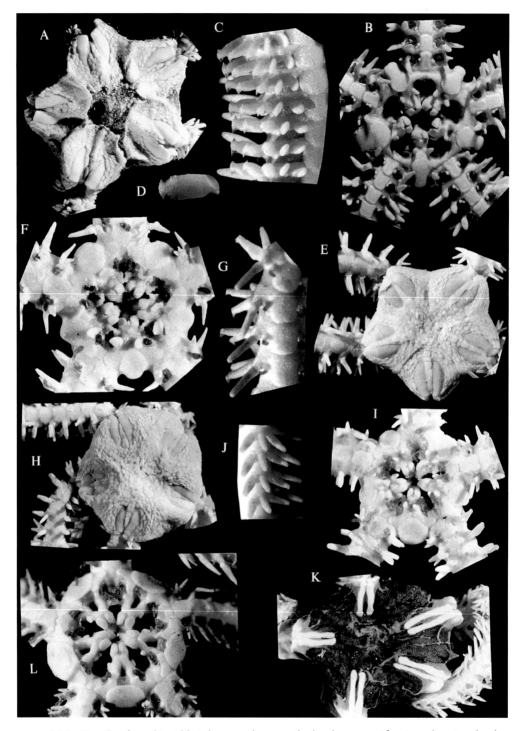

FIGURE 103. Details of amphiurid brittle stars photographed at low magnification, showing the dorsal surface of the disk and arms and the ventral surface of the oral frame and base of the arms of dry preserved specimens. *Amphiura fibulata*: A, dorsal; B, ventral; C, detail of arm spines in dorsal view; D, pickax-shaped arm spine at high magnification; *Amphiura palmeri*: E, dorsal; F, ventral; G, detail of arm spines in ventral view; *Amphiura stimpsonii*: H, dorsal; I, ventral; J, detail of arm spines, dorsal spines on the left; *Amphiura (Ophionema) intricata*: K, dorsal; L, ventral. Photos by G. Hendler.

FIGURE 104. Details of amphiurid brittle stars photographed at low magnification, showing the dorsal surface of the disk and arms and the ventral surface of the oral frame and base of the arms of dry preserved specimens. *Ophiocnida scabriuscula*: A, dorsal; B, ventral; *Ophiostigma isocanthum*: C, dorsal; D, ventral; *Ophiostigma siva*: E, dorsal; F, ventral. Photos by G. Hendler.

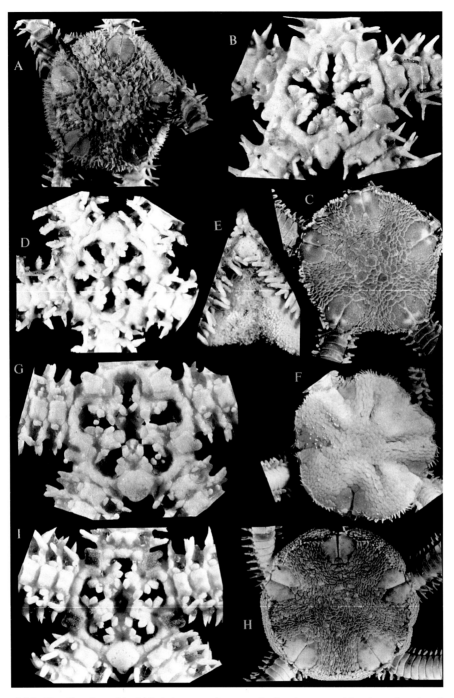

FIGURE 105. Details of amphiurid brittle stars photographed at low magnification, showing the dorsal surface of the disk and arms and the ventral surface of the oral frame and base of the arms of dry preserved specimens. *Ophiophragmus cubanus:* A, dorsal view of a small individual with dorsal and ventral disk papillae and fence papillae; B, ventral; *Ophiophragmus filograneus:* C, dorsal; D, ventral; E, ventral interradius of disk showing granulelike disk papillae; *Ophiophragmus moorei:* F, dorsal; G, ventral; *Ophiophragmus wurdemanii:* H, dorsal; I, ventral. Photos by G. Hendler.

FIGURE 106. Details of amphiurid brittle stars photographed at low magnification, showing the dorsal surface of the disk and arms and the ventral surface of the oral frame and base of the arms of dry preserved specimens. *Ophiophragmus pulcher:* A, dorsal; B, ventral; *Ophiophragmus riisei:* C, dorsal view of a disk with primary plates and well-developed fence papillae (USNM 7583); D, dorsal view of a disk with irregular scalation and few fence papillae (UZM, Copenhagen, type specimen of *Amphiura riisei* Lütken); E, ventral (same specimen as D); *Ophiophragmus septus:* F, dorsal; G, ventral. Photos by G. Hendler.

CLASS ECHINOIDEA
Sea Urchins, Sand Dollars, Heart Urchins

The class Echinoidea comprises about 900 species, including the familiar sand dollars and the spiny, spheroidal sea urchins. Of the 100 or so echinoids known from the Caribbean and Gulf of Mexico, about 30 occur in depths less than 30 m (100 ft). To most people, echinoids are better known for their skeletons, painted as seaside souvenirs, cast as bronze belt buckles, or their gonads served as sushi, than as intricate, beautiful, and bizarre living animals. For a variety of reasons they have been objects of interest since ancient times, but Hyman (1955:413) disdainfully noted that "most zoologists are aware of echinoids chiefly as a source of eggs for experimental purposes." Indeed, they remain of vital interest to cell biologists, but Hyman would have been pleased to consider the wealth of information on echinoids that has been accumulated by physiologists, morphologists, systematists, ecologists, and paleontologists.

FORM AND FUNCTION

Traditionally, the Echinoidea was thought to consist of two extant subclasses, the globose "regular" urchins and the more flattened and bilaterally symmetrical "irregular" urchins. Based on evolutionary relationships within the class, the major division now recognized distinguishes the subclass Cidaroidea from the more diverse subclass Euechinoidea. Within the Euechinoidea, the irregular urchins compose one cohort (with six orders), and the regular urchins consist of three cohorts (and eight orders).

The anatomical structures of echinoids are illustrated in Figures 107–109. Echinoids occur in a surprising variety of shapes and sizes, but all have a hollow shell or test of flattened plates that interlock at their edges (Figures 134–136). There are 20 columns of plates and five paired columns of radially placed ambulacral plates alternating with five paired columns of interradially placed interambulacral plates. The outer surface of the test is equipped with numerous knobs and tubercles that carry movable spines and other specialized structures. The radially symmetrical "regular" echinoids have a test that tends toward a subspherical or hemispherical shape (Figure 107). The mouth is in the middle of the underside of the test and the anus is opposite the mouth, within a centrally placed apical system composed of 10 plates, five genitals alternating with five ocular plates (Figure 108). The genital plates are each penetrated by a pore from which eggs or sperm are released during spawning. One of the genital plates also carries the madreporite of the water vascular system. The ocular plates, so named because they were once thought to contain light-sensitive organs, are perforated by a single tube foot, which is actually the terminal extension of the radial water vessel, equivalent to the plate at the tip of a sea star arm. As a sea urchin grows, new plates are added to the test at the outer edge

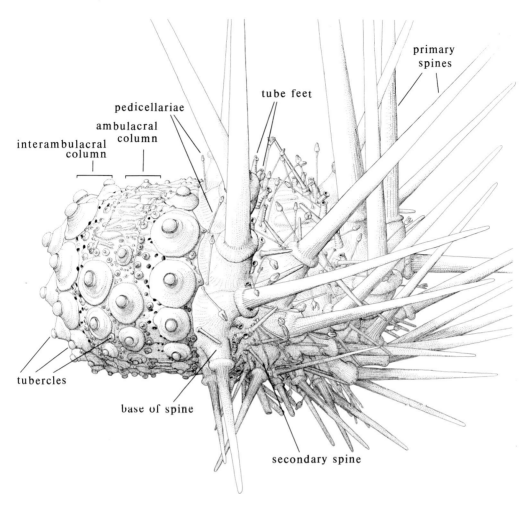

FIGURE 107. Major anatomical features of a regular sea urchin, viewed from the side; half of the body is illustrated without spines and other superficial structures to reveal features of the naked test. Illustration by C. Messing.

of the ocular plates, and the varied rates of plate growth between the apical system and the mouth determine the shape of the test.

In the bilaterally symmetrical, irregular sea urchins the anus migrates posteriorly during development and no longer lies within the apical system in the adult (Figure 109). The test in irregular echinoids ranges from the thick-walled, solid, flattened, discoid shape of sand dollars to the thin, fragile, inflated shape of heart urchins. Some sand dollars have symmetrically arranged slots (lunules) or notches in the test, and spatangoid heart urchins generally have a shovel-shaped lip (labrum) at one edge of the mouth.

In the regular urchins, five bands of tube feet run from the vi-

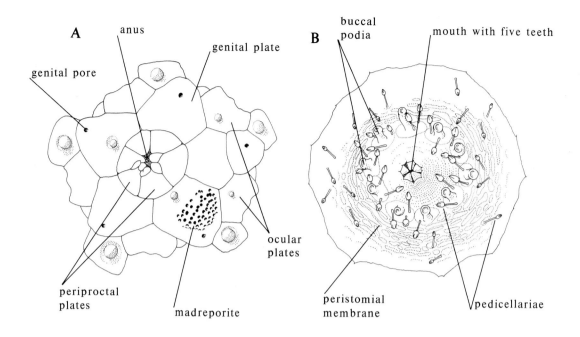

FIGURE 108. Dorsal and ventral regions of a regular sea urchin showing anatomical features of: *A*, the apical system surrounding the anus; *B*, the area around the mouth. Illustration by C. Messing.

cinity of the mouth to the vicinity of the anus. In a living sea urchin the numbers of thin, extensible tube feet are immense, their concentration achieved by a miniaturization and fusion of the perforate, foot-bearing ossicles of the test. The feet are used for locomotion, sensory perception, and for the manipulation of nearby objects. In irregular echinoids, many tube feet are arranged into "petals" on the upper surface that are used for respiration; on the oral surface and elsewhere, numerous scattered tube feet aid in locomotion and feeding. The suckered tip of sea urchin tube feet characteristically has an internal ring of supportive ossicles, and in the heart urchins, the penicillate tube feet that are used to grasp sediment each terminates in a radiating fringe of fingerlike processes with internal supporting ossicles.

The spines usually come in two sizes on a sea urchin; the larger or primary spines are carried on large tubercles, and small tubercles carry the numerous secondary spines. The engine driving each spine is a muscular collar connecting the base of the spine to the glassy smooth tubercle. Beneath the collar lies a layer of tendon known as the "catch apparatus" for its capacity to lock the spine erect. Once the catch apparatus is activated, it is easier to break a spine than to move it. Their spines make sea urchins prickly to touch, and, because some spines easily can penetrate the skin and cause intense pain, sharp-spined sea urchins should be handled with caution.

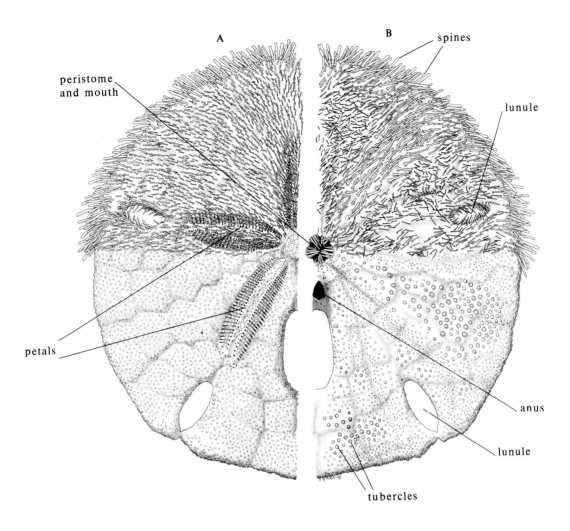

FIGURE 109. Major anatomical features of an irregular sea urchin shown in dorsal (A) and ventral (B) views. Half of each sector is illustrated without spines and other superficial structures to reveal features of the naked test. Illustration by C. Messing.

Spines can be used for defense, as shown in certain deep-sea urchins with poison sacs at the tips of their long, needle-thin spines. However, in the case of sand dollars, their multitudinous spines are so minute and close-set that they give the living animal a velvety appearance and texture. Some burrowing sea biscuits such as cassiduloids and clypeasteroids have spines that insulate the test from surrounding sand and permit oxygen-carrying water to flow around the animal. Similarly, mud-dwelling spatangoid heart urchins use mucus-coated spines as a barrier and have mucus-producing spines concentrated in tracts (fascioles) that are marked by lines of minute tubercles on the bare test. Sand dollars and cidaroids in particular use spines as locomotory appendages; in many other urchins the tube feet play a more substantial role in locomotion.

Also present on the urchin test are numerous, stalked pincerlike organs known as pedicellariae. They have a variety of shapes, sizes, and functions, aiding in removal of debris and in discouraging predators and "fouling" organisms. The globiferous pedicellariae are especially active as defensive structures; their terminal clawlike valves each contains a poison sac. They can pick off unwanted larvae, and they are powerful enough in some species to repel large, predatory sea stars. All urchins but the cidaroids bear microscopic, club-shaped, spinelike structures called sphaeridia that are thought to act as organs of equilibrium.

The test cavity is usually spacious, though less so in the internally buttressed sand dollars, and contains all the major organ systems. The digestive tract consists of a long, tubelike intestine, which makes two loops inside the test, beginning at the mouth and ending at the anus. In most sea urchins the mouth is equipped with five sharp, chisel-like teeth, which act against each other to cut and scrape food material such as seaweeds and seagrasses. The teeth are operated by a complex of 40 ossicles and 60 muscles known as Aristotle's lantern, which is attached by muscle and connective tissue to the wall of the test. The combined activity of teeth and spines enables some urchins to bore into concrete and even into steel. Irregular urchins generally lack a protrusible lantern or entirely lack the lantern complex.

The gonad consists of five interradially placed organs suspended within the body cavity in regular urchins, with three to five in irregular urchins, each of which consists of a clump of anastomosing tubules. A genital duct connects each gonad to a genital plate in the apical system. Because sexual maturity is not attained until openings in the genital plates (the gonopores) develop, it is possible to distinguish juvenile from adult individuals.

GENERAL BIOLOGY

Regular and irregular urchins differ considerably in their ecology. Birkeland (1989) characterized seven "morphological life forms" of regular urchins living on coral reefs, of which four are represented in the Caribbean. The long-spined, large, motile diadematids (e.g., *Diadema antillarum*) are found in coral, seagrass, and mangrove habitats. They are herbivorous, but also ingest encrusting animals, and they even consume living coral when seaweeds are scarce. The short-spined, motile toxopneustids (e.g., *Lytechinus variegatus*) have a delicate test, which limits them to areas with low wave impact. They are found in seagrass, rubble-strewn sand bottom, and coral reef. In contrast to diadematids, they are browsers rather than scrapers, nibbling on plants and detritus. The short-spined, sedentary *Echinometra* species have a thick test and can withstand strong surf. They live in branching corals, under reef rubble, or in burrows. Their food is drifting and attached algae and the microscopic boring algae that live within the reef. By excavating carbonate reef rock for food and shelter, they can create labyrinthine burrow systems. The slate-pencil cidaroid urchins, represented by *Eucidaris tribuloides*, are omnivorous grazers on encrusting organisms, animals in particular. They are usually cryptic, but along with *Echinometra* species they may emerge to feed at night.

Burrowing heart urchins and sand dollars, when present in dense populations, can alter sediment composition by their burrowing activities. Most irregular urchins feed on sediment as they plow through the substratum. Sediment particles are moved to the mouth by the tube feet and spines, and in some cases small animals and plants such as foraminiferan protozoans and algae are selectively ingested. Heart urchins that live just below the surface of the sediment collect particles using penicillate tube feet and pass them to the mouth. Those that burrow deep in soft sediments stabilize their burrows with mucus; their penicillate tube feet can stretch as long as 160 mm (6 in) to reach the surface. They collect particles, which are bound in mucus and passed to the mouth by ciliary tracts. Feces and excavated particles may be passed to a so-called "sanitary drain," a burrow to the rear of the urchin.

A number of regular urchins display a well-known, but poorly understood "covering reaction." They use their tube feet to pass objects from the surrounding substrate to the top of the test and then hold the small collection of debris in place, covering the apical system and plates nearby. Experiments variously have shown that the covering reaction is a response to reduce the intensity of direct sunlight, for camouflage, and for the storage of food items such as seaweed.

Many urchins have annual reproductive cycles, even some species in the relatively aseasonal Tropics and the deep sea. Some sea urchin populations spawn monthly at a certain phase of the moon. In most echinoids, fertilization is external. The embryo grows into a beautiful planktonic echinopluteus larva with a body form that is characteristic of the family to which it belongs. The nearly transparent larvae generally have four to six or more pairs of movable, skeleton-supported larval arms bearing a ciliated band that is used for food collection and locomotion. After a period in the plankton, from 9 days to 15 weeks, the larva sinks to the seafloor and undergoes a radical metamorphosis from a pelagic filter feeder to a benthic omnivore. During this amazing process, the rudimentary urchin adheres with its tube feet, the arms of the larva are bent backward, its skin collapses, the skeleton is shed, and a globose urchin is revealed, all in less than a minute. In a small percentage of sea urchins, particularly in antarctic species, the larval stage is reduced or suppressed, and the slowly developing young are held on, or inside, the body of the female parent (i.e., brooded) until they can crawl away as young sea urchins. Some female brooding echinoids have drastic modifications of the test, such as deep pouches that shelter the developing young. Sexual dimorphism sometimes is shown in non-brooding urchins, usually marked by differently shaped gonopores or genital papillae in males and females.

Sea urchins serve as food for fish and for other animals that are capable of breaching an armored test and spines. On Caribbean reefs, at least 34 species of fish eat sea urchins; they are the principal food source of six species, as well as a dietary component for crustaceans and birds. Intertidal sea urchins suffer heavy mortalities on the occasions that they are exposed to air and sun during extreme midday low tides. Mass mortalities caused by pathogens, affecting extensive regions, have a role in regulating the population growth of some echinoids. However, urchins generally have the capacity to endure moderate physical injury. Lost or damaged spines are repaired by regeneration, and small wounds to the test can heal. More extensive damage to the test usually results in infection and the death of the echinoid. Urchins serve as hosts to a multitude of parasites and benign symbionts. Many species of ciliate protozoans live in the intestines of echinoids; flatworm and crustacean commensals are also common, and one species of crab lives only in the rectum of sea urchins.

ECOLOGICAL AND COMMERCIAL IMPORTANCE

The gonads of regular urchins have been eaten as a delicacy, or out of necessity, for several thousand years. Today, an extensive fishery for echinoids registers an impressive annual world harvest of 47,560 metric tons (reviewed by Sloan 1985). Japan dominates the sea urchin fishery, harvesting 14 native species and importing roe from 13 countries; sizeable exporters include Chile and the United States. These fisheries are poorly regulated, and depletion of the resource is a recurring problem. The European fishery for *Paracentrotus lividus* (Lamarck) has resulted in overfishing, habitat destruction, and drastic population declines of urchins. Overfishing has impoverished the northeastern Pacific fishery for *Strongylocentrotus franciscanus* (A. Agassiz) and the Chilean fishery for *Loxechinus albus* (Molina). Urchins are also fished in the Pacific and Caribbean regions, and in places such as Barbados, urchin populations recently have been devastated.

Amazingly, it appears that echinoids are responsible for more than 90% of the bioerosion in Caribbean waters. Urchins, by their incessant grinding of the substrate, erode carbonate rock, producing quantities of sediment and gradually altering the structure of the reef framework. The precise impact of urchins in complex food webs is far from clear, but their grazing can create extensive "barren grounds." Like a horde of locusts or a column of army ants, fronts of hungry urchins are capable of stripping the seafloor of edible plants and animals. The vast numbers of *Strongylocentrotus droebachiensis* (O. F. Müller), which overgraze algae in eastern Canada and New England, may have an influence on the lobster fishery. Similarly, the food web involving sea otter and sea urchins on the west coast of the United States affects the commercial harvesting of kelp. Sea urchin grazing has a major impact on coral reefs, and their presence unquestionably regulates the diversity and abundance of tropical marine plants and animals. The far-reaching influence of the long-spined *Diadema antillarum* was proved by a five-fold increase in algal biomass on coral reefs after epidemic mortality of the species (see below under *Diadema*).

TERMINOLOGY AND CONVENTIONS

The size of sea urchins is expressed in terms of overall diameter, measured across an imaginary sphere that encloses the longest spines, or in terms of the test (diameter, length, or height) without spines. In cases where it is necessary to denude the test of an urchin to examine the distribution of tubercles, pore pairs, and so on, the bleaching procedure described for sea stars can be used.

IMPORTANT REFERENCES

Mortensen's (1928–1952) classic monograph of the echinoids is the "last word" on these animals up to 1950 or so. Because the echinoids have elicited curiosity since ancient times, the literature on their systematics and biology is voluminous. Harvey (1956) devoted a fine book to a single species, *Arbacia punctulata*. A wealth of information on the systematics and structures of the group is collected in the *Treatise on Invertebrate Paleontology* (Durham et al. 1966), and in a classic monograph by Jackson (1912); an excellent recent summary of echinoid functional morphology and phylogeny is available in Smith (1984a). H. L. Clark (1933) and Serafy (1979) reviewed the common sea urchins in the Caribbean area, and many authors noted below deal with the biology of individual species. A recent review of their reproductive biology was offered by Pearse and Cameron (1991), and an overview of their feeding was presented by Jangoux and Lawrence (1982).

FIGURE 110. *Eucidaris tribuloides.*

FAMILY CIDARIDAE

Eucidaris tribuloides (Lamarck)
Figures 110, 134A

DESCRIPTION: This urchin is easily recognized because of its small number of solid, thick, brown, cylindrical spines arranged in 10 vertical series. The spines are attached to a thick, globular test. Overall diameter, including spines, of this species can reach 130 mm (5 in). Ground color of the test is light brown to red-brown. The muscle bases of all secondary spines are brown to red-brown, the shafts of the spines are dirty white, and the distal ends lightly tinged with light brown. Muscle bases of the primary spines are variegated brown, and the shafts of the spines are variously colored because of epizoic animals and plants; naked primary spines often have broad, dirty white and pinkish brown bands. The oralmost primary spines are often banded dirty white and light pinkish. The tube feet are light brown; aboral feet are very broad at the base, more or less pointed, and very extensile; the oral feet have well-developed dirty white terminal disks. From the adapical to the oral feet, the terminal disks become broader, better developed, and more conspicuous. The spines lack a covering of living tissue and consequently are frequently adorned with algae and other epizoans.

HABITAT: On coral reef in small crevices, in turtle grass beds, or under rocks and rubble in back reef lagoon areas.

DISTRIBUTION: Cape Hatteras, North Carolina, south throughout the Caribbean to Rio de Janeiro, Brazil (Serafy 1979). A variety or subspecies, *E. t. africana* Mortensen, is recorded from the Gulf of Guinea in West Africa and the Cape Verde Islands. DEPTH: 0–800 m (0–2,324 ft), but most commonly in less than 50 m (164 ft) (Serafy 1979).

BIOLOGY: Length of the spines in *E. tribuloides* may be related to habitat: in high-energy areas the spines tend to be short and thick; in low-energy areas they are long and slender. The spines are slow-growing; a new spine takes a year to reach a length of 23 mm (Cutress 1965). In grass beds, *E. tribuloides* is easily seen because it does not cover itself with plant and shell debris as do *Lytechinus variegatus* and *Tripneustes ventricosus*. However, the spines of this species are often overgrown with encrusting animals and plants, presumably attached to dead skeleton, because only the growing, basal region of the spine is covered with living epithelium (Märkel and Röser 1983a). The "dead" distal part of the spine, which has a toughened, polycrystalline calcite cortex, is shed when the spine is damaged, and a new spine tip regenerates. This process, which has been likened to "the annual exchange of a stag's antlers" (Märkel and Röser 1983b:56), results in the distinctive shapes of spines on different parts of the test (Cutress 1965; Märkel et al. 1971; Märkel and Röser 1983a,b). The neurophysiology of the primary spine ligament, or "catch-apparatus," was investigated by Morales et al. (1993). The stiffening and softening responses of the ligament appear to be mediated by a combination of cholinergic, adrenergic, nicotinic, and muscarinic receptors. The changes in plates and spines with growth, the anatomy of Aristotle's lantern, and the process of biomineralization of the tooth have been examined in several studies of *E. tribuloides* (Cutress 1965; Märkel 1979, 1981; Märkel et al. 1986).

According to McPherson (1968a), *E. tribuloides* may leave its protective site at night, presumably for foraging. In intra- and interspecific aggression experiments, Shulman (1990) found that resident *Eucidaris* only occasionally pushed and bit intruding *Eucidaris* and *Echinometra* species. Like other cidaroid sea urchins, *E. tribuloides* is very slow-moving. Individuals in crevices may be so tightly wedged that they cannot be removed without some breakage of spines. The elaborate connective tissue system that enables the spines to lock in position has been explored with electron microscopy (Smith et al. 1990), and the biochemical composition of collagen in the spine ligament has been analyzed (Trotter and Koob 1994).

In 1984–1985, on the northwestern coast of Puerto Rico, a die-off of this species occurred (Williams et al. 1986). Heavy waves cast ashore 6,000–8,000 dead specimens. The cause of the mortality is unknown; it was presumably not storm-related because individuals living in crevices are protected from waves. The tests of dead individuals disarticulate rapidly (Greenstein 1991). Fossil fragments have been found in Plio-Pleistocene sedimentary sequences from Venezuela, Tobago, and Jamaica (Donovan 1993b; Donovan and Gordon 1993).

This species seems to be omnivorous; its stomach contents include pieces of algae and bryozoans (Mortensen 1928), coral fragments, gastropod shells, echinoid spines, sponges, and turtle grass (McPherson 1968a,b). The diet seems to vary with locality (Lawrence 1975; De Ridder and Lawrence 1982). According to McClintock et al. (1982), this species prefers *Donax* (clam) prey models to *Thalassia* (turtle grass) prey models. Using artificial foods (1 or 10% fish meal in agar), Lares and McClintock (1991) found that *E. tribuloides* can respond to a low-quality diet by increasing its feeding rate and absorption efficiency.

In a year-long laboratory study, McClintock and Watts (1990) artificially altered the photoperiod for this species and found that short days and long nights enhance and entrain development of gametes (eggs and sperm). Preliminary results suggest that the sex steroids estradiol and progesterone, present in the gonads of male and female individuals, may modulate seasonal phases of gametogenesis (Hines et al. 1992). Lessios (1987, 1991) found that through time, the size of the eggs released may vary by a factor of two, and individuals in Panama tend to spawn monthly, when the moon is full.

There is evidence, though inconclusive, that gender is regulated by sex chromosomes in this species (Pearse and Cameron 1991). McPherson (1968a) found that individuals of 2 years age are sexually mature, that their gonads are ripe in late summer and early fall, and that the embryo develops from fertilized egg to metamorphosed juvenile in 25 days. However, Schroeder (1981) described relatively slow development from fertilization to the late two-armed larval stage.

More recent research on the species supports the distinction in development between cidaroids and other, more advanced, echinoids (Wray and McClay 1988; Vodicka et al. 1990). Emlet (1988) described the development through metamorphosis of a related species, *E. thouarsii* (Valenciennes), and showed how cidaroid sea urchins have a characteristic larval form and a simpler metamorphosis into juveniles than noncidaroid urchins. He also suggested that the larvae described and illustrated by McPherson (1968a) lack the characteristic, highly developed lobes, and that they were probably larvae of noncidaroid echinoids. Individuals are thought to be slow-growing and long-lived, surviving for over 5 years in an aquarium (Moore 1966).

Parasites of *E. tribuloides* include the gastropod *Sabinella troglodytes* (Thiele), which forms cuplike depressions in deformed primary spines of the urchin, and *Nanobalcis worsfoldi* Warén, which also occurs on the spines (Warén 1984; Warén and Mifsud 1990). Another gastropod, *Trochostilifer eucidaricola* Warén & Moolenbeek, is found near the mouth of *E. tribuloides* (Warén and Moolenbeek 1989).

SELECTED REFERENCES: Lamarck 1816: 56 [as *Cidarite tribuloides*]; Kier and Grant 1965:11, pl. ii, figs. 1–3; Zeiller 1974:109, 1 fig.; Voss 1976:136, 1 fig.; Colin 1978:415, 471 (fig.); Serafy 1979:15, fig. 3 [as *Eucidaris tribuloides tribuloides*]; Kaplan 1982:191, pl. 34, fig. 14; Pawson 1986:533, pl. 176.

FAMILY DIADEMATIDAE

Astropyga magnifica A. H. Clark
Figure 111

DESCRIPTION: This beautiful, brilliantly colored urchin shares many morphological characters with *Diadema antillarum*; both species have a somewhat flattened test with conspicuous perforate and crenulate tubercles, and long fragile spines. Unlike *D. antillarum*, *A. magnifica* has a fairly flexible test, with extensive naked interambulacral areas on the upper surface. The test can reach 20 cm (8 in) in horizontal diameter. Also, the spines of *A. magnifica* are not hollow, but are filled with a calcite meshwork.

The spines are banded in reddish brown and yellowish white. The naked areas of the test are golden yellow, and the brown ambulacra are bordered by single rows of brilliant iridescent blue spots. The inflated anal cone is bluish white, with a brown peripheral ring and a dark brown anus. When viewed from above, the anal cone resembles a bull's-eye.

FIGURE 111. *Astropyga magnifica*.

HABITAT: Kier and Grant (1965) reported this species from a sand terrace on the seaward side of the reef at Key Largo, Florida. Serafy (1979) described specimens from algal sand, carbonate sand, and crushed shell habitats and limestone outcroppings.

DISTRIBUTION: South Carolina and the southeastern Gulf of Mexico through the Greater and Lesser Antilles to Colombia, Venezuela, and Surinam. DEPTH: 11–88 m (36–289 ft) (Serafy 1979).

BIOLOGY: Like *Diadema antillarum*, this species has a patchy distribution; it seems to be gregarious in some Caribbean localities, with groups of specimens clumping together (Kier and Grant 1965). At Galeta, Panama, solitary individuals occasionally occur on an off-reef sand plain at a depth of 12 m (Hendler, previously unpublished). On the northwestern coast of Puerto Rico, Williams et al. (1986) reported a die-off of 68 specimens of *A. magnifica* in a water depth of 0.6–1.2 m. Two aggregations at greater depths of 14 and 21 m were unaffected. No cause was suggested for the die-off.

The blue iridescent spots of this striking species were once thought to be light receptors (Mortensen 1940), but that function appears unlikely (Yoshida 1966). When threatened, or when the incident light intensity changes rapidly, for example with a passing shadow, this urchin groups its spines into five cone-shaped bundles (Kier and Grant 1965). This species is also highly mobile, moving at least 1 m/minute (Kier and Grant 1965). As in *D. antillarum*, the anal cone enables the animal to eject feces away from the body.

Some commensals, especially shrimps, have been observed by Kier and Grant (1965), Gooding (1974), and Criales (1984). Kier and Grant (1965) reported a small species of fish (*Apogon* sp.) swimming among the spines of *A. magnifica*.

SELECTED REFERENCES: A. H. Clark 1934:52; Kier and Grant 1965:15, pl. 1, figs. 1–5; Serafy 1979:27, fig. 7.

FIGURE 112. *Diadema antillarum*.

Diadema antillarum (Philippi)
Figures 112, 136A

DESCRIPTION: This urchin has extremely long, fragile, slender, hollow, black spines equipped with numerous whorls of spinelets. Overall diameter of fully grown individuals can exceed 500 mm (20 in). The spines can be 300–400 mm (12–16 in) long, up to four times the diameter of the test. The test is low and flattened, with the height usually less than 50% of the width. The apical system and central areas of the broad interambulacra are depressed, and the narrow ambulacra are slightly inflated, with the pore pairs in two fairly straight series in each area. The tubercles are perforate and crenulate. The test is so thin and fragile that it remains intact for only a short time after the death of the urchin.

The typical color of test and spines is black, but individuals may have few to many lighter-colored white or gray spines, and some are almost entirely white, as seen in the individual illustrated here. Juveniles always have spines with black and white bands and may be mistaken for a different species. In general, darker animals live in bright light and clear water, and lighter individuals live in darker, more turbid conditions, in crevices (Moore 1966) or deep water.

HABITAT: Commonly on coral reefs, also in turtle grass beds, on sand or rock bottoms, and in mangroves; it prefers "quiet" waters, actively avoiding heavy wave action.

DISTRIBUTION: In the western Atlantic, from the Gulf of Mexico, Bermuda, and from southeastern Florida to Rio de Janeiro, Brazil; in the eastern Atlantic at the Azores, Madeira, the Canaries, the Cape Verdes, Annobón Island, and the Gulf of Guinea (Chesher 1966a; Serafy 1979). DEPTH: 0–400 m

(0–1,310 ft), but usually less than 50 m (165 ft).

BIOLOGY: The literature on the biology of this sea urchin is extensive. Studies before 1978 on diet, growth and reproduction, predators, and commensals were summarized in some detail by Serafy (1979). Those earlier findings, along with some research after 1979, are discussed here.

D. antillarum is a highly active urchin. In response to a disturbance, such as a sudden change in light intensity or unusual water movement, it waves its spines rapidly and directs many spines toward the source of disturbance. It can also move away rapidly. The "dermal light sense" in this species is acute, and it responds to changes in light intensity by altering the color of its test (Millott 1965, 1968; Millott and Coleman 1969).

D. antillarum is gregarious, often occurring in very large numbers in its preferred habitats, where it can reach densities of more than 20 individuals per square meter (Scoffin et al. 1980). It has the ability to reduce its body size (i.e., to shrink its component skeletal plates) in response to food limitations caused by increasing population density (Levitan 1988b, 1989, 1991); conversely, it can take advantage of decreasing population density by increasing in size and fecundity. Populations have been observed to aggregate for the purposes of spawning and for mutual protection (Snyder and Snyder 1970). In contrast, Levitan (1988a) observed that at the U.S. Virgin Islands spawning was not highly synchronized and little aggregation occurred.

Typically, *D. antillarum* remains in crevices during the day, emerging at night to graze, although clumps of individuals can often be seen in the open, for example in turtle grass beds. The "homing" behavior of this species was studied by Lewis (1964), Ogden et al. (1973a,b), Carpenter (1984), and others. Carpenter found that the "homing frequency" varied between 30 and 84% and that homing behavior was positively correlated with abundance of predators. Woodley (1982) showed that in daylight, individuals tend to move toward darker areas (including other specimens of the same species); presumably this is a defensive response. In a study of aggressive behavior in sea urchins, Shulman (1990) found that *D. antillarum* would aggressively bite intruding *Echinometra l. lucunter* and *E. viridis*, and that the *Echinometra* spp. would retreat. Shulman believed that the greater dispersion of *Echinometra* spp. relative to available food resources, and their tendency to inhabit crevices, is caused by *D. antillarum*.

This species is primarily a grazer, feeding on algal turf (Ogden et al. 1973a,b; Hay 1984; Carpenter 1986). Williams and Carpenter (1988) found that algal turfs grazed by *D. antillarum* are two to 10 times more productive than ungrazed turfs; this is because excretion by *D. antillarum* provides supplemental nitrogen to the nutrient-limited algal turfs. Hawkins (1981) found that young urchins ingest far more crustose coralline algae, and larger individuals graze on epipelic and endolithic algae. *D. antillarum* also feeds on young corals (Rylaarsdam 1983) and zoanthids (Karlson 1983). Removal of specimens from feeding areas, and exclusion experiments, have shown that in the absence of *D. antillarum*, the growth of algae is very rapid and detrimental to the well-being of reef corals (Ogden et al. 1973a,b; Sammarco et al. 1974; Carpenter 1981, 1986; Hawkins 1981; Sammarco 1982a,b; Hay 1984; Hay and Taylor 1985; and numerous other studies). In controlling growth of algae, *D. antillarum* also "clears the way" for settlement of coral larvae; when urchins are removed from an area, larval settlement of corals is drastically reduced (Sammarco 1980, 1982a). However, in the course of its grazing, *D. antillarum* also abrades calcium carbonate from the coral reef itself. Scoffin et al. (1980) estimated that at Barbados, where this species averages 23 individuals per square meter, it abrades approximately 9,750 kg of coral skeleton per year; this is almost half of the total estimated annual production of coral.

In 1983 a massive die-off of *D. antillarum* occurred all over the Caribbean, some areas losing more than 97% of their mature individuals. An unidentified host-specific water-borne pathogen is assumed to be responsible (Lessios et al. 1984a,b). Bacterial pathogens, *Clostridium perfringens* (Veillon & Zuber) Haduroy et al. and *C. sordelli* (Hall & Scott) Prévot, were isolated from dead laboratory-reared *D. antillarum* and found to be lethal to *D. antillarum* in laboratory situations (Bauer and Agerter 1987). It is not known whether these bacteria were involved in the mass mortality.

Several follow-up studies were made in various localities (for example, Bak et al. 1984; Lessios 1985a; Lewis and Wainwright 1985; de Ruyter and Bak 1986; Hunte et al. 1986; Liddell and Ohlhorst 1986; de Ruyter and Breeman 1987; Hughes et al. 1987; Carpenter 1988, 1990; Hunte and Younglao 1988; Levitan 1988c). Short-term and chronic adverse effects of the *D. antillarum* die-off on coral reefs include a dramatic increase of algal cover (de Ruyter and Breeman 1987; Hughes et al. 1987; Carpenter 1988; Morrison 1988; Forcucci 1994; Hughes 1994) and a reduced settlement of coral larvae. As *D. antillarum* populations increase, algal cover is greatly reduced. However, Jackson and Kaufmann (1987) found that in cryptic coral reef environments the mass mortality hardly changed community composition, because alternative species failed to become established. Greenstein (1989) found that after the 1983 mass mortality in Bonaire, Netherlands Antilles, the skeletal remains of *D. antillarum* contributed nothing of any significance to the total echinoderm content of surface sediments, so that the sedimentary record could not register the occurrence of this major ecological event. However, fossil material that is attributed to this species has been recovered from the last interglacial Falmouth Formation of Jamaica (Donovan and Gordon 1993).

In Panama, Lessios (1988a) found that for the period 1983–1987, recovery from the 97% die-off of 1983 was very slow. The meager recruitment of juveniles was attributed to an inadequate supply of planktonic larvae, not to increased predation on juveniles or to absence of adults that might provide settlement cues. Levitan (1988a) suggested that fertilization success is poor, and thus recruitment is poor, because spawning is not highly synchronized; at the U.S. Virgin Islands, only 5% spawned spontaneously at any one time, and there was only slight evidence of aggregation. Although there is evidence that some populations are slowly recovering, Karlson and Levitan (1990) determined that local *D. antillarum* populations are recruitment-limited and that there is no indication of recovery toward pre–mass mortality population levels. In the Florida Keys, Jamaica, and elsewhere in the Caribbean, populations of *D. antillarum* are not expected to recover for at least several decades (Forcucci 1994; Hughes 1994). Lessios (1988b) summarized what has been learned from the *D. antillarum* disaster.

More than 20 fish species prey upon *D. antillarum* (Randall et al. 1964; Randall 1967). Hoffman and Robertson (1983) found that two species of toadfish feed primarily on this species and that they remove as many as 20,000 urchins per hectare per year. Robertson (1987) found that in Panama, following the 1983 mass mortality of *D. antillarum,* the two species of toadfish remained at their pre-event abundance and they had changed their diet to include a broader range of mobile benthic invertebrates. He suggested that recovery of *D. an-*

tillarum populations might be hampered by persistence of high population levels of these predators. Among invertebrates, two species of the gastropod mollusk *Cassis* and the spiny lobster *Panulirus argus* (Latreille) are common predators of *D. antillarum* (Schroeder 1962; Snyder and Snyder 1970; Serafy 1979). Based on their *Cassis tuberosa* (Linnaeus) predation studies, Levitan and Genovese (1989) concluded that patch reefs are a refuge from predation for *D. antillarum*. Hendler (1977b) showed that at a reef flat at Galeta, Panama, deaths from predation were trivial in comparison with deaths from physical stress.

The timing of spawning may vary considerably; Serafy (1979) suggested that spawning times change from year to year and differ among populations. Lessios (1981a, 1984, 1987, 1991) found dissimilarities in peak spawning period among different Panamanian populations. He ascertained that egg size fluctuates during the month and found a striking pattern of lunar periodicity in spawning. In Bermuda, *D. antillarum* spawns from early summer to early winter and has a well-defined lunar rhythm (Iliffe and Pearse 1982). In a study of growth rates, Bauer (1982) found that small individuals 5–32 mm in diameter can grow at a rate of 4.5 mm/month, and larger individuals 32–42 mm in diameter grow 1.9 mm/month. The aboral-oral gradient in the growth rate of test plates, responsible for the shape of the test, has also been examined (Märkel 1981). The larva of this species has been described, but its metamorphosis has not (Mortensen 1921; Pearse and Cameron 1991).

The brittle but sharp spines of this species can easily penetrate human flesh, causing intense pain, to which many swimmers and scuba divers can attest! The spines shatter in the wound and are virtually impossible to remove, but eventually they dissolve. They always transfer organic debris and bacteria into a wound, sometimes leading to infection, and they may deposit a mild venom as well, but this has yet to be proven.

Numerous invertebrates and some vertebrates have commensal associations with *D. antillarum*. These include several small crustaceans that may swim among the spines, such as shrimps, mysids, and copepods (Randall et al. 1964; Gooding 1974), and the young of many reef fish (Randall et al. 1964). Recent records include the black shrimp, *Tuleariocaris neglecta* Chace, which positions itself head-down among the urchin's spines (Criales 1984). An internal turbellarian parasite, *Syndisyrinx antillarum* (Stunkard & Corliss), has been found (Hertel et al. 1990), and a parasitic snail (*Pelseneeria* sp.) has been reported from African specimens (Chesher 1966a).

REMARKS: There is a close resemblance, and a great biochemical similarity, between *D. antillarum* and its eastern Pacific geminate congener, *D. mexicanum* (Lessios 1981b, 1984, 1990).

SELECTED REFERENCES: Philippi 1845: 355 [as *Cidaris* (*Diadema*) *antillarum*]; Lewis 1964:549; Randall et al. 1964:421; Kier and Grant 1965:13, pl. 2, figs. 4–7; Zeiller 1974;109, 1 fig.; Voss 1976:136, 1 fig.; Colin 1978:423, 425 (fig.), 428 (fig.); Serafy 1979:24, fig. 6; Kaplan 1982:190, fig. 8, pl. 6, pl. 17; Pawson 1986:533, pl. 176, pl. 14.1.

FIGURE 113. *Arbacia punctulata*.

FAMILY ARBACIIDAE

Arbacia punctulata (Lamarck)
Figures 113, 134B

DESCRIPTION: This dark brown to purple urchin has long, slender spines with sharp tips and reaches a total diameter of about 100 mm (4 in); diameter of the test reaches approximately 50 mm (2 in). The area of the apical system is naked, and the periproct is covered by four conspicuous plates. On the upper surface of the test, the interambulacral areas are also naked and very conspicuous. Color in life varies from reddish gray, to reddish, to purplish, to brownish, to almost black (Harvey 1956). The spines are usually of the same color as the test, and the muscle bases are dirty white with numerous brown to purple spots. Tube feet are inconspicuous aborally, very light olive drab overall. Pedicellariae aborally have the same color stalks as muscle bases. The oral side of the test is light brown to light purple, the peristome area is white, the numerous circumoral feet are silvery white, and their terminal disks are white. Oral primaries are lighter than aboral primaries, sometimes almost banded, white and pinkish. Pore pairs are arranged in a simple vertical series. The interambulacral tubercles are of equal size.

HABITAT: In turtle grass beds, under pieces of rubble on coral reefs, and singly or aggregated on rocky, sandy, or shelly bottoms (Harvey 1956; Sharp and Gray 1962; Kier and Grant 1965; Kier 1975; Serafy 1979).

DISTRIBUTION: Cape Cod, Massachusetts, southward to Cuba and the Yucatán Peninsula; Texas to Florida in the Gulf of Mexico; Panama to French Guiana and north to Barbados (Serafy 1979). DEPTH: 0–225 m (0–738 ft) but most common in depths of less than 50 m (164 ft) (Serafy 1979).

BIOLOGY: A thorough treatment of the biology of this species was given by Harvey (1956). Serafy (1979) summarized aspects of growth, recruitment, and population dynamics, based upon the material available to him.

This species is primarily an herbivore, but Harvey (1956) noted that it also eats sponges, coral polyps, sand dollars, and even other *A. punctulata*. Fell et al. (1984) found that it eats *Cliona* sponges in Connecticut. Hay et al. (1986) found that the chemical attraction of *A. punctulata* to its preferred algal food increased at night; they suggested that increased foraging at night decreases its susceptibility to predators. The species prefers to graze on *Gracilaria tikvahiae* McLachlan rather than *Sargassum filipendula* C. Agardh, and thus it can affect the relative abundance of these two seaweed species (Pfister and Hay 1988). Renaud et al. (1990) showed that after desiccation, the less palatable seaweed *Padina gymnospora* (Kützing) Vickers is preferred over the more palatable *G. tikvahiae* by *A. punctulata*; they suggested that *P. gymnospora* loses some chemical defenses during desiccation and that the history of physical or biological stress can affect a plant's susceptibility to herbivory. Boolootian and Cantor (1965) reported on the feeding preferences and the salinity tolerance of this urchin in the Cape Cod region.

Serafy (1979) noted a disjunct distribution pattern for this species (there are "north" and "south" American populations), and he recommended further study. Offspring of laboratory-reared *A. punctulata* from two widely separated localities, Massachusetts and the northeastern Gulf of Mexico, had spines of different lengths, and "hybrids" had spines of intermediate length (Marcus 1980). Thus, considerable morphological variation, under genetic control, occurs over the wide geographic range of this species.

A. punctulata has a long reproductive season in warm waters such as in Florida, but it spawns in the spring and summer near the northern limits of its range (Harvey 1956; Serafy 1979). The eggs have been used extensively in biological research, to the point that *A. punctulata* is virtually extinct near major northeastern United States marine laboratories, probably because of over-collecting (Harvey 1956).

Bell and McClintock (1982) found that up to 145 different organisms, especially copepods, are associated as commensals with this echinoid.

SELECTED REFERENCES: Lamarck 1816: 47 [as *Echinus punctulatus*]; Harvey 1956:1–298; Kier and Grant 1965:17, pl. 2, figs. 8, 9, pl. 10, fig. 5; Zeiller 1974:110, 1 fig.; Kier 1975:17; Voss 1976:137, 1 fig.; Gosner 1979:257, pl. 62; Serafy 1979:30, fig. 9; Kaplan 1982:189; Serafy and Fell 1985:21; Ruppert and Fox 1988:73, fig. 114.

FIGURE 114. *Lytechinus variegatus*.

FAMILY TOXOPNEUSTIDAE

Lytechinus variegatus (Lamarck)
Figures 114, 134D

DESCRIPTION: This urchin has short spines, and it reaches a total diameter of about 110 mm (4 in); diameter of the test reaches about 85 mm (3 in). The test is hemispherical with smoothly curving sides and carries many small tubercles. Tuberculation is poorly developed in aboral portions of ambulacra and interambulacra, where conspicuous naked areas occur. The globiferous pedicellariae, visible to the naked eye, are very numerous on living specimens, appearing as stalked, almost spherical, white or pink structures. Color of test and spines may vary and has been used as a basis for distinguishing subspecies (Serafy 1973). In the Florida Keys, two so-called subspecies occur: *L. variegatus variegatus* (Lamarck) usually has a greenish test and green spines, and *L. variegatus carolinus* Agassiz usually has a light red test and spines. However, color may be highly variable within a population.

HABITAT: Most common in quiet water, especially in turtle grass beds; also occurs on rock or sand. Moore et al. (1963a) noted that this species is intolerant of suspended silt and abandons areas where the water is turbid.

DISTRIBUTION: Beaufort, North Carolina, southward throughout the Caribbean to Santos, Brazil. Also occurs in Bermuda (*L. variegatus atlanticus* Agassiz). DEPTH: 0–250 m (0–820 ft), but the species is most common in less than 50 m (164 ft) (Serafy 1979).

BIOLOGY: Moore (1965) found a correlation between symmetry, spination, and color pattern in this species, based upon his study of deformed specimens. Allain (1978) suggested that deformations of the test in specimens from Colombia may be linked with pollution. Test deformations have also been noted in Bermuda populations in areas where suspended sediments have been increased by earthworks nearby (Pawson, previously unpublished). The production of hermaphroditic individuals in this normally gonochoric species was related to abnormally low water temperatures by Moore et al. (1963a). There is also evidence, though inconclusive, that males and females of this species have different numbers of chromosomes (Pearse and Cameron 1991).

In its preferred turtle grass habitat, this species can reach high population densities, sometimes extensively overgrazing the turtle grass (Valentine and Heck 1991). Camp et al. (1973) reported an astonishing elevated density of up to 636 individuals per square meter. In a seagrass habitat in Nicaragua, Vadas et al. (1982) showed that this species consumes 0.6 g dry weight of *Thalassia* per day and that dead *Thalassia* leaves and sediment are the major food items; there, the effect on live *Thalassia* is minimal.

L. variegatus exhibits a covering reaction, piling debris from the surrounding seafloor on the upper surface of the test and holding it there with its tube feet (Mortenson 1943a; Millott 1956; Sharp and Gray 1962: Kier and Grant 1965). In his detailed discussion of this phenomenon, Millott (1956) showed that, in this species, covering is related to light and diurnal light changes and suggested that the cover acts as a screen against strong light. Not surprisingly, the righting response time of this urchin is directly related to temperature (Kleitman 1941).

Moore and McPherson (1965) found that *L. variegatus* has a higher growth rate in winter than in summer and that its assimilation efficiency was about 50–60%. Empirical and experimental studies on its feeding are numerous (see Lawrence 1975; De Ridder and Lawrence 1982; Montague et al. 1991). McClintock et al. (1982) found that *L. variegatus* consumed more *Donax variabilis* Say (a bivalve mollusk) artificial prey models than *Thalassia testudinum* prey models, although the urchin probably never encounters this bivalve species in the wild, and they showed that prey manipulation is an important determinant of feeding rates. *L. variegatus* is inefficient at digesting eelgrass, *Zostera marina* Linnaeus, ejecting 1- to 2-mm particles in its feces (Drifmeyer 1981). The gut tissue of this species, the stomach wall in particular, seems to function as a nutrient storage organ (Klinger et al. 1988; Bishop and Watts 1992). Bishop and Watts (1994) found that after food is provided to previously starved individuals of *L. variegatus*, the urchins respond with gamete growth, followed by the production of new gametes.

The spawning period of *L. variegatus* is shorter in northern localities, such as Bermuda, than at lower latitudes (Moore et al. 1963a). In Panama, it is sexually ripe throughout the year (Lessios 1985b, 1991), but at Tarpon Springs, Florida, substantial variations in reproductive patterns prevailed among subpopulations exposed to differing environmental regimes (Ernest and Blake 1981). Mazur and Miller (1971) maintained Florida animals in aquaria and reported a reproductive season of March to October. In Puerto Rico, Cameron (1986) found a broad, loosely synchronous reproductive season, with the least gonad development in the spring. Spawning in Panama, and possibly elsewhere, follows a semilunar rhythm; individuals tend to spawn at the new and full moon (Lessios 1991).

The processes of fertilization and early development in this species, studied by numerous authors, were reviewed in Pearse and Cameron (1991). In artificial seawater in the laboratory, development from zygote to metamorphosis takes 33–43 days (Mazur and Miller 1971). The rate of development is accelerated by increased temperature (Petersen and Almeida 1976). Salinities below 35 ppt result in decreased larval survival and rates of development. These effects are not mitigated by the acclimation of adult sea urchins to low salinities before spawning and fertilization (Roller and Stickle 1993). Boidron-Metairon (1988) found that as available food diminishes, larvae of *L. variegatus* can increase the size of the ciliated bands, thereby increasing feeding efficiency. After metamorphosis, the juveniles grow to about 15 mm horizontal test diameter in 6 months and to 25 mm in 9 months (Pawson and Miller 1982; Michel 1984).

Keller (1983) studied competition and predation in this species and in *Tripneustes ventricosus*, which co-occur in *Thalassia* beds at Discovery Bay, Jamaica. Randall (1967) recorded several species of fish as predators. Shorebirds can eat large numbers of these urchins when they are exposed by unusually low tides. Ruddy turnstones peck holes in their tests (Hendler 1977b), and herring gulls carry urchins into the air and drop them onto wet sand, breaking them open (Moore et al. 1963a). Hendler (1977b) reported predation by *Cypraecassis testiculus* (Linnaeus) snails in Panama. Engstrom (1982) found that in Puerto Rico, predation by the helmet shell *Cassis tuberosa* was significant (0.8 *L. variegatus* per square meter per year), but that populations of *L. variegatus* did not decline, because population density was maintained by immigration. He suggested that recruitment was highly localized, based on differences between the size frequency distributions of two populations under study. Alarm responses to extracts of conspecifics were minimal for specimens living in turtle grass beds (Parker and Shulman 1986).

A high incidence of infestation by the turbellarian *Syndisyrinx collongistyla* Hertel et al., 1990 in the body cavity and intestines of *L. variegatus* has been observed (Nappi and Crawford 1984; Hertel et al. 1990), and the digestion and chemical composition of the parasite have been studied (Mettrick and Jennings 1969). Protozoan parasites of this urchin are reported as well (Mortensen 1943a), and the polychaete *Podarke obscura* Verrill is an occasional associate (Ruppert and Fox 1988).

REMARKS: The status of the so-called subspecies of *L. variegatus* requires further investigation. Pawson and Miller (1982) used laboratory experiments to demonstrate a genetic basis for morphological differences of Bermuda *L. variegatus* from those in Florida. Rosenberg and Wain (1982) studied isozyme variation in individuals from Florida and Bermuda and found little justification for distinguishing subspecies of this sea urchin.

SELECTED REFERENCES: Lamarck 1816: 48 [as *Echinus variegatus*]; Kier and Grant 1965:21, pl. 3, fig. 1; Serafy 1973:525; 1979: 49, fig. 21 [as *Lytechinus variegatus carolinus*]; Zeiller 1974:110, 2 figs.; Voss 1976: 137, 1 fig.; Colin 1978:415, 420 (fig.); Kaplan 1982:192, pl. 32, fig. 7; Pawson 1986:533, pl. 176; Ruppert and Fox 1988: 74, pl. A31.

Lytechinus williamsi Chesher
Figures 115, 134C

DESCRIPTION: This is a small species, total diameter reaching 50 mm (2 in), test diameter rarely exceeding 30 mm (1 in). Its spines are usually deep green, sometimes white, and the large, conspicuous globiferous pedicellariae are purple. The test is beige, with a purple-brown stripe running down the median longitudinal suture in each interambulacrum. On the naked test, two rows of large tubercles are present on each interam-

FIGURE 115. *Lytechinus williamsi.*

bulacrum, and two rows of smaller tubercles in each ambulacrum.

L. williamsi is similar in general appearance to *L. variegatus,* but differs in having purple globiferous pedicellariae; in *L. variegatus* they are white or pink. Also, *L. williamsi* has far fewer plates in each column of the test and fewer wedges in the spines (12 versus 24) than *L. variegatus.*

HABITAT: Crevices in coral reefs and rocky areas, or on exposed surfaces of corals such as *Millepora* or *Agaricia* spp., depending on locality.

DISTRIBUTION: The Dry Tortugas and the Florida Keys, Colombia, Panama, and Belize. DEPTH: 5–92 m (16–302 ft) (Serafy 1979; Gallo 1988).

BIOLOGY: Although adults are often rare, Henkel (1982) noted that this species was "very common" on the branches of living *Acropora cervicornis* and in *Acropora* spp. rubble at depths of 3–8 m at Enmedio Reef, southwestern Gulf of Mexico. Hendler (previously unpublished) noted that this species was common in similar habitats at Bocas del Toro, Panama. On the Belize Barrier Reef juveniles were found in *Halimeda* and larger individuals in finger-coral rubble (Miller and Hendler, previously unpublished). Those on living coral are thought to graze on the coral polyps (Caycedo 1979).

Lessios (1985b, 1991) reported that in Panama, *L. williamsi* has an annual reproductive cycle, with an extended spawning period during the rainy season; its spawning behavior is not related to lunar phases. In tests of escape responses, Parker and Shulman (1986) noted that this species, when exposed to chemical extracts from conspecifics, relaxed its hold on the substratum and dropped into the crevices of the coral reef. In studies of intra- and interspecific aggression, Shulman (1990) found that intruding individuals of this species were very seldom attacked by resident individuals of *Diadema antillarum.*

Hertel et al. (1990) found a previously undescribed species of umagillid turbellarian, *Syndisyrinx collongistyla,* in the intestine of individuals from Jamaica.

SELECTED REFERENCES: Chesher 1968b: 1, figs. 1–5; Colin 1978:418, 420 (fig.), 421 (fig.); Humann 1992:290, 291 (fig.).

Tripneustes ventricosus (Lamarck)
Figures 116, 134E

DESCRIPTION: This is a large urchin that reaches a diameter of up to 150 mm (6 in). The test is hemispherical, with ambulacral pores in three vertical series, and with numerous small tubercles, tending to form horizontal series of five or six on each interambulacral plate. In living individuals, the test appears uniform dark brown, including the basal muscles of primary spines. The short primary and secondary spines are a showy and contrasting white to dirty white. Aboral tube feet are uniformly brown, including the terminal disks; near the disk is a conspicuous white nerve ganglion visible as a double bump. In midradii and interradii, and around the periproct, are bands of hundreds of globiferous pedicellariae with calcareous stalks. They are covered with a thick layer of dark brown tissue, with the valves darker brown than the stalk; points of the valves are conspicuously white. The oral side of the test is lighter brown; the spines and feet are as on the aboral side. The peristome is light brown with scattered olive green patches marking the presence of small clusters of spinelets and pedicellariae. The tissue around the jaws is dark brown.

HABITAT: Grassy areas on sand bottoms; also on reefs and among rocks and rubble.

FIGURE 116. *Tripneustes ventricosus.*

DISTRIBUTION: Florida and Bermuda south to Brazil; also off West Africa, Ascension, Fernando de Noronha, and Trindade islands. DEPTH: 0–55 m (0–180 ft) (Mortensen 1943a; Chesher 1966a; Serafy 1979).

BIOLOGY: Its tenacious grip on the substratum enables this species to live in relatively high-energy areas such as reef-crest bedrock and on *Sargassum*. In turtle grass beds, *T. ventricosus* lives on the sandy bottom among the grass clumps, but it also can climb blades of the grass (Kier and Grant 1965). It often co-occurs with *Lytechinus variegatus*, and it seems able to tolerate more turbid conditions than the latter species (Pawson, previously unpublished). However, it rapidly succumbs when subjected to extreme physical stress caused by reef-flat exposure such as that described by Hendler (1977b). This species may cover part of its upper test with fragments of grass, shell, and rubble, but the tendency is much less strongly developed than in *Lytechinus variegatus*.

Its intestinal contents are diverse, and may include turtle grass *(Thalassia)* and assorted detrital material (Mortensen 1943a). Tertschnig (1984) discussed feeding propensities of *T. ventricosus* in turtle grass *(Thalassia)* beds in Bermuda, concluding that it has a negligible effect on living grass, preferring to ingest dead grass blades, algal crusts, and sediment adhering to grass blades. But in a later paper, he noted that at St. Croix, Virgin Islands, this species feeds day and night on *Thalassia*, consuming approximately 3.6% of the daily production of the turtle grass (Tertschnig 1989). Moore and McPherson (1965) found that this urchin has a higher growth rate in winter than in summer, and the assimilation efficiency year-round is 50–60%. The rate of body growth is faster outside the reproductive season (Lewis 1958).

Keller (1983) studied competition and predation in this species and in *L. variegatus*, which co-occur in turtle grass beds at Discovery Bay, Jamaica. He noted that juvenile individuals are preyed upon by fish. At St. Croix, Virgin Islands, high rates of locomotion at night and low activity during the day were suggested to be related to predation by night-active helmet conchs (*Cassis* spp.) (Moore, 1956; Tertschnig 1989).

At Barbados a centuries-old artisanal fishery of this species collapsed because of overfishing in the late 1970s and early 1980s (Scheibling and Mladenov 1987). Some populations still flourish in remote and inaccessible areas around Barbados, and in time the commercial populations could be reestablished.

Kier and Grant (1965) found numerous immature individuals about 25 mm in diameter near a grass patch, clustered together with immature *Lytechinus variegatus* in clumps of algae, broken shells, and sponge fragments. According to McPherson (1965), juveniles of this species grow rapidly during the first year, with a mean test diameter of 75 mm being reached by the following summer.

In Florida, in the autumn, the gonads first develop in young specimens with a minimum test diameter of 35–45 mm, and the spawning season is in spring and summer (Moore et al. 1963b; McPherson 1965). At one locality, spawning patterns may vary from year to year (McPherson 1965). In Panama, the species remains ripe throughout the year, and spawning occurs at random during the lunar cycle (Lessios 1985b, 1991). In Puerto Rico, Cameron (1986) reported that it has a broad, loosely synchronous reproductive season, with a resting phase apparently occurring in the autumn.

The development of the echinopluteus larva, its metamorphosis, and the characteristics of settling juveniles have been described (Mortensen 1921; Lewis 1958). Moore et al. (1963b) noted that the young are common in the intertidal, but adults are generally restricted to the subtidal. They also concluded that instances of abnormal, hermaphroditic individuals are related to unusually low temperatures in the field. Evidence that the gender of this species is chromosomally determined is inconclusive (Pearse and Cameron 1991).

Associated animals include the cryptically colored shrimp *Gnathophylloides mineri* Schmitt, which lives among the spines and feeds on the epidermis of the spines (Patton et al. 1985). Criales (1984) reported that this shrimp is commonly located near the urchin's mouth. Protozoan parasites have also been reported (Mortensen 1943a).

SELECTED REFERENCES: Lamarck 1816: 44 [as *Echinus ventricosus*]; Kier and Grant 1965:24, pl. 3, fig. 2; Voss 1976:138, 1 fig.; Colin 1978:419, 421 (fig.); Kaplan 1982: 193, pl. 34, fig. 8, pl. 37, fig. 8; Pawson 1986: 535, pl. 176.

FAMILY ECHINOMETRIDAE

Echinometra lucunter lucunter
(Linnaeus)
Figures 117, 134F

DESCRIPTION: This urchin reaches a maximum diameter of 150 mm (6 in), but most individuals are half that size. It has an elongate oval test with two rows of large tubercles along the ambulacra and interambulacra, pore pairs in arcs of six, and a large peristome. The spines are long (although shorter than those of *E. viridis*), thick at the base, slender elsewhere, and sharply pointed.

Aborally, the primary and secondary spines are very dark olive green, greenish violet to purple at the tips; superficially, the general color of the spines is blackish, although some specimens are reddish dorsally, as may be seen in the individual illustrated here. Test and muscle bases of the spines are light to dark red-brown. Aboral tube feet are light brown, and the terminal disks dark brown to blackish. Oral spines are lighter colored than aboral spines, light olive green, changing to violet distally. The test and peristome are variegated creamy brown. Near the mouth the tube feet are translucent, almost colorless, their terminal disks about twice as large as disks on the aboral feet, creamy white, the edge of the disk with a very narrow, dark brown band.

HABITAT: Limestone reef rock, often in the surf zone.

DISTRIBUTION: Beaufort, North Carolina, and Bermuda, southward throughout the Caribbean and eastern Central America to Desterra, Brazil; also West Africa. Another subspecies, *Echinometra lucunter polypora* Pawson, is common at Ascension and St. Helena islands (Pawson 1978). DEPTH: 0–45 m (0–148 ft) (Serafy 1979).

BIOLOGY: On hard substrates, this species may be very common, often in exposed, shallow fore reef habitats, including the midlittoral yellow beachrock zone (Bratström 1992; McGehee 1992). It usually occupies depressions or shallow burrows abraded by action of the animal's spines and teeth on the rock substrate. In Panama it occurs in dense concentrations on the reef crest, in rough surf (Hendler, previously unpublished). Similar assemblages occur elsewhere, including West Africa (Mortensen 1943b; Stephenson and Stephenson 1972). Pompa et al. (1990) reported an average density of almost 11 indi-

FIGURE 117. *Echinometra lucunter lucunter*.

viduals per square meter in the Gulf of Cariaco, Venezuela, and noted the presence of a mixed size class. Up to 129 individuals have been counted in a square meter quadrat (Greenstein 1993). This species is somewhat resistant to physical stress caused by such factors as increased temperature and salinity (Hendler 1977b).

Because of its burrowing ability, the density of *E. l. lucunter* is a significant factor in the mechanical breakdown (bioerosion) of the fabric of coral reefs. Ogden (1977) found that the rate of erosion of coral substrate at the Virgin Islands attributable to this species is 3.9 kg/m² per year; at Bermuda it is 7.0 kg/m² per year; at Barbados it is 24 g per urchin per year. Hoskin and Reed (1985) calculated an erosion rate of 3.2 kg/m² per year at Little Bahama Bank; they estimated that burrows were excavated in an average of 2.9 years.

This species feeds mostly at night; McPherson (1969) and Ogden (1976) noted that its primary food is drift algae. Hendler (previously unpublished) noted that in Panama, individuals' feeding activities clear the area around the burrow of all but encrusting calcareous algae. McClintock et al. (1982) found that *Donax variabilis* (a small bivalve mollusk) prey models are preferred to *Thalassia testudinum* (turtle grass) prey models by this species, although the urchin probably never encounters this particular bivalve species in the wild.

A few individuals move out of their holes to feed, but most remain in them (McPherson 1969; Ogden 1976). In a detailed study of activity patterns, behavior, and feeding in this species, Abbott et al. (1974) found that urchins that moved out to feed at night usually returned to the same hole. Social interactions in the form of agonistic behavior were observed by Abbott et al. (1974) and Grunbaum et al. (1978). Conspecific intruders actively were discouraged from entering occupied burrows; resident urchins pushed and bit intruders, and such encounters were usually won by the original occupants of burrows. Escape (flight) responses were elicited when this species was exposed to chemical extracts from conspecifics and congeners

(Parker and Shulman 1986). In a later study, Shulman (1990) confirmed earlier observations and found that *Diadema antillarum* actively pushed and bit intruding *E. l. lucunter* (see under *Diadema antillarum* above).

Lewis and Storey (1984) compared *Echinometra l. lucunter* from high-energy habitats and from low wave-energy habitats. Specimens from the high-energy area had thicker, smaller, flatter, and narrower tests, a distinctive pattern of insertion of ocular plates in the apical system, and they spawned once a year. Specimens from the low-energy area spawned twice a year.

In the Florida Keys in 1965, this species showed an annual reproductive cycle, spawning in late summer (McPherson 1969). At Puerto Rico, Cameron (1986) found that it, along with *E. viridis*, has a short, well-defined reproductive season, which peaks in the fall when water temperature is highest. At Panama, its reproductive periodicity is markedly less well defined than in *E. viridis* (Lessios 1981a); the capability to spawn fluctuates between 20 and 90% of the population of *E. l. lucunter* from month to month (Lessios 1985b). Information on the reproductive periodicity of this species is summarized in Pearse and Cameron (1991). In the course of a study of the cytology and histochemistry of the gonads of *E. l. lucunter*, Tennent et al. (1931) noted that spawning takes about 15 minutes for an individual.

The larva of *E. l. lucunter* has been reared through metamorphosis (Mortensen 1921). In a study of gametic incompatability, Lessios and Cunningham (1990) found that this species and *E. viridis* maintain their genetic integrity despite the fact that they are incompletely isolated gametically. Not surprisingly, the processes of fertilization and development are adversely affected by ferrous sulfate and by sodium hypochlorite (Muñoz and Ellies 1982) and by reduced salinity (Petersen and Almeida 1976).

Ablanedo et al. (1990) found that individuals of this species accumulate certain heavy metals in the gonads, test, spines, and lantern, reflecting the degree of environmental pollution to which they are exposed. Little information is available on the physiology of this *Echinometra* species, but the rate of oxygen uptake of its lantern muscle tissue has been studied (Bianconcini et al. 1985).

Predators (Abbott et al. 1974) include birds such as ruddy turnstones, conchs, and fish, including triggerfish, grunts, jacks, and wrasses. Tidal exposure of this species at Galeta Island, Panama (Hendler 1977b; Schneider 1985), resulted in heavy predation by ruddy turnstones; in contrast, no mortality was noted in areas protected from predation by birds. At the same locality it is also subject to predation by a snail, *Cypraecassis testiculus* (Linnaeus), which bores a hole through the urchin's test (Hendler 1977b).

Warén and Moolenbeek (1989) noted that the eulimid gastropod *Monogamus minibulla* (Olsson & McGinty) is parasitic on this echinoid. Hertel et al. (1990) described a new umagillid turbellarian, *Syndisyrinx collongistyla,* from the intestine of Jamaican specimens; a related worm, *S. evelinae* (Marcus), had previously been noted in specimens from St. Barthélemy. Protozoans infest the species as well (Mortensen 1943b). Several associates, including a gobiesocid fish, a porcellanid crab, and an ophiotrichid brittle star, have been found to live in the burrows of Colombian *E. l. lucunter* (Schoppe 1991).

Fossil specimens of this species from the late Tertiary of Florida have been reported (Kier 1963), and there are records for Oligocene and later deposits from elsewhere in the Caribbean and Angola (Gordon 1991; Donovan 1993b; Donovan and Gordon 1993; Donovan and Jones 1994). However, the species is so infrequently fossilized that these records probably do not reflect its past geographic range (Donovan and Jones 1994).

REMARKS: *E. l. lucunter* differs from *E. viridis* in having a darker colored test and spines, and in having dark milled rings near the bases of the spines. The number of pore pairs is a characteristic than can be used to distinguish between the denuded tests of the two species (Mortensen 1943b).

SELECTED REFERENCES: Linnaeus 1758: 665 [as *Echinus lucunter*]; Kier and Grant 1965:18, pl. 16, figs. 1–4; McLean 1967: 586; McPherson 1969:194; Abbott et al. 1974:1–111; Zeiller 1974:111, 1 fig.; Voss 1976:138, 1 fig.; Colin 1978:419, 424 (fig.); Kaplan 1982:190, pl. 32, fig. 6, pl. 34, fig. 7; Pawson 1986:535, pl. 176.

Echinometra viridis A. Agassiz
Figures 118, 134G

DESCRIPTION: This species is similar in general shape and size to *Echinometra l. lucunter*, although the test is more nearly circular in outline. There are two rows of large tubercles in each ambulacrum and interambulacrum, pore pairs in arcs of five, and spines strong and sharply pointed. In living specimens, the general test color is reddish brown; basal muscles of the spines are also reddish brown. The conspicuous milled rings of the spines are white. The spines are brownish green, with the distal 10% uniformly olive green, but tipped with purple. The aboral feet are as long as primary spines, light brown overall, including disks. Color of the oral surface is lighter than the aboral surface; the test is light brown, the peristome dark red-brown. All spines on the oral surface are lighter colored, but otherwise not conspicuously different from aboral spines. Feet of the oral surface are similar in color to aboral feet, but the disks are larger.

HABITAT: In coral reefs and rocky areas it can be found in crevices or under rubble, in coral debris, on branching corals, or in sandy areas in association with shell or rock. Dense populations on some Belizean mangrove cays aggregate among mangrove prop roots. In Panama, they can also be abundant in protected habitats on the branches of *Acropora cervicornis* (staghorn coral) (Miller and Hendler, previously unpublished).

DISTRIBUTION: Southern Florida and throughout the West Indies to Venezuela.
DEPTH: 0–40 m (0–131 ft).

BIOLOGY: *E. viridis* is not a rock-borer, and although it usually remains hidden on reefs, it can occur in the open in sheltered habitats, on mangrove roots or branching coral. Thus, its cryptic behavior may be a response to water turbulence as well as to the presence of predators (Hendler, previously unpublished). McGehee (1992) found it most common where water currents are reduced, and McClanahan (1992) monitored its susceptibility to predators. Lessios (1985b) found densities of almost 50 individuals per square meter in corals on a Panamanian reef. McPherson (1969) reported densities of up to 21 individuals per square meter in staghorn coral in the Florida Keys. There, its intestinal contents consisted mainly of algae and a significant fraction of carbonate sediment. Sammarco (1982a) noted high densities of 50 individuals per square meter on reefs at Discovery Bay, Jamaica. Here, the species grazed locally on algae, producing numerous algae-free patches on the reef. At the same locality, removal of *Diadema antillarum* from an area allowed an increase by migration of the population of *E. viridis*, and removal of damselfish increased the presence of *E. viridis* in 3 days (Williams 1981). This species is relatively resistant to physical stress caused by such factors as increased temperature and salinity (Hendler 1977b).

In studies of alarm responses, Parker and Shulman (1986) found that *E. viridis*

FIGURE 118. *Echinometra viridis*.

moved toward the nearest refuge upon exposure to chemical extracts from conspecifics and from some other urchin species. In a later study, Shulman (1990) found that resident individuals reacted aggressively against intruding conspecifics and congeners, the interaction usually taking the form of pushing and, rarely, biting. *Diadema antillarum* exhibited biting behavior against this species.

According to Mortensen (1943b), the genital pores appear in individuals only 12 mm long, indicating sexual maturity at that size. In 1965 in the Florida Keys, this species showed an annual reproductive cycle, spawning in early fall (McPherson 1969). In Panama, the reproductive cycle is annual (April–December), during the rainy season, population-wide, and not correlated with the lunar cycle (Lessios 1981a, 1985b, 1991). At Puerto Rico, Cameron (1986) found that the reproductive cycle is annual and the breeding season is short, peaking in the fall when water temperature is highest. A description of the larval development of this species has not been published (Pearse and Cameron 1991).

Hertel et al. (1990) described a new umagillid turbellarian, *Syndisyrinx collongistyla*, from the intestine of this echinoid.

Donovan and Jones (1994) reported fossil specimens of *E. viridis* from Pleistocene formations of Jamaica.

REMARKS: This species differs from *E. l. lucunter* in having a lighter colored test and spines with purple tips and white milled rings.

SELECTED REFERENCES: A. Agassiz 1863:22; Kier and Grant 1965:20, pl. 10, fig. 6; McPherson 1969:194; Kier 1975:17; Voss 1976:138, 1 fig.; Colin 1978:422, 424 (fig.); Kaplan 1982:190, pl. 32, fig. 5, pl. 34, fig. 5.

FIGURE 119. *Echinoneus cyclostomus.*

FAMILY ECHINONEIDAE

Echinoneus cyclòstomus Leske
Figures 119, 134H

DESCRIPTION: This urchin is small, up to 30 mm (1 in) long, elongate oval, approximately cylindrical in cross section, with rounded ends. The very short spines are whitish to light brown. The tube feet are conspicuously red, in narrow bands in the ambulacra, and may give the animal a reddish cast. The test is white to light brown, with an ovoidal mouth in the center of the oral surface of the test and a large anus just posterior to the mouth.

HABITAT: Under rock slabs or pieces of coral rubble, living in or on coarse sand.

DISTRIBUTION: One of the few echinoderm species reported to be "tropicopolitan," it occurs throughout the Tropics of the world. DEPTH: 5–570 m (16–1,870 ft) (Serafy 1979), but most commonly occurs in shallow depths.

BIOLOGY: This is not strictly a burrowing species, but a secretive one, never found in the open. It leads a sedentary existence and moves very slowly, even when disturbed. Individuals often cling to the undersurfaces of rock slabs that rest on sand, and they swallow pieces of organic material adhering to coarse sand grains and shell fragments (H. L. Clark in Westergren 1911; Mortensen 1948a). The species is common on reef flats and in other reef habitats such as sand channels where debris collects. The numbers of recently dead individuals collected during periods of midday exposures of a Panamanian reef flat suggest that as many as one to two individuals per square meter occurred on the reef. Most of them were killed by heat stress, but some had been preyed on by the boring snail *Cypraecassis testiculus* (Hendler 1977b).

The lantern and teeth are not found in adult individuals of this and other species in the family Echinoneidae. Westergren (1911) discussed at length, and illustrated, the anatomy of the test and internal organs and indicated that a lantern, present in the smallest individuals, is resorbed by the time they grow to 5 mm in length. Jensen (1981) described in detail the lantern structure of juvenile *E. cyclostomus*. Donovan (1993b) confirmed that there are four genital plates, one a madreporite, in the apical system. He also reviewed the lower Miocene and Pleistocene record of the species in the Caribbean region.

SELECTED REFERENCES: Leske 1778: 173, pl. 37, figs. 4, 5; Kier and Grant 1965: 25, pl. 15, fig. 1; Voss 1976:140, 1 fig.; Kaplan 1982:191; Pawson 1986:535, pl. 177, pl. 14.2.

FAMILY CLYPEASTERIDAE

Clypeaster luetkeni Mortensen
Figure 120A,B

DESCRIPTION: This is a medium-sized flattened urchin up to 110 mm (4.3 in) long, with five petals of more or less equal size, the petaloid area occupying slightly more than half the length of the test. The oral surface is slightly concave. The upper surface of the preserved test is light pinkish brown, with darker pigmentation at plate boundaries, so that the mosaic structure of the test is apparent. The oral surface is also light brown, with a blackish tinge in the ambulacral areas because of presence of very numerous black tube feet. Hopkins (1988:339) noted that a specimen he examined "seems to have been rose purple."

HABITAT: At Looe Key, Florida, on a sand plain.

DISTRIBUTION: Known only from off Haiti, St. Thomas (Mortensen 1948b), Montego Bay, Jamaica (Hopkins 1988), and Looe Key, Florida Keys (this report). DEPTH: 9 m (30 ft) (Haiti and Looe Key).

BIOLOGY: Mortensen (1948b) provided a description of this species, based upon two small but probably sexually mature specimens 18 and 22 mm long. The specimen described by Hopkins (1988) was 47.3 mm long, and the Looe Key specimen is approximately 110 mm long, indicating that this species can reach a considerable size. The life history of this handsome urchin is virtually unknown. Presumably, in a similar fashion to *Clypeaster subdepressus*, it covers itself with a layer of sediment, but remains more or less epifaunal.

SELECTED REFERENCES: Mortensen 1948b:121, pl. 25, figs. 3–5, pl. 27, fig. 3, pl. 48, figs. 21, 22 [as *Clypeaster (Stolonoclypus) Lütkeni*].

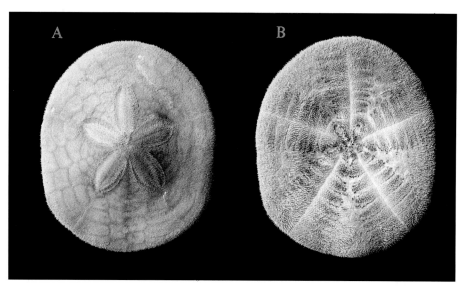

FIGURE 120. *Clypeaster luetkeni*: A, dorsal view; B, ventral view.

Clypeaster rosaceus (Linnaeus)
Figures 121, 134I

DESCRIPTION: This sea biscuit is elongate, up to 200 mm (8 in) long, dark brown, and inflated, with a strongly convex upper surface and a concave lower (oral) surface. The five subequal ambulacral petals are inflated. The naked test is white, and, because it is thick and strong, complete dead tests are encountered frequently. In life, the test is covered with a carpet of short spines.

HABITAT: Turtle grass beds and sand fields bordering turtle grass beds.

DISTRIBUTION: According to Serafy (1979), the range extends from South Carolina to Barbados; also known from Venezuela, Colombia, Panama, Belize, and Texas. DEPTH: 0–285 m (935 ft), but most common at shallow depths. Hopkins (1988) suggested a more limited geographic range (South Florida, southern Gulf of Mexico and Caribbean) and a bathymetric range of 1–50 m (3–164 ft).

BIOLOGY: This urchin lives on the surface of the sand and does not burrow, even when it is on bare sand (Kier and Grant 1965; but see Moore 1966). It frequently covers the upper surface of its test with fragments of turtle grass, shell, and rock. Kier and Grant (1965) discussed the covering reaction and other aspects of its behavior. Hammond (1982a) noted that this species has distinctly nocturnal patterns of feeding and activity. Preferred food appears to be dead turtle grass, collected with the aid of oral surface accessory podia with terminal suckers (Telford et al. 1987), but in areas where *Halimeda* is abundant, segments of the dead alga are the major source of food (Kampfer and Tertschnig 1992). Within the urchin's teeth, deposits of polycrystalline calcite have been identified, which may strengthen the structure (Märkel et al. 1986), and the biting force of the teeth in this species has been analyzed (Ellers and

FIGURE 121. *Clypeaster rosaceus*.

Telford 1991). The grinding of large sediment particles by *C. rosaceus* may play a major role in the production of fine sediment on reefs. In 1 year, a single individual can transform 5.5 kg of coarse sand to fine sand (Kampfer and Tertschnig 1992).

Surface ciliary currents are not associated with feeding processes (Telford et al. 1987). Mooi (1986) described the structure and function of the miliary spines in this species. They help create ciliary currents to move unwanted fine particles from the animal, and they assist in preventing larger particles from falling between the spines toward the test.

In Panama, *C. rosaceus* has an annual reproductive cycle; spawning occurs during the rainy season, at random with respect to the lunar cycle (Lessios 1985b, 1991). In another study in Panama, Emlet (1986) found that the egg and larva of this species contain a considerable amount of yolk. During the brief period of development the larva may or may not feed on plankton; metamorphosis occurs after 5–7 days at an average temperature of 27°C.

Griffith (1987) and Werding and Sanchez (1989) reported that a small pinnotherid crab, a *Dissodactylus* species, is associated with this echinoid.

Based on his study of a range of fossils, Poddobiuk (1985) concluded that the occurrence of *C. rosaceus* extends back to the Miocene era. In the Caribbean region, fossilized specimens have been reported from Jamaica, Puerto Rico, and Cuba (Donovan 1993b).

REMARKS: For a discussion of the correct scientific name of this species, see A. M. Clark (1964).

SELECTED REFERENCES: Linnaeus 1758: 665 [as *Echinus rosaceus*]; Kier and Grant 1965:26, pl. 4, figs. 1–7, pl. 6, fig. 7; Zeiller 1974:112, 2 figs.; Kier 1975:18; Voss 1976: 138, 1 fig.; Kaplan 1982:190, pl. 35, figs. 7, 8.

Clypeaster subdepressus (Gray)
Figures 122, 134J

DESCRIPTION: This large urchin is up to 300 mm (12 in) long, flattened, and conspicuously elongate, with five petals of equal size. The aboral surface is flattened toward the edges and rises to form a low hump at the center. The oral surface is quite flat, with a slight concavity at the mouth. The test is yellow to dark tan and covered with short, densely distributed spines.

HABITAT: Sand fields or shelly sediment, where there is little or no seagrass.

DISTRIBUTION: North Carolina, southward through the Caribbean including the east coast of Central America to Rio de Janeiro, Brazil (Serafy 1979). DEPTH: Commonly 5–50 m (16–164 ft), but reported to 210 m (689 ft). Hopkins (1988) questioned the reliability of the greater depth records.

BIOLOGY: Serafy (1979) discussed the relationship between shape, meristic characteristics, and size in *S. subdepressus*. Kier and Grant (1965) described the burrowing behavior and righting reaction in this species. Individuals occur just beneath the surface of the sand, and when excavated they can rebury in 6–12 minutes. The tendency of covering individuals to convey sediment onto ambulacral sectors of the test is shown in Figure 122. When large individuals are turned upside down, they right themselves in approximately 1 hour. This species occurs on coarse biogenic sands and selects the upper size fractions for ingestion, using oral surface accessory podia with terminal suckers (Telford et

FIGURE 122. *Clypeaster subdepressus*.

al. 1987). Gladfelter (1978) found that the helmet conch *Cassis tuberosa* is a major predator of *C. subdepressus*.

In Panama, this species is ripe throughout the year. At least 60% of individuals examined at any given time were capable of spawning (Lessios 1985b, 1987, 1991). Emlet (1986) found that it has a planktotrophic larva that metamorphoses in 16–28 days at an average temperature of 27°C.

No fewer than five species of pinnotherid crabs of the genus *Dissodactylus* are associated with this echinoid (Griffith 1987; Werding and Sanchez 1989). The polychaete *Ophiodromus obscurus* Verrill is another common commensal (Werding and Sanchez 1989).

Fossils of this echinoid have been reported from late Tertiary deposits in Florida (Kier 1963).

SELECTED REFERENCES: Gray 1825:427 [as *Echinanthus subdepressus*]; Mortensen 1948b:112, pl. 23, figs. 1–3, pl. 24, fig. 3, pl. 25, fig. 6, pl. 26, figs. 1, 6, pl. 27, fig. 4, pl. 65, figs. 4, 11, 14, 15 [as *Clypeaster (Stolonoclypus) subdepressus*]; Kier and Grant 1965:28, pl. 5, figs. 1–6, pl. 6, figs. 1–10, pl. 15, fig. 8; Serafy 1970:672; 1979:65, fig. 28; Zeiller 1974:111, 1 fig.; Kaplan 1982:190, pl. 35, figs. 5, 6.

FAMILY MELLITIDAE

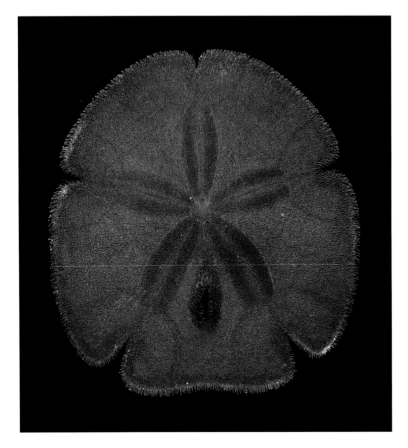

FIGURE 123. *Encope aberrans*. Photo by W. Lee.

Encope aberrans Martens
Figures 123, 136C

DESCRIPTION: The test is strongly constructed, almost circular, the oral surface quite flat. Maximum length is approximately 150 mm (6 in). There are two distinct but shallow posterior notches and three indistinct anterior indentations. The posterior interambulacral lunule is short. There are five long petals, the posterior pair being slightly longer than the others. Specimens preserved in alcohol are dark purplish to dark brown.

HABITAT: Sand plains, on crushed shell and quartz particles.

DISTRIBUTION: From Cape Hatteras, North Carolina, southward to the Bahama Islands, and throughout the Gulf of Mexico. DEPTH: 12–90 m (39–295 ft) (Serafy 1979).

BIOLOGY: In the past, this species has been confused with *Encope michelini* L. Agassiz, but Phelan (1972) established the validity of *E. aberrans*. Virtually nothing is known of its life history, and the distribution pattern has yet to be confirmed. The size and composition of the eggs shed by this species were analyzed by Turner and Lawrence (1979). It is known as a fossil from the Miocene of Venezuela (Cooke 1961) and from the late Miocene of Florida (Kier 1963), according to Phelan (1972).

SELECTED REFERENCES: Martens 1867a: 112; Phelan 1972:125, figs. 2c, 8–10; Serafy 1979:76, fig. 33.

Encope michelini L. Agassiz
Figures 124, 135A

DESCRIPTION: This is a flat, more or less circular, dark brown or deep purple sand dol-

FIGURE 124. *Encope michelini*.

lar of moderate size, up to 140 mm (5.5 in) long, and of robust construction. There is one slit in the test, between the posterior pair of petals, and five conspicuous, nearly equal-sized indentations in the test margin. There are five long petals, the posterior pair being slightly longer than the anterior three.

HABITAT: Fine to coarse deep sand with little or no turtle grass or filamentous algae.

DISTRIBUTION: Cape Hatteras, North Carolina, to the southern tip of Florida, through the Gulf of Mexico to Cozumel, Mexico, but not known from the Bahama Islands. DEPTH: 3–90 m (10–295 ft) (Phelan 1972).

BIOLOGY: In the preferred habitat of this species, the surface of the sand is usually marked by deep ripples, indicating strong water currents. It can burrow to a depth of 10 cm (Kier and Grant 1965), but it is also frequently found at the sediment surface covered with only a thin veneer of sediment. Telford and Mooi (1986) reported that *E. michelini* is found on both biogenic carbonate and siliceous terrigenous substrates. In deeper water, individuals occur in areas of crushed shell and sand (Serafy 1979). A mean population density of 1.3 individuals per square meter has been reported (McNulty et al. 1962).

Serafy (1979) documented meristic and morphometric relationships in this species. According to Telford and Mooi (1986), it is a podial particle picker, living on substrates where approximately 90% of the particles are in the 0.10–0.40 mm range. Only particles larger than 0.10 mm in size are ingested. Deposits of polycrystalline calcite have been identified in the teeth of this species, similar to those in *Clypeaster rosaceus* (Märkel et al. 1971). In the Gulf of Mexico, the gray triggerfish feeds on this species and two sympatric sand dollars, *Mellita tenuis* H. L. Clark and *Leodia sexiesperforata*. The fish excavates and overturns the urchins, breaks into the oral region, and removes soft tissues from the test (Frazer et al. 1991).

Little is known about other aspects of the biology of *E. michelini*. However, fossils of the species have been found in Pliocene and Pleistocene deposits of the eastern United States (Mortensen 1948b). A subspecies, *E. m. imperforata* Kier, has been found in late Tertiary deposits in Florida (Kier 1963).

SELECTED REFERENCES: L. Agassiz 1841:58; Kier and Grant 1965:33, pl. 5, fig. 7, pl. 7, figs. 1–8, pl. 15, fig. 7; Phelan 1972: 124; Zeiller 1974:113, 2 figs.; Serafy 1979: 80, fig. 35.

Leodia sexiesperforata (Leske)
Figures 125, 135B

DESCRIPTION: This is a yellow to light brown subcircular sand dollar, up to 100 mm (4 in) long, with a very flat, thin test and thin margin. The test has six slotlike holes (lunules) and five short, equal petals.

HABITAT: Open sand areas where seagrasses and filamentous algae are rare or absent.

DISTRIBUTION: From Cape Hatteras, North Carolina, southward around the tip of

FIGURE 125. *Leodia sexiesperforata.*

Florida into the Gulf of Mexico, north to Sanibel Island, Florida. Also the Greater and Lesser Antilles, and from the Yucatán Peninsula along the coast of northern South America to Uruguay. DEPTH: 0–60 m (0–197 ft), most common in less than 25 m (82 ft) (Serafy 1979).

BIOLOGY: Kier and Grant (1965) described burrowing activity and the righting reaction in this species. An individual can burrow into sand in 5–7 minutes, usually to a depth of 20–30 mm. An inverted individual rights itself by burrowing vertically into the sand and eventually tilting over, onto its oral surface. Telford (1981) noted that inverted individuals are usually "flipped over" by wave action. The only indication of the presence of this species in an area is the occasional dead test found lying on the sediment surface. Living individuals can be found by "raking" one's fingers through the sediment, although sometimes buried individuals can be detected by six distinctive depressions, corresponding to the six lunules. Telford and Mooi (1986) suggested that this species is restricted entirely to biogenic carbonate sediments, and the five-holed sand dollar *Mellita quinquiesperforata* (Leske) (now *M. isometra*) is found only on siliceous terrigenous substrates. In the Gulf of Mexico the population density of *L. sexiesperforata* was found to increase with the distance from the reef; the trend was attributed to near-reef predation by the gray triggerfish, *Balistes capriscus* Gmelin (Frazer et al. 1991). At San Salvador, Bahamas, a boring snail (the helmet conch *Cassis tuberosa*) is a serious predator on this sand dollar (McClintock and Marion 1993).

Goodbody (1960) and Telford (1981) noted that *L. sexiesperforata* is a ciliary-mucus feeder, transporting particles of 1–20 μm diameter from the aboral surface down the test to the mouth. The particles include algal cells, detritus, and sand grains. Telford and Mooi (1986) reported instead that this sand dollar, like other mellitids, is a podial particle-picker, selecting particles sized 0.05–0.20 mm; this species shuns particles larger than 0.20 mm. Based on a computer simulation, Telford (1990) suggested that this species would draw more than 90% of its diet from the 0.10–0.40 mm particle size class.

Serafy (1979) presented some morphometric and meristic data for a growth series of 23 specimens of this species. Growth to the greatest diameter has been estimated to take 5 years (Crozier 1920). In Panama, individuals are ripe throughout most of the year, but spawning occurs predominantly during the rainy season (Lessios 1985b, 1991). The yolky eggs are generally about 0.26 mm in diameter, but the size of eggs released by females may vary by a factor of three (Mortensen 1921; Lessios 1987). Metamorphosis of the echinopluteus larva occurs 5–8 days after fertilization, when the larva has been presented with suitable substrate (Mortensen 1921).

Polson et al. (1993) succeeded in isolating intraskeletal matrix protein from fossilized *L. sexiesperforata* of early Pleistocene age. The molecular weights of eight fractions were determined as a first step in characterizing the proteins for phylogenetic analysis.

Telford (1978) found that *L. sexiesperforata* is host to the crab *Dissodactylus crinitichelis* Moreira, and Griffith (1987) added a second species, *D. latus* Griffith.

SELECTED REFERENCES: Leske 1778: 199, pl. L, figs. 3, 4 [as *Echinodiscus sexies perforatus*]; Mortensen 1948b:422, pl. 58, fig. 4, pl. 61, fig. 7, pl. 72, fig. 19 [as *Mellita sexiesperforata*]; Kier and Grant 1965:31, pl. 7, figs. 6–8; Voss 1976:140, 1 fig. [as *Mellita sexiesperforata*]; Serafy 1979:74, fig. 32; Kaplan 1982:191, fig. 59, pl. 34, fig. 2; Pawson 1986:536, pl. 177.

Mellita isometra Harold & Telford
Figures 126, 136B

DESCRIPTION: This is a frequently depicted sand dollar, with a thin, flattened test more or less pentagonal (five-sided) in outline, only very slightly wider than long, covered with a dense fur of short spines. The length or diameter of the test seldom exceeds 100 mm (4 in). There are five slotlike holes (lunules) in the test and five short, equal-sized petals on the upper surface. When the animal is viewed from the side, the highest part of the test is slightly forward of the point where the petals meet.

HABITAT: Shallow-water siliceous sediments.

DISTRIBUTION: Cape Cod to Florida, and the Bahama Islands. Not reported from the Florida Keys. DEPTH: 1–50 m (3–164 ft) (Weihe and Gray 1968; Harold and Telford 1990).

BIOLOGY: Until recently, it was believed that a single species of the common five-holed sand dollar—known as *Mellita quinquiesperforata*—existed in the western Atlantic, ranging from Cape Cod, Massachusetts, to Florida, throughout the Gulf of Mexico and the West Indies, and along the Central and South American coasts to southern Brazil. Harold and Telford (1990) made an analysis of collections from numerous parts of the range and concluded that three distinct species of *Mellita* are represented: *M. isometra* from the east coast of the United States and the Bahama Islands, *M. tenuis* (with circular outline and highest point posterior to the point where the petals meet) from the eastern Gulf of Mexico, and *M. quinquiesperforata* (with test much wider than long) from the Caribbean, Central America, and Brazil.

Serafy (1979) summarized the scientific literature on the biology of the species *Mellita quinquiesperforata* considered in the broad sense. Although some generalizations concerning aspects of biology might be applicable to all three species, in the interests of accuracy the summary of biology given here covers only *M. isometra*.

Populations of *M. isometra* may be dense, and individuals tend to aggregate into "clumps," with upwards of 17 animals per square meter reported (Weihe and Gray 1968). Injuries are common and may be caused by fish, by the blue crab *Callinectes sapidus* Rathbun, and by wave action during storms (Weihe and Gray 1968).

These animals generally move forward, as fast as 3 cm/second, along the axis of bilateral symmetry. They also can rotate, which allows for a change in direction, and this commonly occurs when the animal encounters an obstacle or an unfavorable stimulus during forward progress. Burrowing activity involves the removal of sand particles by the spines and feet, and penetration of the animal into the substratum. Normally, individuals burrow to a depth of several centimeters, but during storms they can reach depths of 20–25 cm for protection (Weihe and Gray 1968). In a study of burrowing and feeding activities, Ghiold (1979) identified seven morphologically distinct groups of spines, and he suggested that this species is a selective deposit feeder and that feeding and burrowing can occur simultaneously. Telford et al. (1985) found that feeding is intermittent, alternating with hours of quiescence. The complex process of picking up of particles, passing them between podia, incorporating them into mucus cords, crushing them in the Aristotle's lantern, and ultimately passing them through the intestine was described in detail by Telford et al. (1985). They disagreed with many details of earlier published accounts of feeding mechanisms. In a later pa-

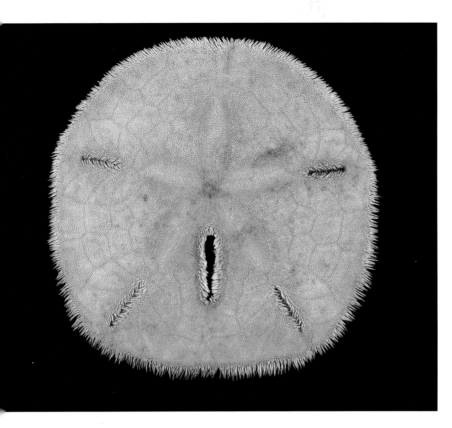

FIGURE 126. *Mellita isometra*. Photo by W. Lee.

per, Telford and Mooi (1986) showed that *Mellita isometra* occurs only on siliceous substrates derived from nearby land, and that this species is a nonselective particle-picker. However, the size range of sediment particles is important for this species and other sand dollars. They suffocate in fine silts, and entire populations can be exterminated by settling dredge spoils (Weihe and Gray 1968).

The righting behavior of upside-down animals is slow and methodical. The posterior edge of the body is moved into the sediment as the anterior end is elevated, until the animal becomes vertical and, often assisted by wave action, it falls over with its oral surface down. Other sand dollar species have been found to right themselves by burying the anterior end (Cabanac et al. 1993).

Migration experiments conducted by Weihe and Gray (1968) showed that individuals travel only short distances (15–30 cm) during 12- or 24-hour periods. Over the course of a week, animals may move an average of 2 m.

Several authors have investigated the function of the slots or lunules. Hyman (1955) and Ghiold (1979) noted that the lunules are not used in burrowing, and Weihe and Gray (1968) found that the lunules are not involved in the process of righting when the animal is inverted. Bell and Frey (1969) reported movement of feces upward by movement of sand through the posterior lunule. Ghiold (1979) and Alexander and Ghiold (1980) concluded that the lunules are primarily food-gathering devices. At about the same time, Telford (1981) showed that the lunules can help to keep the animal in place in strong currents by providing channels for bleeding off excess pressure from the oral surface. Telford ascribed lesser importance to other functions that have been suggested for the lunules, such as assistance with feeding, burrowing, righting, and strengthening the skeleton. In young individuals, the lunules are open and continuous with the edge of the test (Weihe and Gray 1968).

Individuals of *M. isometra* may harbor up to eight individuals of a small commensal crab, *Dissodactylus mellitae* Rathbun. The crab appears to be attracted to water-borne chemicals emitted by the urchins (Gray et al. 1968b).

SELECTED REFERENCES: Telford et al. 1985:431, figs. 1–5; Harold and Telford 1990:987, figs. 11–13.

FAMILY SCHIZASTERIDAE

Moira atropos (Lamarck)
Figures 127, 135C

DESCRIPTION: This is a small, light brown or yellowish to white burrowing heart urchin, subspherical, usually less than 50 mm (2 in) long, with distinctive narrow, deeply sunken petals and a well-developed frontal notch. The test is covered with short spines that become longer on the anterolateral margins. On many individuals a horseshoe-shaped stripe of dark red-brown pigment partially encircles the apical system.

HABITAT: Mud or mud-sand bottoms.

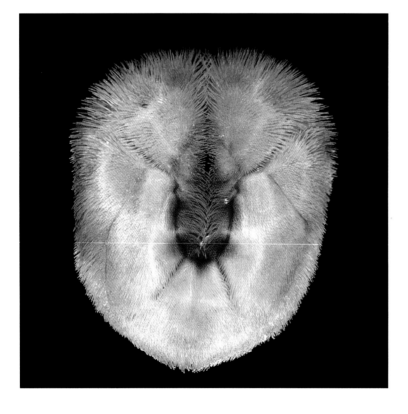

FIGURE 127. *Moira atropos.*

DISTRIBUTION: From Cape Hatteras and Bermuda, southward throughout the Caribbean to São Paulo, Brazil. DEPTH: 0–455 m (0–1,493 ft) but most common in less than 50 m (160 ft) (Serafy 1979).

BIOLOGY: Individuals burrow to a depth of 15 cm in soft sediments (Serafy 1979). Chesher (1963) reported concentrations of up to 32 individuals per square meter in very soft mud, and highest abundances occur in sediments ranging from 75 to 90% sand (McNulty et al. 1962; Moore and Lopez 1966). It is often associated with *Paraster* spp. and *Brissopsis elongata elongata*. Serafy (1979) summarized information on population dynamics and presented some morphological and meristic data for this species.

The structure and function of the frontal ambulacrum was described in detail by Chesher (1963). He showed that this species feeds on detritus, extending tube feet from the frontal ambulacrum through a tunnel in the sediment to the seafloor surface and using the sticky penicillate feet to transfer detritus down through the tunnel toward the mouth. Moore and Lopez (1966) found that in Biscayne Bay, Florida, this species reached a length of 38 mm at the age of 3 years, and Serafy (1979) suggested that specimens of 58 mm length were probably older than 3 years. Predators on this species include the asteroid *Luidia clathrata* (Serafy 1979) and the margate fish *Haemulon album* Cuvier & Valenciennes (Randall 1967).

Moore and Lopez (1966) reported that spawning of *M. atropos* occurred in Biscayne Bay, Florida, in April 1964 and March 1965, and that there was a significant correlation between spawning and lunar phases, with peak spawning occurring immediately after a full moon. The larva has been reared through metamorphosis, but a detailed description was never published (Grave 1902a,b).

Tertiary fossils of this species have been reported from the eastern United States (Mortensen 1951).

SELECTED REFERENCES: Lamarck 1816: 32 [as *Spatangus atropos*]; Chesher 1963: 549, figs. 1–11; Kier 1975:15; Voss 1976: 140, 141 (fig.) [as *Moira atropus*]; Gosner 1979:258, pl. 62; Serafy 1979:91, fig. 39; Kaplan 1982:192, pl. 34, fig. 6; Pawson 1986:536, pl. 177; Ruppert and Fox 1988: 76, pl. A32.

Paraster doederleini Chesher
Figures 128, 135F

DESCRIPTION: This is a medium-sized, fragile, yellowish tan or light brown urchin up to 75 mm (3 in) long, with a distinctly heart-shaped, approximately spherical test, covered with fine, fragile spines. There are four sunken petals, the anterior pair curved and much longer than the posterior pair. A distinctive, dark brown peripetalous fasciole is present. The tube feet in the anterior groove are bright red.

HABITAT: Mud fields.

DISTRIBUTION: The Florida Keys, Colombia and Belize (Chesher 1972; Kier 1975), and the Dry Tortugas (Kier and Hendler, previously unpublished). DEPTH: 12–220 m (39–722 ft).

BIOLOGY: These urchins live buried, 20–100 mm below the surface, in densities of one to two individuals per square meter (Kier 1975); the only evidence of their presence in the area is dead tests at the sediment surface. Kier (1975) performed some morphometric analyses of samples of specimens from Belize, Florida, and Colombia and concluded that perhaps more than one taxon was represented in these populations, but that more

FIGURE 128. *Paraster doederleini*.

study was required of representatives from Colombia.

Fossil specimens of this species have been reported from Pleistocene deposits in Jamaica (Donovan et al. 1994).

SELECTED REFERENCES: Chesher 1972: 10, figs. 1–9; Kier 1975:9.

Paraster floridiensis (Kier & Grant)
Figures 129, 135D

DESCRIPTION: This is a small, fragile, heart-shaped urchin up to 12 mm (0.5 in) long, with a light pink to yellowish test and anterior petals much longer than the posterior petals. The fascioles and tube feet of the petals are red, and the tube feet of the anterior groove are yellow-orange.

HABITAT: Mud or sand fields (Chesher 1966b).

DISTRIBUTION: The Florida Keys, the Bahama Islands, Dominica, and the San Blas Islands of Panama. DEPTH: 2–25 m (7–82 ft).

BIOLOGY: This species burrows to depths of up to 250 mm (Chesher 1966b), and there is no evidence of its presence in an area, apart from dead tests on the surface of the sediment. It frequently occurs in association with *P. doederleini* and other fragile burrowing spatangoids.

Chesher (1966b) noted that a specimen was found in the stomach of a black margate fish, *Anisotremus surinamensis* (Bloch).

SELECTED REFERENCES: Kier and Grant 1965:50, pl. 13, figs. 4–6, pl. 14, figs. 1–9 [as *Paraster (Schizaster) floridiensis*]; Chesher 1966b:1, figs. 1–6.

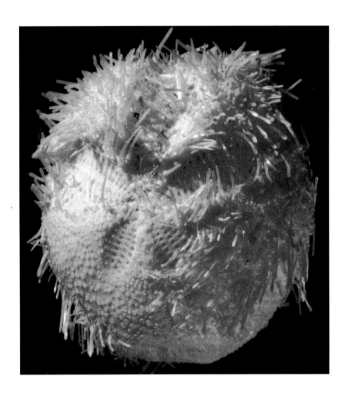

FIGURE 129. *Paraster floridiensis.* Photo by G. Hendler.

FAMILY BRISSIDAE

Brissopsis elongata elongata
Mortensen
Figures 130, 135G

DESCRIPTION: This is a small urchin, less than 100 mm (4 in) long, elongate with rounded ends. There are four petals of more or less equal length, the posterior pair almost contiguous for much of their length, diverging distally. The test is whitish, with a covering of small tan spines.

HABITAT: Mud fields.

DISTRIBUTION: Published records for Belize (Kier 1975) and from Panama to Venezuela (Chesher 1968a; Kier 1975). Museum collections (USNM) include specimens from the Florida Keys, the Dominican Republic, Puerto Rico, and Dominica. DEPTH: 13–72 m (43–236 ft).

BIOLOGY: This species burrows to a depth of 4–10 cm in mud fields (Kier 1975), where it is often found in association with *Paraster* spp. and *Moira atropos*. Chesher (1968a) pointed out that this is probably the only species of *Brissopsis* that could be studied easily in the field; regrettably, even today little is known of its biology. Like all spatangoids, it lacks a lantern and is a sediment swallower. Kier (1975) found that *B. elongata* burrowed beneath the seafloor surface in 10 minutes and illustrated the process with photographs taken in situ.

REMARKS: Turner and Norlund (1988) reviewed the status of western Atlantic *Bris-*

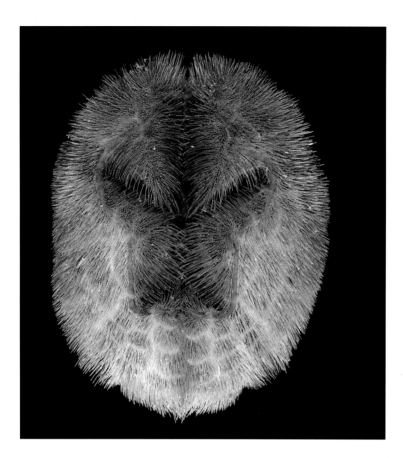

FIGURE 130. *Brissopsis elongata elongata*.

sopsis species and identified reliable systematic characters for the group. *B. elongata jarlii* Mortensen, which is known only from the Gulf of Guinea, Africa, differs from *B. elongata elongata* in the relative density of tube feet in different regions of the body and in some features of the pedicellariae (Chesher 1968a).

SELECTED REFERENCES: Mortensen 1907:163, pl. 3, figs. 4, 14, 15, 19, pl. 4, figs. 1, 4, 13, 18, pl. 8, figs. 2, 15–17, 21, 28, pl. 19, figs. 12, 17; Chesher 1968a:63, figs. 7, 16, 17, pls. 13, 15–18; Kier 1975:16, pl. 9, figs. 1–6; Turner and Norlund 1988:890.

Brissus unicolor (Leske)
Figures 131, 135E

DESCRIPTION: This heart urchin seldom exceeds 50 mm (2 in) in length, but can reach a length of more than 127 mm (5 in). It is elongated, inflated posteriorly, with a blunt anterior end and a bluntly pointed posterior end. The test and its covering of short spines are light tan to whitish, and there are some conspicuous black pedicellariae. There are four petals; the anterior pair is shorter than the posterior pair. The anterior petals lie at an angle of 180° to one another, and the posterior petals lie at an acute angle of less than 45°.

HABITAT: Under rocks, on or in sand (Mortensen 1951; Kier and Grant 1965).

DISTRIBUTION: Bermuda, and Florida through the Caribbean to Brazil. In the eastern Atlantic, in the Mediterranean as far east as the Adriatic Sea, also the Cape Verde Is-

FIGURE 131. *Brissus unicolor.*

lands and the Azores. DEPTH: 0–240 m (0–787 m).

BIOLOGY: This species is apparently never very abundant, but it is known from numerous localities. Densities can approach 0.3 per square meter, judging from the number of dead individuals collected after a low-tide mortality event on a Panamanian reef flat (Hendler 1977b). It lives under rock slabs, buried in the sand, or buried in sand-filled pockets in reef rock (Kier and Grant 1975; Hendler, previously unpublished). Individuals may be preyed upon by the boring snail *Cy praecassis testiculus* (Hendler 1977b).

SELECTED REFERENCES: Leske 1778: 248, pl. 26B,C [as *Spatangus Brissus unicolor*]; Mortensen 1951:509, pl. 38, fig. 10; Kier and Grant 1965:38, pl. 15, figs. 4, 5; Kaplan 1982:190.

Meoma ventricosa ventricosa
(Lamarck)
Figures 132, 135H

DESCRIPTION: This is the most common large heart urchin in the Florida Keys. It reaches a length of 200 mm (8 in). The test is high and covered with short spines. All spines are red-brown; the ground color of the test and of the muscle bases of spines is a darker red-brown than the spines. On the oral surface, the overall color is generally lighter, the spines often tending toward olive green. The cleaned, naked test is white, as it is in other burrowing urchins, with four conspicuous sunken petals of more or less equal size.

243

FIGURE 132. *Meoma ventricosa ventricosa*.

Chesher (1970) provided persuasive evidence for regarding the western Atlantic and eastern Pacific populations as subspecies rather than as distinct species. *Meoma ventricosa grandis* Gray occurs along the Pacific coast of Central and South America, from the Gulf of California to Colombia and the Galápagos Islands (Serafy 1979).

HABITAT: Areas of coarse sand and shell fragments, either associated with turtle grass beds and patch reef, in pockets on reef flats, or deep-water sandy areas.

DISTRIBUTION: Fort Lauderdale, Florida, and the Bahama Islands, southward through the Greater and Lesser Antilles; in the Gulf of Mexico, from southern Florida westward to Central America; its southern limit is the Orinoco River. DEPTH: 2–200 m (7–760 ft).

BIOLOGY: Serafy (1979) summarized information on diet, growth, and reproduction and population dynamics of this species. Kier and Grant (1965) discussed it in detail, including notes on burrowing and righting behavior. Kier (1975) described the morphology of juvenile specimens. Chesher (1969) and Serafy (1979) commented on the diversity of habitats occupied by this species, with sediments ranging from fine silty sand to coral rubble. It is frequently common in large numbers in sand fields adjacent to turtle grass beds or in small sand patches within the turtle grass. The echinoid's presence is usually indicated by a slight mound in the sand at the head of a distinct trail and a concentration of coarser sediments or shell fragments over the top of the apical system. It occurs with other irregular urchins such as *Paraster floridiensis, Brissus unicolor, Plagiobrissus grandis, Clypeaster rosaceus,* and *Echinoneus cyclostomus* (Chesher 1969).

Meoma v. ventricosa may form large feeding aggregations (e.g., Scheibling 1982b); it ingests material from the seafloor, such as sand and crushed shell (Serafy 1979). The grain size of its intestine contents does not differ from that of surrounding sediments (Hammond 1982b), but there is some evidence to suggest that nutrient-rich fragments are ingested selectively (Hammond 1983). Over 3,000 specimens representing 137 species of mollusks were found in the intestines of one urchin; whether they were alive when ingested is not known (Serafy 1979). This species has an alkaline foregut and does not dissolve calcium carbonate in the intestine (Hammond 1981).

Chesher (1969) suggested that the nocturnal emergence from burial of *M. v. ventricosa* is necessitated by a nightly decrease in interstitial oxygen concentration; the animal moves upward to obtain more oxygen. It is most active nocturnally, but the gut is filled with sediment at night and during the day (Hammond 1982a,b). Young individuals may be more common under coral slabs in shallow reef areas, and those buried in the sand do not emerge at night as do the adults (Chesher 1969).

Within 1 year, individuals can reach a length of 88 mm, and they can grow to 134 mm in 4 years. Those from open sand areas or deeper water grow to a larger size, up to a known maximum of 200 mm (Chesher 1969). Individuals 2 years old are sexually mature. Spawning occurs from August to February in Florida (Chesher 1969). Chesher (1969) observed isolated males ejecting columns of sperm up to 20 cm into the water. Kier (previously unpublished) photographed males and females spawning simultaneously at Looe Key, Florida; the eggs and sperm were released into the water above the seafloor. A description of the larva of this subspecies has not been published, but *M. v. grandis* larvae have been studied (Mortensen 1921).

Predators on *Meoma v. ventricosa* include loggerhead turtles, stingrays, other fish, helmet conches (Chesher 1969; Gladfelter 1978), and the sea star *Oreaster reticulatus* (Kier and Grant 1965). The small crab *Dissodactylus primitivus* Bouvier infests 80–100% of *Meoma v. ventricosa* specimens; they prefer to live near the mouth or inside the esophagus (Chesher 1969; Telford 1978; Griffith 1987). Werding and Sanchez (1989) recorded a second *Dissodactylus* species from *Meoma v. ventricosa* in Colombia and the association of *Ophiodromus obscurus,* a commensal polychaete. A small bivalve mollusk, *Neaeromya* sp. (identified by Paul H. Scott, personal communication), was found crawling among the short spines on the oral surface of this species at Looe Key, Florida.

Fossil specimens of *M. ventricosa* have been reported from Pleistocene formations on Jamaica and Barbados (Donovan and Jones 1994).

SELECTED REFERENCES: Lamarck 1816: 29 [as *Spatangus ventricosus*]; Kier and Grant 1965:38, pl. 3, figs. 4, 5, pl. 9, figs. 1–4, pl. 10, figs. 1–4, pl. 11, figs. 1–6, pl. 12, figs. 1–4, pl. 13, figs. 1–3, pl. 15, fig. 6, pl. 16, figs. 5, 6; Chesher 1969:72, figs. 1–14; 1970:731; Kier 1975:7, fig. 6, pl. 4, figs. 3, 4, pl. 5, pl. 6, figs. 1–3; Voss 1976:142, 1 fig.; Colin 1978:427, 428 (fig.); Kaplan 1982: 192, pl. 35, figs. 3, 4; Pawson 1986:537, pl. 177.

Plagiobrissus grandis (Gmelin)
Figures 133, 135I

DESCRIPTION: This is a large, elongate oval, irregular urchin, with a fragile, tan test, reaching a length of 220 mm (8 in). Its spines are short, except on the upper surface where conspicuous long, needlelike spines are directed posteriorly. Visible on the bare test are four petals (the two anterior shorter) and a slight anterior notch; conspicuous larger tubercles for carrying large spines are restricted to the area between the petals.

Kier (1975) described in detail the morphology of immature specimens of this species.

FIGURE 133. *Plagiobrissus grandis*, showing a sequence of burrowing behavior.

HABITAT: Sandy areas where seagrass and algae are sparse or absent.

DISTRIBUTION: Cape Canaveral, Florida, southward through the Caribbean and the Central and South American coasts to São Paulo, Brazil. DEPTH: 1–210 m (3–689 ft), but most common in less than 50 m (164 ft) (Serafy 1979).

BIOLOGY: This burrowing species leaves scant evidence of its presence visible on the surface, in the form of a projecting cluster of translucent spines. The spines are generally erected only when an individual is disturbed, presumably as a defense against predators. It typically lives buried about 50 mm deep in the sediment (Kier and Grant 1965). Occurrence of dead tests on the surface can provide another indication of presence nearby of live animals, because the delicate test probably does not persist for long. Serafy (1979) reported specimens from areas of crushed shell and quartz sand. McNulty et al. (1962) found it living in the same habitat as *Clypeaster subdepressus* and *Encope michelini* off Key Largo, Florida.

When brought to the surface alive, *P. grandis* moves rapidly across the substratum, at rates of up to 10 cm/second (Kier and Grant 1965). It can bury itself completely within 10 minutes. When moving under the sand, it may leave a weakly defined trail at the surface.

The intestine of this species is full both night and day (Hammond 1982a). Individuals selectively ingest nutrient-rich grains of sediment; they do not select particles according to their size (Hammond 1982b, 1983). The foregut is alkaline and does not dissolve calcium carbonate sand (Hammond 1981).

The helmet conchs *Cassis tuberosa* and *C. madagascariensis spinella* Clench are major predators (Moore 1956; Gladfelter 1978), and most attacks are made on the anterior end of the urchin (Chesher in Serafy 1979). According to Moore (1956:74) "One urchin that had escaped was seen trundling along on its secondary spines with a large *Cassis* in hot pursuit." Associated animals include the pinnotherid crab *Dissodactylus primitivus* (Griffith 1987).

SELECTED REFERENCES: Gmelin 1791: 3200 [as *Echinus grandis*]; Kier and Grant 1965:36, pl. 8, figs. 1–6, pl. 15, figs. 2, 3; Kier 1975:3, figs. 2–5, pls. 1–3, pl. 4, figs. 1, 2; Voss 1976:142, 1 fig.; Serafy 1979:98, fig. 43; Kaplan 1982:193, pl. 35, figs. 1, 2.

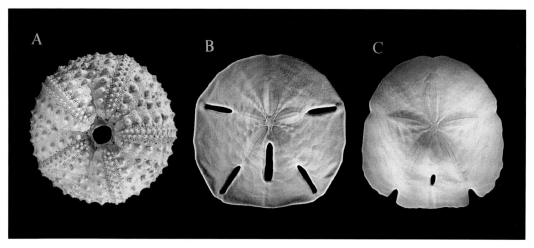

FIGURE 136. The distinctive skeletons of sea urchins in dorsal view, revealed by removing soft tissue and spines. The overall shape and fine structure of the naked test is characteristic for each species shown. A, *Diadema antillarum*; B, *Mellita isometra*; C, *Encope aberrans*. Photos by D. Meier.

FIGURES 134 and 135 are shown overleaf (pp. 248, 249).

CLASS ECHINOIDEA

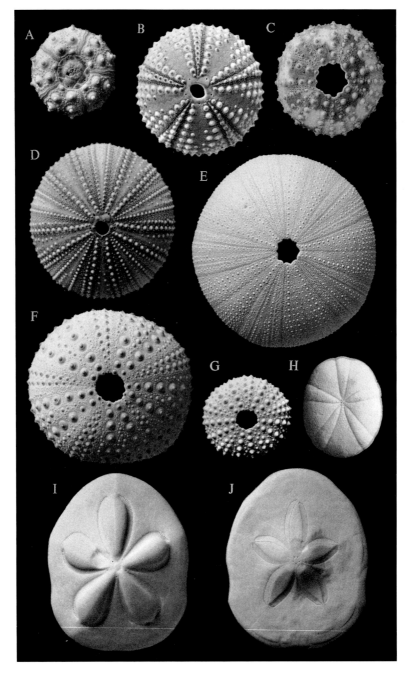

FIGURE 134. The distinctive skeletons of sea urchins in dorsal view, revealed by removing soft tissue and spines. The overall shape and fine structure of the naked test is characteristic for each species shown. A, *Eucidaris tribuloides*; B, *Arbacia punctulata*; C, *Lytechinus williamsi*; D, *Lytechinus variegatus*; E, *Tripneustes ventricosus*; F, *Echinometra lucunter lucunter*; G, *Echinometra viridis*; H, *Echinoneus cyclostomus*; I, *Clypeaster rosaceus*; J, *Clypeaster subdepressus*. Photos by T. Smoyer.

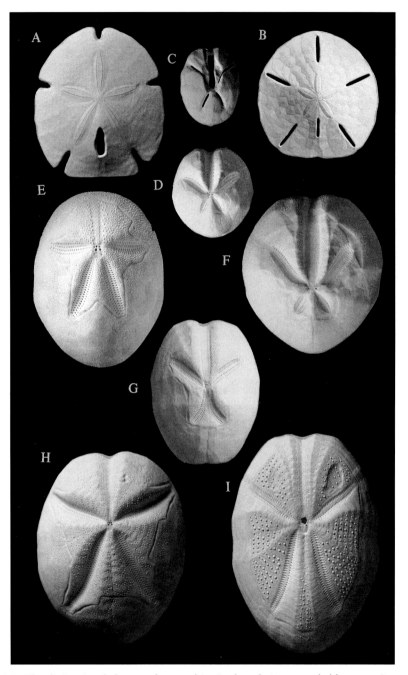

FIGURE 135. The distinctive skeletons of sea urchins in dorsal view, revealed by removing soft tissue and spines. The overall shape and fine structure of the naked test is characteristic for each species shown. *A, Encope michelini; B, Leodia sexiesperforata; C, Moira atropos; D, Paraster floridiensis; E, Brissus unicolor; F, Paraster doederleini; G, Brissopsis elongata elongata; H, Meoma ventricosa ventricosa; I, Plagiobrissus grandis.* Photos by T. Smoyer.

CLASS HOLOTHUROIDEA
Sea Cucumbers

Holothuroids are too often viewed as unattractive and sluggish, and Hyman (1955:122) remarked that "interest in this group appears relegated to taxonomic specialists." Certainly, the sea cucumbers differ more than superficially from other echinoderms (Figure 2), for they usually do not possess a conspicuous skeleton, and their body wall is leathery or soft. All the same, a close inspection of sea cucumbers reveals features that link them with their more heavily calcified sister classes, such as the microscopic ossicles, skeletal structures embedded in the body wall. Depending on the species, the ossicles constitute 3% to over 70% of the dry weight of the holothuroid body wall. One might also consider that in the deep sea, holothuroids can compose up to 90% of the ecosystem biomass, and because deep waters cover over 70% of the surface of the earth, holothurians are among the dominant organisms on our planet. In the tropical Indo-Pacific, large sea cucumbers throng quiet lagoons and back reef areas, but in the Atlantic they are usually less conspicuous and fewer in number. Indeed, less than 100 of the 1,250 known species of sea cucumbers occur in the Florida Keys and the Bahama Islands, and at diving depths fewer than 60 species occur.

FORM AND FUNCTION

The anatomical structures of sea cucumbers are depicted in Figures 137–139. Sea cucumbers range from a few centimeters (about an inch) in length to at least 2 m (7 ft)! The body is approximately

FIGURE 137. Major anatomical features of a sea cucumber, in a dendrochirote species. Illustration by C. Messing.

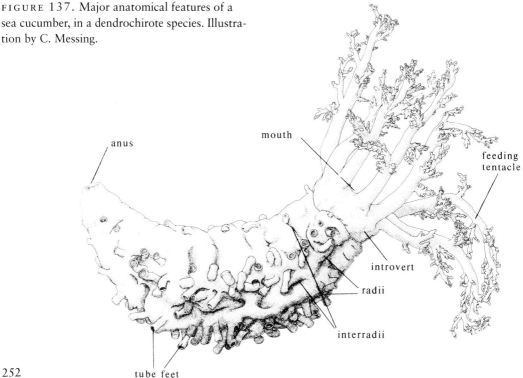

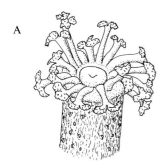 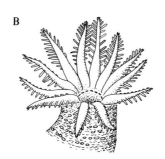 

FIGURE 138. Three of the four distinctive forms of sea cucumber tentacles. *A,* peltate: stalks with shield-shaped tips; *B,* pinnate: featherlike; *C,* digitate: simple, fingerlike. Illustration by P. Mikkelsen. The fourth form, dendritic or branching tentacles, can be seen in Figure 137.

cylindrical, with a mouth at one end and an anus at the other. The mouth is surrounded by a ring of 10–30 retractile feeding tentacles, which are actually modified tube feet arising from the water vascular system (Figure 138). The five radii run along the body from mouth to anus. Tube feet are usually present—but absent from so-called "apodous" sea cucumbers—and their arrangement on the body wall can vary considerably. Tube feet may be present on all radii, sometimes forming conspicuous bands running from mouth to anus. They are often more numerous ventrally than dorsally and can be scattered in interradii as well as in the radii. Dorsal tube feet may be modified to form papillae. Holothuroid tube feet usually have a terminal skeletal disk, somewhat like echinoid tube feet.

Within the dermis of the body wall, the sea cucumber skeleton takes the form of vast numbers of microscopic ossicles. They display a profusion of beautiful geometric shapes, though the form of ossicles in any species is limited and taxonomically characteristic. They are denoted by descriptive terms such as buttons, cups, tables, plates, rods, anchors, and wheels (Figure 139). In some sea cucumbers the presence of ossicles renders the body wall stiff and rough to touch.

The interior of the body wall is lined with circular muscles overlain by five radial, longitudinal muscles. The esophagus is surrounded by a calcareous ring comprising 10 or more sizable ossicles. This ring serves to support the esophagus and is an attachment point for the five longitudinal muscles, which are used to contract the body, and, when present, for the pharyngeal retractor muscles used to withdraw the tentacles. The stomach-intestine runs posteriorly, then anteriorly, then posteriorly again to terminate at the anus. In this class there is no madreporite to be seen on the body wall; it is internal, with a stone canal leading to the water vascular ring that lies just behind the posterior margin of the calcareous ring. One or more thin-walled, saclike Polian vesicles also attach to the water

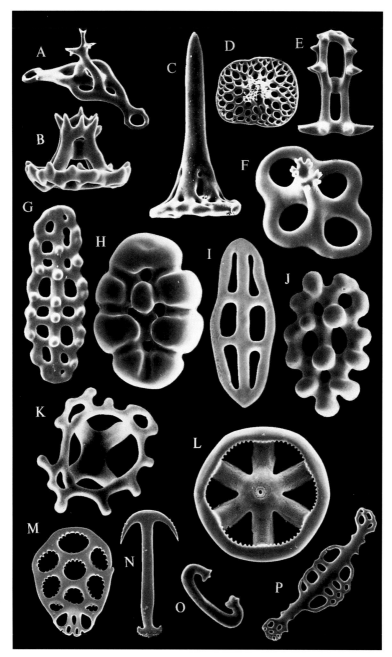

FIGURE 139. Representative types of holothuroid ossicles. Structures were extracted from the body wall of various species and photographed at high magnification using scanning electron microscopy. *A*, table, dorsal view; *B*, table, lateral view; *C*, tacklike table; *D*, complex table, dorsal view; *E*, table, lateral view; *F*, table, dorsal view; *G, H*, knobbed buttons; *I*, smooth button; *J*, knobbed button; *K*, basket; *L*, wheel; *M*, anchor plate; *N*, anchor; *O*, C-shaped element; *P*, rod.

ring. The gonad comprises one or two tufts attached to the dorsal body wall, which are usually composed of numerous branched or unbranched tubules. A single genital duct, running anteriorly in the dorsal mesentery, leads to an opening immediately posterior to the ring of tentacles, in the middorsal interradius.

In most large and thick-bodied species, branching respiratory trees occupy the posterior part of the coelomic cavity. They are composed of paired arborescent systems of tubes that are connected by means of a short duct to the terminal, enlarged part of the large intestine, the cloaca. The respiratory trees are aerated by pumping movements of the cloaca.

There are six orders in the class Holothuroidea, three of which are represented in the shallow waters of this region. The Dendrochirotida have a substantial body wall that may be soft or firm. They have an introvert, a collar of flexible tissue behind the tentacles that is pulled into the body by retractor muscles. The tentacles are branched or dendritic, and the most ventral pair is usually smallest (Figures 137, 138). Species with many tentacles have small tentacles either interspersed with the larger ones or forming an inner circlet. The calcareous ring is simple or with well-developed posterior processes. Two separate gonadal tufts also characterize the order. As one might surmise, based on the net formed by their tentacles, dendrochirotids are suspension feeders, ingesting microscopic plants such as diatoms and unicellular algae, animals such as protozoans, nematodes, ostracods, copepods, jellyfish, and larvae, and microscopic particles of detritus. Their food is trapped in, or adheres to, mucus on the tentacle branches. Representatives of the order are uncommon on reefs, presumably because of the relatively low concentrations of plankton there.

Aspidochirotida have a thick body wall and clear bilateral symmetry. The ventral surface bears numerous locomotory tube feet and in some species is flattened into a sole. The tube feet of the dorsal surface are often modified to form papillae. Aspidochirotids have 10–30 (usually 20) tentacles that usually are peltate (i.e., each bearing a shield-shaped group of branches arising from a short central stalk), although there are exceptional species, such as *Holothuria glaberrima* (compare Figures 138 and 165), that have dendritically branched tentacles. A collar (rim composed of fused tube feet) surrounds the tentacles and closes over them when they are retracted, serving the same protective function as the introvert of the Dendrochirotida. The calcareous ring is simple. The Aspidochirotida are deposit feeders, using short tentacles to pass food to the mouth, and most occur in low-energy environments where rich sediments accumulate. They characteristically select for organic content rather than for the size of their food particles. They, and other deposit-feeding holothurians, consume organic and inorganic detritus and associated microorganisms, protozoans, and meiofauna.

The Apodida have a thin body wall and 10–20 digitate or pinnate tentacles (Figure 138). They characteristically lack tube feet and respiratory trees. Apodids possess easily recognized anchor or wheel ossicles, which often may be seen within their translucent body wall. They are generally nocturnal deposit feeders, collecting particles by the mechanical operations of the tentacles or by the adhesiveness of the tentacles. Discrete sensory organs are more evident in the family Synaptidae in this order than in other sea cucumbers, perhaps related to their delicate construction and mobility. Some species have pigmented photoreceptors at the base of the tentacles and some also have statocysts, organs of balance.

GENERAL BIOLOGY

The slow-moving ways of sea cucumbers are legendary. One *Cucumaria* individual stayed in the same spot in an aquarium for 2 years. Some sea cucumbers can move by using their podia; others produce locomotory waves of the body wall, conjuring up the image of a giant caterpillar. Surprisingly, certain species can swim. For example, a *Leptosynapta* species moves through the water by alternately flexing in a U-shape and extending its body.

The sedentary behavior and soft body wall of most sea cucumbers place them at the mercy of carnivorous fish and crustaceans. It is perhaps not surprising then that there is evidence of toxic compounds sequestered in the bodies of coral reef holothuroids, substances that deter attacks by predators. Also in the defensive arsenal of sea cucumbers are the Cuvierian tubules attached to respiratory trees in some Aspidochirotida. They are very elastic white, pink, or red tubules that are ejected, blind-end first, through the anus and after discharge, stretch up to 20 times their original length. They are not only readily expelled but rapidly regenerated. Whether cast out singly or as an entire tuft, the tubules swell into sticky threads. The tubules appear to be toxic, and they are extremely adhesive, instantly fouling any object they contact. Some species undergo autoevisceration (ejection of the intestine and related structures) if stressed, but the self-inflicted damage can be repaired by subsequent regeneration. Large aspidochirotids (e.g., *Holothuria, Stichopus, Actinopyga*) rupture at the cloaca and contract the body wall to expel the respiratory trees, gonads, and gut through the anus. Dendrochirotids (e.g., *Thyone, Phyllophorus*) rupture at the introvert and shed the aquapharyngeal bulb and attached structures.

There are separate males and females in most sea cucumber species with the exception of 12, six of which are synaptids. Spawning individuals usually stretch the anterior extremity off the bottom and into the water. Aspidochirotids characteristically sway to and fro in a "cobra-like" fashion, and dendrochirotids wave their tentacles, dispersing and mixing sex cells. These spawning events may be brief, lasting only seconds, or may last for several hours. Cases are reported of spawning aggregations and of "pseudocopulation"—the pairing of mating individuals with fore-ends intertwined.

Indirectly developing species in the Aspidochirotida and Apodida have a pelagic larva, an auricularia with a looped ciliary band that passes over the oral and anal hoods. Some auricularia larvae have internal ossicles or clear spherules. During metamorphosis, the ciliary band breaks and reforms as rings, and the viscera are modified, but no torsion occurs because the adult body has the same orientation as the larva. The majority of holothuroids develop from a yolky vitellaria (also called "doliolaria") larva that has a series of transverse ciliary bands similarly arranged as in the postmetamorphic pelagic juvenile. The planktonic life of these larvae usually spans 2 weeks to 2 months, after which the settling juveniles are able to attach using their five buccal tube feet.

At least 41 species of sea cucumbers worldwide brood; a majority of these are in the Cucumariidae. Brooders deposit eggs among their tentacles, or they use the tentacles to position eggs underneath the body or in special incubatory pockets in the body wall. There are 14 species that retain developing young in the gonad or coelom. Only nine species of cucumarid dendrochirotids and holothurid aspidochirotids reproduce asexually; two, *Holothuria parvula* and *Holothuria surinamensis*, occur in the Caribbean. The process of asexual transverse fission takes 14 hours to 5 days and is followed by regeneration of structures missing from the severed anterior and posterior halves of the animal.

Sea cucumbers host a wide assortment of commensals and parasites, some highly modified, including protozoans, flatworms, polychaetes, copepods, crabs, pycnogonids, clams, and snails. One of the latter produces fine tubes that resemble holothuroid gonadal tubules, which were long mistaken as part of the host until it was discovered that small snails inside the tubes were actually the developing young of a gastropod parasite. Possibly the most interesting commensal is the pearlfish, which inhabits the posterior intestinal tract or respiratory tree.

ECOLOGICAL AND COMMERCIAL IMPORTANCE

At least 13,371 metric tons of sea cucumbers are landed every year, mostly from the Japanese and Korean fishery for *Stichopus japonicus* Selenka, and are consumed raw and undried (Sloan [1985] reviewed the sea cucumber fishery). The species is not only heavily fished, but it is also reared in aquaculture. There is an additional harvest of over 625 metric tons per year of the North American *Parastichopus* spp., but only the muscle bands of these thin-bodied species are used. Dried sea cucumbers, known as trepang (Malaysian) or bêche-de-mer (French rendering of the Portuguese bicho do mar), are gourmet fare on some tables and have been considered to possess aphrodisiac or curative properties. At least nine shallow-water tropical species are harvested for human consumption and exported primarily to Hong Kong, Singapore, and Malaysia. These are *Holothuria, Thelonota,* and *Actinopyga* species, referred to by names such as "prickly-red fish," "mammy fish," and "sand fish." After collection they are cleaned, gutted, boiled, and dried, often without removing ossicles from the body wall. Recipes calling for trepang use the rehydrated, cooked body wall.

Ecological effects of sea cucumbers on marine communities are not well documented. *Holothuria arenicola,* a species that lives on sand bottoms, dredges 47 kg/m^2 per year (10 lb/ft^2 per year) of dry-weight sediment from 15–20 cm (6–8 in) below the surface, and this must impact fauna living near the small volcanoes of sediment that mark the sea cucumber's numerous burrows. As well, on Indo-Pacific reef flats, where sea cucumbers may occur in excess of 35 per square meter (3.3 per square foot) and where individuals process 80 g (2.6 oz) dry weight of sediment per day, there must be some effect on ecological processes. However, sediment is not markedly altered by sea cucumbers; often, microscopic plants and animals are unaffected even when ingested, and cucumber feces do not appreciably enrich the sediment.

TERMINOLOGY AND CONVENTIONS

The size of sea cucumbers is given as a total length from the anterior end of the body to the posterior end. The size can change dramatically as the animal contracts and expands, so that body measurements alone are an unreliable indicator of the identity of a species. Dissections to study internal anatomy are made by cutting longitudinally on the right or left side of the dorsal body wall.

The all-important ossicles in the body wall, tube feet, and tentacles can be studied only with a high-power ("compound") microscope. If a small piece of tissue, cut from the specimen with fine scissors or a razor blade, is placed on a glass microscope slide and then covered with several drops of liquid household bleach, the bleach dissolves away the soft tissues. The ossicles remain in the liquid, which can then be studied under the microscope. Permanent slide preparations can be made by carefully replacing the bleach with water several times, using an eye dropper, allowing the preparation to dry thoroughly, and adding a commercial mounting medium and coverslip to the slide.

The ossicles of adult specimens are used to characterize sea cucumbers taxonomically. However, the ossicles of juvenile specimens may differ from those of adults in the relative abundance and precise shape of the various ossicle forms (Cutress, in press).

IMPORTANT REFERENCES

A classic reference to the fauna of this region is Deichmann's (1930) study of the western Atlantic sea cucumbers. Miller and Pawson's (1984) report on the species of the Gulf of Mexico is a very helpful adjunct, containing photographs of many species and information on species distributions. Cutress's (in press) study of changes in the ossicles during body growth may facilitate the identification of juvenile individuals. A general reference to the limited fossil history of the group is the treatment by Frizzell and Exline (1966) in the *Treatise on Invertebrate Paleontology*. More recent studies of the fossil holothuroid fauna have treated deposits of ossicles and the structures of exquisitely preserved, complete specimens (Smith and Gallemí 1991; Gilliland 1992). The reproductive biology of holothuroids was recently reviewed by Smiley et al. (1991), and their feeding by Jangoux and Lawrence (1982).

FAMILY CUCUMARIIDAE

Duasmodactyla seguroensis
(Deichmann)
Figures 140, 178A,B

DESCRIPTION: This small holothuroid reaches a length of approximately 10 cm (4 in). The stout body is curved to U-shaped, swollen in the middle, and tapered toward both ends; the tapering is more pronounced at the posterior end. The body wall is thin, soft, flexible, and slightly gritty to the touch. The mouth is surrounded by 20 dendritic tentacles, five large pairs alternating with five smaller pairs. Side branches arising from the tentacle stalks are not very bushy. The cylindrical tube feet are scattered, except on the introvert where they form five distinct rows. They are especially numerous ventrally and are so crowded that the anterior and posterior ends of the body appear to be entirely

FIGURE 140. *Duasmodactyla seguroensis*.

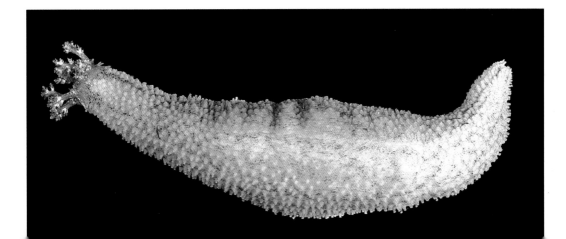

composed of tube feet. The fully extended feet appear as narrow, transparent threads, emerging from holes in the body wall. The tip of each tube foot carries a conspicuous circular disk supported by terminal ossicles. Surrounding the anus are five clusters of conical papillae, each with five papillae per cluster.

All the body wall ossicles (Figure 178A,B) are tables, with undulating or scalloped margins perforated by about 12 peripheral and four central holes. The short spires are composed of four pillars that terminate in 15–16 small spines.

This species varies from dark greenish brown to a mottled brown or dirty white with darker flecking. There is a tendency for the flecking or mottling to produce faint longitudinal bands along the body wall. The tentacle fronds are white and yellow-olive, the stems lighter.

HABITAT: Coastal areas including seagrass beds, near the low-tide mark.

DISTRIBUTION: Florida (Biscayne Bay), and the Dry Tortugas, Jamaica, Puerto Rico (Miller, previously unpublished), Venezuela (Martínez 1991a), and Brazil. DEPTH: Less than 1–5.5 m (less than 3–18 ft).

BIOLOGY: *D. seguroensis* is an uncommon, and virtually unstudied, sea cucumber. Individuals may burrow among seagrass rhizomes or live attached to the underside of rocks.

This species may be confused with *Phyllophorus (Urodemella) occidentalis*, which is sometimes found living with it in the same habitat. They can be difficult to distinguish without an examination of the ossicles, but the body wall of *P. occidentalis* is smooth and the terminal disks of the tube feet are characteristically surrounded by a dark ring of pigment. In contrast, the body wall of *D. seguroensis* is slighty gritty to the touch, and the tube feet disks lack a ring of dark pigment.

SELECTED REFERENCES: Deichmann 1930:141, pl. 17, figs. 10–13 [as *Phyllophorus seguroensis*]; Tommasi 1969:9, fig. 9.

Ocnus surinamensis (Semper)
Figures 141, 178C,D

DESCRIPTION: This species can reach a length of 10 cm (4 in), but most individuals are 5–7 cm (2–3 in) long. The body is elongate and cylindrical, tapering toward the ends, particularly the posterior end. The body can be slightly curved, especially in preserved material. Because of a dense concentration of ossicles, the thin body wall is rigid and slightly gritty to the touch. In small individuals, the tube feet are aligned in double or triple rows along the radii, and scattered in the interradii. As the animal grows, the numbers of interradial tube feet increase disproportionately, so that the adult appears to have a uniform covering of feet. The mouth is surrounded by 10 richly branched tentacles of equal length. An introvert is not conspicuous.

The body wall ossicles (Figure 178C,D) are knobbed buttons and baskets. Some buttons have large, swollen knobs and small perforations; others have smaller knobs and larger holes. The baskets are nearly as large as the buttons and have a deep concavity defined by four spokes. The spokes attach to a rim that bears knobs on its inner and outer margins.

O. surinamensis varies in color from light to dark brown or gray to purple. The numerous white tube feet, which are flecked with dark brown or black pigment spots, contrast with the dark body wall.

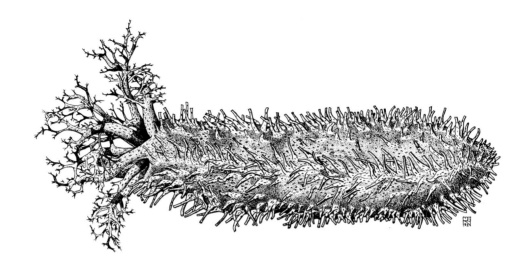

FIGURE 141. *Ocnus surinamensis*. Illustration by M. Nelson-Poole.

HABITAT: Near shore beneath rocks, in crevices, and among seagrass.

DISTRIBUTION: Bermuda to Colombia, Venezuela, and Surinam, including Florida (Biscayne Bay), Cuba, Puerto Rico, Jamaica, and several islands of the Lesser Antilles. DEPTH: Less than 1 m (less than 3 ft).

BIOLOGY: Individuals cover themselves with detritus and shell or rock fragments, grasping the material with their powerful tube feet. So tenaciously does it cling to the rocks or rubble under which it lives, that it is impossible to remove an individual without tearing off some of its tube feet.

REMARKS: *O. surinamensis* is similar to *O. suspectus*, but the body wall of *O. suspectus* is smooth and pliable, with few ossicles present, whereas in *O. surinamensis* the body wall is rough and stiff because of the dense concentration of ossicles. The body wall ossicles of these species differ markedly in shape (compare Figure 178C,D and E–J). In addition, the body of *O. suspectus* is short and inflated, whereas *O. surinamensis* is elongate and cylindrical. The suggestion (Massin 1993) that both these species be united in the genus *Parathyone* Deichmann, 1957, is regarded as premature pending the completion of major revisions of the dendrochirotes, currently in progress.

Another related species, *O. pygmaeus* (Théel 1886:83, pl. 4, fig. 9 [as *Colochirus pygmaeus*]), has been collected from the east and west coasts of Florida (but not the Florida Keys and the Dry Tortugas) (Miller and Pawson 1984). It has large tube feet, most of which are confined to the radii, a chocolate brown pigmentation, and five conspicuous, flaplike oral valves that are capable of covering the mouth. Like *O. suspectus* and *O. surinamensis*, *O. pygmaeus* has been found intertidally, but it usually occurs subtidally on sand and shell bottoms. Individuals have also been found living deep inside sponges (Miller and Pawson 1984).

SELECTED REFERENCES: Semper 1868: 65, pl. 15, fig. 15 [as *Thyone surinamensis*]; Deichmann 1926:25, pl. 3, fig. 1a–e [as *Thyone surinamensis*]; Pawson 1986:540, pls. 14.9, 178.

Ocnus suspectus (Ludwig)
Figures 142, 178E,F,G,H,I,J

DESCRIPTION: This is a small species that can attain a length of approximately 6–7 cm (2–3 in), although most individuals are generally less than half that size. The squat, curved body is swollen ventrally, with mouth and anus turned upward. Near the extreme posterior end, the body tapers abruptly to form a short cone. The mouth is surrounded by 10 equally short, highly branched tentacles that are carried on a naked, narrow introvert. The body wall is thin and soft to the touch, because of the relatively small number of embedded ossicles. Tube feet are scattered over the body, but are most numerous ventrally. They tend to form discrete rows along the radii, especially in small individuals.

Body wall ossicles (Figure 178E–J) are not numerous in this species; most occur near the anus. They consist of buttons, baskets, and slender, perforate rods. The buttons have four large holes and knobbed margins; the baskets are shallow and composed of four spokes connected to a rim bearing approximately 12 knobs.

Minute brown pigment spots are scattered over the body and tube feet, giving *O. suspectus* a brownish hue. However, the smaller the specimen, the more concentrated are the pigment spots. Thus, young individuals are almost black, intermediate-sized individuals are dark brown, and fully grown specimens are a mottled brown. Dark brown or black blotches tend to be concentrated in the interradii, thereby forming longitudinal stripes of darker color along the body. The tentacle stems are variegated black and white, or brown and white. The bushy side branches are white, because of a dense concentration of skeletal rods.

HABITAT: Among rocks in the intertidal zone and in shallow seagrass beds.

FIGURE 142. *Ocnus suspectus.*

DISTRIBUTION: Jamaica, St. Martin, St. Kitts, Barbados, and Colombia. Recently found in Florida (Biscayne Bay) and Puerto Rico (Miller, previously unpublished). Reports from Brazil are questionable, because *O. suspectus* has been confused with *O. braziliensis* (Verrill) (see e.g., Theél [1886] and an opposing view [Deichmann 1930]). DEPTH: Near the low-tide mark to 60 m (197 ft) (Martínez 1991a).

BIOLOGY: Most distributional records are based on collections of single individuals. However, on rocky coastlines along southwestern Puerto Rico, this species is quite common near Cabo Rojo. At one site, more than 25 individuals were found under rocks and rubble along 20 m (66 ft) of shoreline (Miller, previously unpublished). It is interesting that Biscayne Bay, as noted above, is the only locality in Florida from which *O. suspectus* has been confirmed, and a similar congener, *O. surinamensis*, has been found there as well. Near La Parguera, Puerto Rico, these species co-occur at the rocky shoreline (Miller, previously unpublished).

REMARKS: As noted under the previous species, Massin's (1993) placement of *O. surinamensis* and *O. suspectus* in the genus *Parathyone* should be considered only after a major revision of *Ocnus* and related dendrochirote genera.

SELECTED REFERENCES: Ludwig 1875: 92, pl. 6, fig. 19a–c [as *Thyone suspecta*]; Deichmann 1926:23, pl. 3, fig. 2a–d [as *Thyone suspecta*]; Pawson and Miller 1981:393.

Thyonella gemmata (Pourtalès)
Figures 143, 179A,B,C,D

DESCRIPTION: This is a medium-sized, burrowing species that can attain a length of 15 cm (6 in). In the field, freshly excavated individuals are usually distinctly U-shaped. The body tapers toward the bluntly rounded mouth and narrows more drastically toward the stiff anal cone. Because of the relatively thick body wall and a dense layer of ossicles, the skin is rigid and rough. Cylindrical tube feet are found along the midbody, and conical papillae near the ends. The tube feet are arranged in two distinct rows along the radii and scattered over the interradial areas, but they are largest and most numerous on the ventral surface. The mouth is surrounded by 10 stout, bushy tentacles, of which the ventral two are the smallest. Surrounding the anus are five calcareous plates, one at the end of each radius; anterior to each plate are one to five large, conical papillae.

The numerous body wall ossicles (Figure 179A–D) consist of buttons, baskets, rods, and perforated plates. The heavily knobbed buttons are irregular and variable in length. The tiny baskets are shallow and carry seven to nine marginal teeth. Large perforated plates and rods are found in the tube feet.

In both the anterior and posterior regions, there are numerous, large, perforated plates with a smooth to heavily knobbed surface. They are seldom found in the middle portion of the body, where buttons and baskets prevail. Unfortunately, the characteristic baskets, which are invaluable for unequivocal identification, may be completely lacking in some specimens, apparently resorbed with age.

Most individuals are a mottled gray, brown, or olive green, but some are nearly uniformly tan or black. The radial areas usually are lighter than the interradial areas, giving specimens a striped appearance. In live *T. gemmata*, the tips of the tube feet may appear red, because of the presence of hemoglobin-containing cells in the water vascular system.

HABITAT: Muddy or sandy areas, often associated with the seagrasses *Halodule* or *Thalassia*; rarely, beneath rocks.

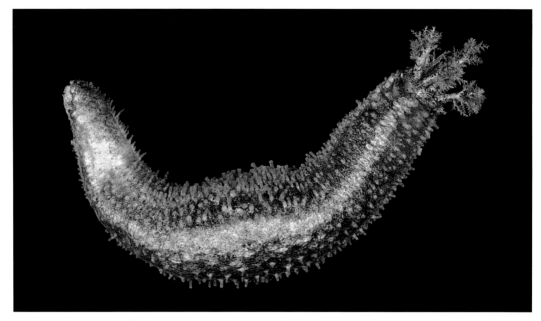

FIGURE 143. *Thyonella gemmata*.

DISTRIBUTION: New England to Florida, and along the Gulf of Mexico coastline to Texas, Cuba, and the Yucatán Peninsula. Around Florida, from the Indian River (east-central Florida), Biscayne Bay (Miller, previously unpublished), Marathon and Conch keys (Miller, previously unpublished), and the entire West Florida coast. DEPTH: Low-tide mark to 6 m (20 ft), but specimens dredged near Daytona Beach were found at depths of 15–20 m (50–66 ft) (Florida Department of Natural Resources, SeaMap Program).

BIOLOGY: This is one of the most common holothurians in shallow waters of the eastern Gulf of Mexico. Occasionally, large numbers are found washed ashore after storms. Adults almost always burrow in mud or sand and rarely attach below rocks; juveniles have been found in algae collected near Conch Key, Florida (M. Byrne, personal communication).

Manwell and Baker (1963) studied two populations of *T. gemmata* off the west coast of Florida that differed slightly in morphological and behavioral characteristics. Chemical analyses of the hemoglobin from their water vascular system and body cavity indicated that the populations of "thin" and "stout" individuals might be sibling species. Differences in the composition of hemoglobin in the coelomic and water vascular erythrocytes were also described (Manwell 1966). The structure and function of hemoglobin in this species has continued to attract the attention of physiologists (Terwilliger and Terwilliger 1988).

SELECTED REFERENCES: Pourtalès 1851: 11 [as *Colochirus gemmatus*]; Miller and Pawson 1984:20, figs. 12, 13; Ruppert and Fox 1988:77, fig. 117.

Thyonella pervicax (Théel)
Figures 144, 179E,F,G,H

DESCRIPTION: This is a small species that usually reaches 4–8 cm (1.5–3 in) in length.

The body is cylindrical, elongate, and tapered only at the extreme anterior and posterior ends, which are bluntly rounded. Usually the mouth and anus are turned slightly upward, but specimens may be distinctly U-shaped. The entire body wall is covered with tube feet in the form of low yet conspicuous conical warts, which are most numerous and crowded on the lower surface. The skin is moderately thick and very stiff and rough because of dense layers of skeletal ossicles.

The body wall ossicles (Figure 179E–H) consist of heavily knobbed, four-holed buttons of two distinct sizes and shallowly to moderately concave baskets with dentate margins.

T. pervicax is generally mottled white or tan and light to dark brown, with conspicuous, scattered patches of brown. The upper surface is generally darker than the lower. Some individuals may be almost uniformly white, but others from the same locality are dark brown.

HABITAT: Soft, sandy sediments, usually at some distance from shore. Off the west coast of Florida, it occurs on shell and quartz sand bottom covered with algae and seagrass.

DISTRIBUTION: Massachusetts southward to East Florida and the Dry Tortugas, and west to the western coast of Florida and Texas in the Gulf of Mexico. It also occurs in Panama (based on USNM specimens) and off Bahia, Brazil. DEPTH: 6–70 m (20–230 ft).

BIOLOGY: This species burrows in soft sediment and may occur in extensive aggregations, judging from the large numbers of individuals collected by dredging.

SELECTED REFERENCES: Théel 1886:93, pl. 5, fig. 9, pl. 12, fig. 3 [as *Thyone pervicax*]; Pawson 1977:6; Miller and Pawson 1984:23, figs. 15, 16.

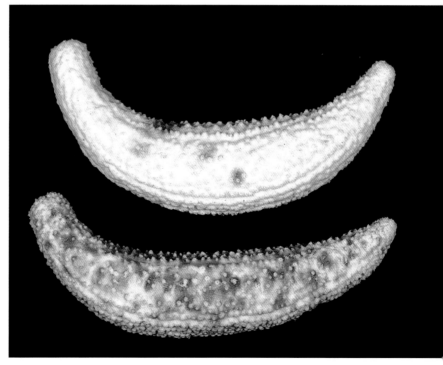

FIGURE 144. *Thyonella pervicax.*

FAMILY SCLERODACTYLIDAE

Euthyonidiella destichada
(Deichmann)
Figures 145, 181D,E,F

DESCRIPTION: This is a small species, reaching a maximum length of 10 cm (4 in); individuals of less than half that size are usually collected. The body is stout, barrel-shaped, and slightly curved dorsally near the anterior and posterior ends. Near the mouth and anus, the body tapers to bluntly rounded ends. The numerous tube feet are irregularly scattered over the body; they are slightly more numerous ventrally. The skin is thin, flexible, and relatively smooth, because of the large numbers of soft tube feet. The feet terminate with a prominent circular calcite disk. The 15–20 tentacles, usually contracted deep within the mouth, are unequal in size, half of them being about twice the length of the others.

The body wall ossicles (Figure 181D–F) are tables having a smooth, oblong disk with eight holes (usually four large and four small). The table spires are two stout pillars that each terminate in five or six tiny, conical spines.

The dark brown or brownish purple body wall pigmentation contrasts with the white-tipped tube feet in this species.

HABITAT: Sand and mud areas associated with turtle grass and among wave-washed rocks near the shore.

DISTRIBUTION: Previously known from Florida (Biscayne Bay and the Dry Tortugas) and Venezuela (Martínez 1991b), and here reported from Puerto Rico (La Parguera), the southern tip of Martinique, Belize (Carrie Bow Cay), and Panama. DEPTH: Low-tide mark to about 4 m (13 ft).

REMARKS: The species can burrow in soft sediments or attach to the underside of rocks. Its natural history is otherwise unknown.

SELECTED REFERENCES: Deichmann

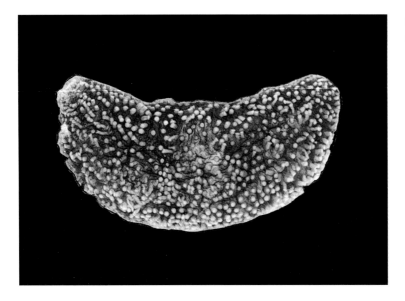

FIGURE 145. *Euthyonidiella destichada*.

1930:146, pl. 18, fig. 3 [as *Phyllophorus destichadus*]; Heding and Panning 1954:118, fig. 47.

Euthyonidiella trita (Sluiter)
Figures 146, 181G,H

DESCRIPTION: This tiny species only grows to a length of 3 cm (1 in). Typically, the slender, cylindrical body is slightly tapered and turned upward near the posterior end, but when the tentacles are retracted, the anterior end is truncated. Numerous cylindrical tube feet are scattered over the entire body, except on the introvert where they are arranged in two or three rows in each radius. Frequently the tube feet are fully contracted into shallow pits. The body wall is smooth, thin, and distinctly wrinkled, yet rigid. Eighteen tentacles of two distinct sizes surround the mouth. Generally one or two smaller tentacles alternate with pairs of larger tentacles. The tentacles have short, thick trunks and small side branches. Surrounding the anus are five inconspicuous, radial, calcareous teeth.

The body wall ossicles (Figure 181G,H) are oblong tables with irregular margins. Most of the tables have four large disk perforations, but a few have one to four smaller additional holes. Their short spires are composed of two pillars, each ending in several small spines.

E. trita varies from violet to gray. The disks of the tube feet are yellow or white, and the tentacles have beige stalks with light to dark brown branches.

HABITAT: Near the shoreline, under rocks and in tide pools.

DISTRIBUTION: Previously reported from the Dry Tortugas, Puerto Rico, the Virgin Islands, St. Martin, Antigua, and Venezuela (Martínez 1991b). Also known from Jamaica and Stuart, Florida (Miller, previously unpublished). DEPTH: Low-tide mark to 3.6 m (12 ft).

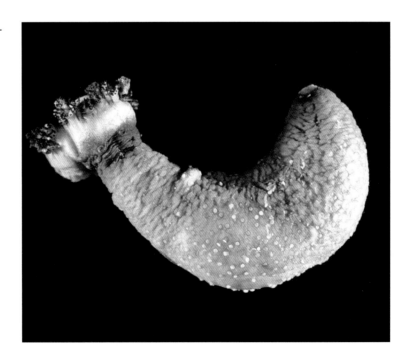

FIGURE 146. *Euthyonidiella trita*. Photo by B. Cutress.

BIOLOGY: Dissection of a specimen collected in May, off Stuart, Florida, revealed that the ovary was well developed. The large eggs (0.5 mm in diameter) and the gonadal tubules were a striking lime green (Miller, previously unpublished).

SELECTED REFERENCES: Sluiter 1910: 339, fig. Ea,b,c [as *Thyone trita*]; Deichmann 1930:147, pl. 18, figs. 4–8 [*Phyllophorus tritus*]; Heding and Panning 1954:116, fig. 45a–g.

Pseudothyone belli (Ludwig)
Figures 147, 180A,B,C

DESCRIPTION: This is a small sea cucumber, and although there are reports of specimens 5 cm (2 in) in length, most individuals are considerably less than half that size. The body is usually curved, somewhat swollen at the middle, and slightly tapered at the ends. The thin body wall is very stiff, because of a profusion of skeletal ossicles. Numerous long, cylindrical tube feet are scattered over the body wall, and there is some tendency for the tube feet to be aligned in rows near the mouth and anus. On the ventral surface of the body they are most numerous, longest, and hairlike. They appear incapable of full retraction, probably because of the dense layer of ossicles in the tube foot wall. In juveniles 0.5 cm (0.25 in) long, the tube feet are in double rows along each radius. Surrounding the anus are five small, radial, calcareous teeth, and two small papillae are situated above each tooth. The mouth is surrounded by eight long, slender, abundantly branched tentacles and two shorter ventral ones about one-third the length of their neighbors.

The body wall ossicles (Figure 180A,B) are knobbed buttons with four perforations, two large central knobs, and 9–12 marginal knobs. The tables from the tube feet have slender, curved disks and short, robust spires terminating in several small teeth (Figure 180C).

P. belli is dirty white with flecks of brown or maroon. In very small specimens, the flecks are gray and quite dense along the upper surface, especially near the ends of the body. These small individuals appear black

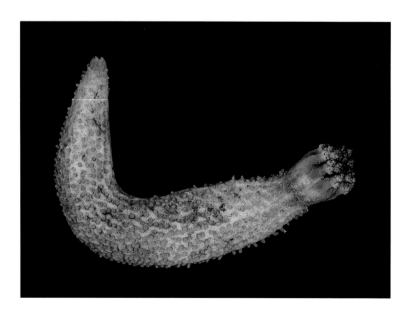

FIGURE 147. *Pseudothyone belli.*

to the naked eye. The tips of the tube feet are bordered by a brown ring. The tentacles have yellowish tan stalks and orange side branches with black fronds.

HABITAT: Off the west coast of Florida, it occurs in sandy carbonate sediments (Miller and Pawson 1984), and off the east coast of Florida, at least to Stuart, it occurs in high densities on coquinoid limestone ledges (Miller, previously unpublished). Juveniles were found in the sediments of a back reef seagrass bed at Looe Key, Florida (Miller, previously unpublished).

DISTRIBUTION: Bermuda (Pawson, previously unpublished), Florida (Vero Beach to Tampa), the Dry Tortugas, Panama, Puerto Rico, Trinidad, Tobago, and Brazil. DEPTH: Low-tide mark to 37 m (121 ft).

BIOLOGY: Typically, the species burrows in soft sediments, and a few individuals have been found clinging to the underside of rock slabs. Off East Florida, *P. belli* apparently settles on the coquinoid reefs, recruiting to crevices just large enough to contain the adult. Individuals can be located when their tentacles are extended for feeding. The tube feet of *P. belli* provide a tenacious grip on hard substrates. Live specimens cannot be dislodged from a smooth surface without tearing off a considerable number of tube feet.

SELECTED REFERENCES: Ludwig 1886: 21, pl. 1, fig. 6 [as *Thyone belli*]; Miller and Pawson 1984:27, figs. 19, 20.

Sclerodactyla briareus (Lesueur)
Figures 148, 180*D,E,F*

DESCRIPTION: This is a medium-sized sea cucumber that can reach a length of 12–15 cm (5–6 in). The body is stout and robust, swollen near the middle, and gently tapered at each end. Generally the mouth and anus are curved slightly upward. Specimens excavated from soft sediment contort their bodies into a spherical shape about the size of a golf ball by bringing mouth and anus into close proximity. The body wall is thin and very soft, because of the small number of skeletal ossicles and the numerous hairlike tube feet scattered over the entire body. The tube feet are cylindrical, though somewhat tapered near their sucking disks, and they are longest and most numerous on the ventral surface. The mouth is surrounded by eight large and bushy tentacles and two ventral tentacles about one-fourth the size of their neighbors. The anus is surrounded by five large, triangular, radial teeth, each of which is overlain by two pairs of papillae, the inner pair twice the length of the outer pair.

The number of ossicles is greatly reduced in large, old individuals, although some usually remain in the feet or in the body wall surrounding the mouth and anus. The body wall ossicles (Figure 180*E,F*) are tables with a flat, squarish disk; in the feet the tables (Figure 180*D*) have an elongate curved disk. Both types of tables have tall spires composed of four pillars.

In life, *S. briareus* is green or brown, although some individuals appear to be almost black. Usually, conspicuous darker patches of pigment cover the mouth and anus and are especially noticeable when the body wall is contracted into a spherical shape. In some individuals, the tube feet are brownish orange. The introvert is gray to black, the tentacle stems black, and the tentacle branches gray.

HABITAT: Soft, muddy substrates, often associated with seagrass beds.

DISTRIBUTION: Nova Scotia, southward along the eastern seaboard of the United States and around the Gulf of Mexico coastline to Texas. It is also reported from Venezu-

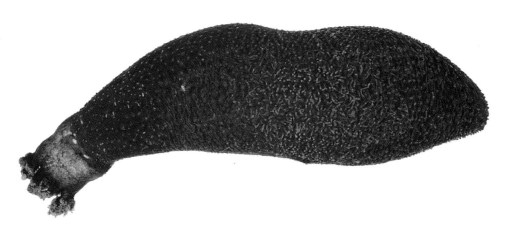

FIGURE 148. *Sclerodactyla briareus*.

ela (Martínez 1991b). In southern Florida, it has been found in Biscayne Bay and off Key West. DEPTH: Previously reported from 0 to 66 m (217 ft); a museum (USNM) specimen from off Georgia was collected at 183 m (600 ft) (Miller, previously unpublished).

BIOLOGY: Juveniles have been found clinging to seagrass blades. The adults burrow just below the sediment, holding the mouth and anus above the substratum (Deichmann 1954). At the northern end of its range, individuals burrow deeper in the sediment during the winter (Edwards 1909). At that time, they do not feed and appear to hibernate (Farmanfarmaian 1969a). The adults have a higher tolerance than most echinoderms for brackish environments (Pearse 1908), and *S. briareus* has been characterized as a low-salinity species (Stickle and Diehl 1987). In suitable mud flats and seagrass meadows, it may occur in great numbers. Often, after a major storm, the beaches adjacent to these habitats are littered with numerous decomposing sea cucumbers that have been dislodged by wave action and washed ashore, their tube feet "sanded off" in the surf zone (Ruppert and Fox 1988).

The species has been widely used as a laboratory animal in biology classes around the United States, and it has been intensively studied by scientists interested in the reproduction, behavior, and physiology of sea cucumbers. Pearse (1908) described a selective feeding behavior of this species, wherein sand particles are rejected in favor of detritus and small organisms. Farmanfarmaian (1969a,b) found that particles of a seagrass, *Zostera*, may compose a significant part of the diet during the summer. He showed that the intestine is morphologically well adapted to digestion and absorption of *Zostera* carbohydrates and that nutrients are transported to outlying tissues by the perivisceral fluid. Ferguson (1982b) documented the modest uptake, presumably through the skin, of free amino acids from the surrounding seawater; presumably the other layers of the body derive some benefit from this supplementary source of nutriment.

Pearse (1908) noted that respiration involves a pulsation of the entire body, by which means water is drawn into the respiratory trees and then forcibly expelled. When placed upon sand or mud, individuals burrow by peristaltic contraction and extension of their bodies, pushing the sand or mud aside and gradually sinking beneath the surface. The animal can advance with the anterior, posterior, or even one side of the body in the lead, and its rate of burrowing is enhanced if the tube feet can find purchase on hard objects (Pearse 1908; Stier 1933). The physiology of smooth muscle contraction was examined by Prosser et al. (1965), who showed that impulses travel between muscle fibers of this species, independent of the nerves.

Individuals in the sediment elevate and expose more and more of the body surface as the concentration of oxygen in the surrounding medium falls (Brown and Shick 1979). The species can survive exposure to anoxic or hypoxic conditions for more than 2 days (Ellington 1976); this capability is strengthened by lactate dehydrogenase in the longitudinal muscles, which helps maintain glycolytic flux. Hemoglobin in the water vascular system of this species may also contribute to resistance to hypoxia; its molecular structure was analyzed and compared with hemoglobin of other holothuroid species by Roberts et al. (1984). Additional research on sea cucumbers has confirmed the high oxygen affinity of their hemoglobin and its adaptive value for low-oxygen habitats (Baker and Terwilliger 1993). Literature related to respiration in this species was reviewed by Shick (1983).

Menton and Eisen (1970) studied the structure of the integument, and in a later paper (1973) they examined the process of wound healing. Pearse (1908) found that autotomy of the anterior end of the body and viscera occurs in response to fouling of the water and to toxic chemicals. Kille (1935) described in detail the regeneration of this species following autotomy. Complete regeneration takes about 37 days. Smith and Greenberg (1973) identified an "evisceration factor" of unknown composition in the coelomic fluid of autoeviscerating individuals. This factor induces autotomy when injected into intact animals.

Ruppert and Fox (1988) noted that *S. briareus* is preyed upon by the fairly undiscriminating loggerhead turtle, *Caretta caretta* (Linnaeus). In a laboratory setting, de Vore and Brodie (1982) tested various parts of the holothurian's body for palatability using the killifish, *Fundulus diaphanus* (Le Sueur). The integument and gonads were the least palatable, and de Vore and Brodie suggested that the distastefulness of these tissues could enhance survival of the holothurian. The species has a "shadow response"; both the posterior and anterior parts of the body will withdraw into the sediment if light impinging on the body is eclipsed (Pearse 1908).

Ultrastructure of the developing oocytes has been described (Kessel 1964). The gonads reach maturity in June in Woods Hole, Massachusetts (Ohshima 1925). The release of gametes occurs late in the day in animals brought into the laboratory, and males always spawn first. The spawning process, which lasts from 15 minutes to 4 hours, is accompanied by elongation of the body and waving of the tentacles (Colwin 1948). The species lacks a free-swimming larval stage; about 4 days after zygote formation, a "metadoliolaria" larva escapes from the egg membrane and crawls on the bottom, soon forming a typical pentactula stage.

REMARKS: Although it is often referred to as *Thyone briareus,* Panning's (1949) placement of this species in the genus *Sclerodactyla* is currrently accepted.

SELECTED REFERENCES: Lesueur 1824: 161 [as *Holothuria briareus*]; Gosner 1979: 254, pl. 46; Miller and Pawson 1984:30, figs. 22, 23; Ruppert and Fox 1988:77, fig. 116.

FAMILY PHYLLOPHORIDAE

Neothyonidium parvum (Ludwig)
Figures 149, 179I,J

DESCRIPTION: This fairly small species reaches a length of approximately 8 cm (3 in). The crescent-shaped body is cylindrical in cross section and slightly tapered toward the bluntly rounded ends. The body wall is thin, wrinkled, and soft, although firm in strongly contracted individuals. Well-developed tube feet are scattered over the body but are more numerous on the ventral surface. The 18–20 tentacles are arranged in indistinct circlets surrounding the mouth. The longest tentacles are four or five times as long as the shortest, and they are bushy, somewhat resembling the tips of celery stalks.

Body wall ossicles (Figure 179I,J) are uniform tables with oblong disks, perforated by 10–20 holes of variable size. Spires are composed of two pillars, terminating in 8–12 bluntly rounded spines.

In life, the body wall along the midsection of *N. parvum* is pigmented bright red or brownish red; toward the mouth and anus the skin is white or pink, with flecks of red pigment. The tips of the tube feet are white, and the tentacles are violet to dark brown. Specimens preserved in alcohol quickly lose their color, becoming pale beige or yellow.

HABITAT: In seagrass areas near the low-tide mark.

DISTRIBUTION: Puerto Rico (Miller, previously unpublished), Florida (Biscayne Bay), Antigua, and Brazil. DEPTH: Less than 2–3 m (less than 7–10 ft).

BIOLOGY: This burrowing species may be associated with the rhizomes of seagrasses. It is reportedly common in Brazil (Deichmann 1926), but has only been collected in small numbers elsewhere.

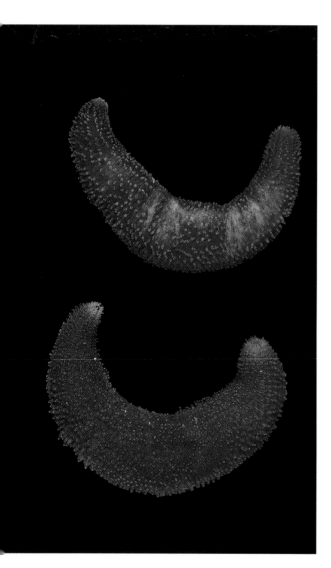

FIGURE 149. *Neothyonidium parvum*.

REMARKS: The bright red coloration of living individuals serves to separate this species from all other cylindrical dendrochirotes inhabiting the tropical western Atlantic.

SELECTED REFERENCES: Ludwig 1881: 54, pl. 3, figs. 16–18 [as *Thyonidium parvum*]; Deichmann 1939:133, figs. 19–21 [as *Phyllophorus parvum*]; Heding and Panning 1954:198, fig. 99a–c.

Phyllophorus (Urodemella) arenicola Pawson & Miller
Figures 150, 182A,B,C

DESCRIPTION: This is a moderate-sized sea cucumber that exceeds 30 cm (12 in) in length when expanded; contracted animals are usually less than 10 cm (4 in) long. The body is approximately cylindrical, more or less U-shaped, and tapers gently toward the anterior and posterior ends. The body wall is thin and soft, becoming firmer and thicker in contracted individuals. The small, soft tube feet are scattered over most of the body, tending to form double rows in the radii; the double rows are most evident on the introvert. There are 20 tentacles; an outer ring of five large pairs alternate with five small pairs forming an inner ring.

The body wall ossicles (Figure 182A–C) are squarish tables with scalloped margins and four or more, commonly eight, perforations. The spire is reduced to form four discrete short, bluntly pointed, vertical projections. The introvert contains tables and rosettes.

FIGURE 150. *Phyllophorus (Urodemella) arenicola*.

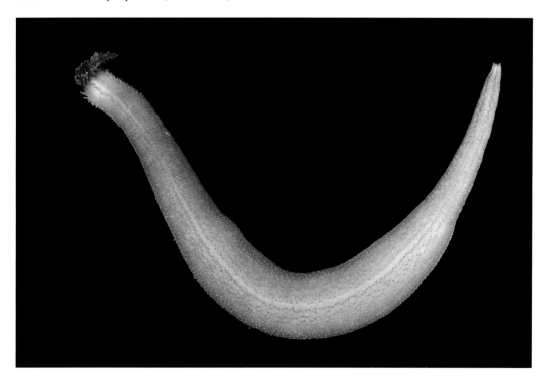

The color in life is uniform to variegated light reddish brown.

HABITAT: Offshore soft sediments, such as unconsolidated sand and shell hash.

DISTRIBUTION: Georgia and eastern Florida. DEPTH: 6–158 m (6–514 ft).

BIOLOGY: Individuals were observed in situ at depths of 9–15 m off Fort Pierce, Florida, in coarse, ripple-marked sediment. Other echinoderms in this habitat include the sand dollar *Encope aberrans* and burrowing amphiurid brittle stars. Their feeding tentacles are sometimes extended into the water column; however, most individuals remain hidden beneath the ripple crests. When disturbed, they quickly withdraw their tentacles and introvert, as deep as 10 cm below the sediment surface. When tightly grasped and roughly pulled from the sediment by hand, their bodies are quite long (more than 30 cm) and contorted. Within 10–15 seconds after removal from the sediment, they contract into a uniform U-shape, less than 10 cm in length, and resemble individuals gently excavated from below. Evidently, "ballooning" the body is a defensive mechanism that serves to obstruct removal from the sediment (Miller and Hendler, previously unpublished).

Specimens have most commonly been collected by dredge, a process that results in a characteristic pattern of damage. The dredged individuals separate (autotomize) into two pieces, the main body from just behind the introvert to the anus and a much-shorter piece consisting of the introvert, tentacles, calcareous ring, and associated structures.

SELECTED REFERENCES: Miller and Pawson 1984:36, figs. 27, 28 [as *Phyllophorus (Urodemella) occidentalis*]; Pawson and Miller 1992:483, figs. 1–4.

Phyllophorus (Urodemella) occidentalis (Ludwig)
Figures 151, 181A,B,C

DESCRIPTION: This is a curved to U-shaped species that can reach a maximum length of 10 cm (4 in), but most individuals are half that size. The body is swollen near the middle and tapered at both ends. In some specimens, the posterior end is drawn out into a short tail that lacks tube feet. The skin is soft, thin, and smooth, and in live individuals it is somewhat slimy. Short, cylindrical tube feet are set on low, conical warts and are somewhat scattered. However, there is a tendency, which is most pronounced on the introvert, for the tube feet to be aligned in double rows along the radii, the rows separated by a narrow strip of naked body wall. The tube feet have well-developed terminal suction disks, and they can retract completely below the surface of the body wall. Dorsally, the tube feet are widely spaced, and ventrally they are close set and numerous. The mouth is surrounded by 20 tentacles, an outer whorl of five large pairs and an inner whorl of five smaller pairs. Between each adjacent pair of large tentacles in the outer whorl, there is a smaller pair situated closer to the mouth. The larger tentacles are four or five times the length of the smaller ones and more richly branched. The anus is surrounded by 10 small papillae, two in each radius.

The body wall ossicles (Figure 181A–C) are tables that have four to eight large perforations and, occasionally, one to six smaller accessory perforations. The tables are delicate, with flat, thin disks and conspicuously dentate margins. The reduced spires of two short pillars terminate in one to three short, acute spines.

The coloration of *P. occidentalis* varies from a uniform light or dark brown to yellow or golden brown. The tube feet are transparent, except for a dark ring of pigment surrounding the white suction disk. In many in-

SEA CUCUMBERS

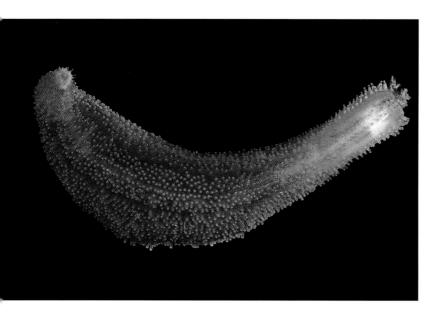

FIGURE 151. *Phyllophorus* (*Urodemella*) *occidentalis*.

dividuals, there are patches of dark brown flecks on the body wall, adjacent to the tube feet. The dark brown or black tentacle stems contrast with the lighter branches, which have heavy accumulations of ossicles. When fully extended, the introvert is transparent, and the internal calcareous ring is visible.

HABITAT: Near the low-tide mark in shallow water and among the roots of seagrasses.

DISTRIBUTION: Florida (the Dry Tortugas), Puerto Rico, Antigua, Barbados, Grenada, Aruba, Trinidad, Surinam, and Brazil. DEPTH: 1–2 m (3–6 ft).

BIOLOGY: *P. occidentalis* burrows in soft sediments and beneath seagrass rhizomes and has been found clinging beneath rocks.

SELECTED REFERENCES: Ludwig 1875: 119 [as *Thyonidium occidentale*]; Deichmann 1954:402 [as *Trachythyonidium occidentale*]; Heding and Panning 1954:164; Tikasingh 1963:96 [as *Trachythyonidium occidentale*].

Stolus cognatus (Lampert)
Figures 152, 182*D,E,F,G*

DESCRIPTION: This small to moderate-sized species attains a maximum length of 15 cm (6 in); most individuals are 5–10 cm (2–4 in) long. The body is strongly curved and spindle-shaped; tapering is most pronounced at the posterior end, where the body wall forms a rigid anal cone. The small and inconspicuous tube feet are arranged in five bands along the radii and are scattered in the interradii. The body wall is thin, stiff, and slightly gritty. The small tube feet and the numerous ossicles in the body wall give the skin a very smooth appearance. The mouth is surrounded by 10 small tentacles of nearly equal length.

The body wall ossicles (Figure 182*D,E*) are large, elongate plates with uneven margins and usually two rows of perforations. The feet contain slightly curved rods with perforations at each end and an expanded portion near the middle (Figure 182*F,G*).

The body wall is white or grayish with a peppering of brown or black pigment. In liv-

275

CLASS HOLOTHUROIDEA

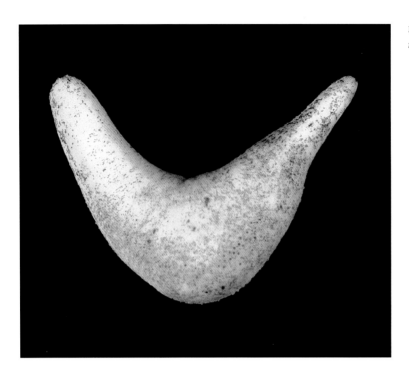

FIGURE 152. *Stolus cognatus*.

ing animals, a few of the tube feet along the lateral margin are bright red or yellow. The tentacles are grayish brown. Deichmann (1939) reported that several individuals collected in Biscayne Bay, off Miami, Florida, were a mottled reddish brown.

HABITAT: Seagrass beds among the roots of *Syringodium* or *Thalassia*.

DISTRIBUTION: Off Florida, at Key Largo (Martin, personal communication), Key West, and reportedly common south of Cape Florida (Biscayne Bay), and around the Dry Tortugas. Also known from Cuba, the Yucatán, Venezuela, Aruba, and Brazil. DEPTH: Less than 3 m (less than 12 ft).

BIOLOGY: *S. cognatus* lives in a shallow, U-shaped burrow, with its mouth and anus slightly extended above the substrate (Miller, previously unpublished).

SELECTED REFERENCES: Lampert 1885: 251 [as *Semperia cognata*], fig. 51 [as *Cucumaria cognata*]; Panning 1949:462 [as *Stolus cognitus*]; Caycedo 1978:165, pl. 4, figs. a, b, text fig. 3 [as *Thyoneria cognata*].

Thyone deichmannae Madsen
Figures 153, 182H,I,J

DESCRIPTION: This species is small, reaching a maximum length of 7 cm (2.8 in). The body is stout, cylindrical, slightly swollen at the middle, and bluntly rounded at both ends. The body wall is thick, smooth, and soft. The numerous tube feet, all with terminal suckers, are scattered over the entire body and lack an ordered arrangement along the radii. Most of the feet remain extended, even in preserved material, giving the species a furry appearance. There are 10 tentacles, of which the ventral pair is smallest.

The body wall ossicles (Figure 182*I,J*)

SEA CUCUMBERS

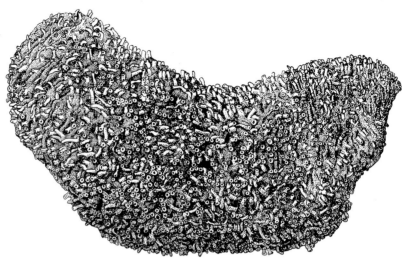

FIGURE 153. *Thyone deichmannae*. Illustration by C. Messing.

are large tables with squarish to rectangular disks, usually perforated with four to eight large holes and up to 10 smaller ones. At the center of the disk there is a spire composed of two pillars and ending in a few small spines. The supporting ossicles of the tube feet are large tables with elongate curved disks and robust spires (Figure 182*H*).

T. deichmannae is drab, with a grayish brown body wall and light gray to light brown tube feet.

HABITAT: Soft sediments, such as crushed shell, quartz sand, and calcareous silt.

DISTRIBUTION: Previously reported (as *T. inermis* Heller) from Florida (Jacksonville, St. Augustine, Tampa Bay, Captiva Key), the Florida Keys, Cuba, and Tobago. Material from North Carolina, the northeastern Gulf of Mexico, and Panama is represented in museum (USNM) collections (Miller, previously unpublished). DEPTH: 6–366 m (20–1,200 ft).

BIOLOGY: *T. deichmannae* is probably a burrowing species and is known to ingest surface deposits (Miller and Pawson 1984).

Its stomach contents include foraminiferans, molluscan shell fragments, ostracod valves, echinoid spines, diatom frustules, and some amorphous material. That is unexpected, because the majority of dendrochirote holothurians feed on suspended particles, using their arborescent feeding tentacles. Specimens measuring 16–42 mm can have gonads with mature sperm and eggs. The large, yolky eggs indicate that development is direct, as is common for dendrochirotids (Miller and Pawson 1984).

SELECTED REFERENCES: Madsen 1941: 26; Miller and Pawson 1984:40, figs. 32, 33 [as *T. inermis*].

Thyone pseudofusus Deichmann
Figures 154, 182*K,L,M*

DESCRIPTION: This is one of the smallest sea cucumbers in the Florida Keys. Most individuals are less than 2 cm (1 in) long; exceptional specimens reach 4–5 cm (2 in) in length. The body form is cylindrical, slightly swollen in the middle, and gently tapering toward the upturned mouth and anus. The body wall is moderately thin and pliable; it is rough because of a dense layer of ossicles. Tube feet are in double rows along the radii, with fewer, scattered feet in the interradial regions. They can retract completely, forming pits in the body wall, but when extended they are long and hairlike, especially those on the ventral surface. The mouth is located in the center of a conical introvert, surrounded by 10 tentacles of which the lowermost (ventral) pair is one-half the size of adjacent pairs. When fully extended, the introvert composes one-third to one-half the length of the body. The anus is surrounded by a ring of five small radial anal papillae.

The body wall ossicles (Figure 182*K,L*) are tables that have thick oval disks with four perforations, which are partially obscured by a robust spire. The short arch-shaped spires terminate in 6–10 small spines and are opposed on the underside of the disk by a distinct loop. Tables with curved, elongate disks and tapering spires of two or three pillars are abundant in the tube feet (Figure 182*M*).

According to H. L. Clark (1933:114), this species is "dirty whitish or very pale brown" when alive. In preserved material, the disks of the tube feet are darker than the surrounding body wall tissue.

HABITAT: Calcareous or quartz sand, covered with a layer of the green alga *Caulerpa* and the seagrass *Halophila* (Miller and Pawson 1984), among coralline algae (H. L. Clark 1933), and beneath rubble at the base of limestone ledges on the east coast of Florida (Miller, previously unpublished).

DISTRIBUTION: North Carolina, East and West Florida, Key West, the Dry Tortugas, Texas, the Yucatán Peninsula, Panama, Colombia, Tobago, and Brazil. DEPTH: 6–46 m (20–151 ft).

BIOLOGY: Like *T. deichmannae*, *T. pseudofusus* appears to be an exceptional deposit-feeding dendrochirotid. Its gut contents consist of 95% calcareous remains and 5% quartz sand (Miller and Pawson 1984), and it is reported to feed on at least 25 species of diatoms (Martínez 1989).

Randall (1967) reported that *T. pseudofusus* has been found in stomach contents of the white grunt, *Haemulon plumieri* (Lacepède).

SELECTED REFERENCES: Deichmann 1930:168, pl. 14, figs. 6–9; Miller and Pawson 1984:43, figs. 36, 37.

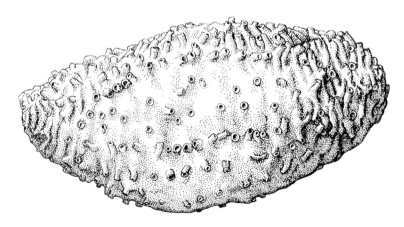

FIGURE 154. *Thyone pseudofusus*. Illustration by C. Messing.

FIGURE 155. *Astichopus multifidus*.

FAMILY STICHOPODIDAE

Astichopus multifidus (Sluiter)
Figures 155, 187D,E,F

DESCRIPTION: The largest individuals are approximately 50 cm (20 in) long and weigh as much as 2.5 kg (5.5 lb). The upper and lateral surfaces are covered with hundreds of conical papillae 3–5 mm (0.12–0.20 in) high. These tiny papillae give the living sea cucumber a hairy or furry appearance. The papillae found at the lateral margin are the same size as, or only slightly larger than, those of the dorsal surface. The ventral surface is flattened, wider than the dorsal surface, and covered with a dense layer of cylindrical tube feet. The body wall is fairly thin and soft. Unlike many sea cucumber species, *A. multifidus* does not become rigid and hard when disturbed, but the body wall is stiff and thickened in contracted, preserved specimens. The tentacles, which have large knoblike disks, number about 20. They project from the ventrally positioned mouth and are surrounded by a narrow, yet distinct, tentacular collar.

The abundant body wall ossicles (Figure 187D–F) are irregular C-, O-, and S-shaped elements; miliary grains are very numerous.

The upper surface of most individuals is mottled brown and white, with translucent white papillae. Some specimens are completely white, except for dark brown rings surrounding the bases of the papillae. Generally, the lower surface is white with sparse flecks of brown pigment between the tube feet. However, Glynn (1965) reported that the ventral surface of two Puerto Rican specimens was chocolate brown.

HABITAT: On muddy bottoms near Campeche Bank in the Gulf of Mexico (Glynn 1965), but elsewhere this species lives on sandy substrates in back reef areas, in or around seagrass beds of *Thalassia*, *Syringodium*, or *Halophila*. It tends to occur in deep, calm reef habitats that are not subject to constant wave action or strong bottom surge.

DISTRIBUTION: The Bahama Islands, Florida (Key Biscayne, Fort Myers), the Florida Keys, the Dry Tortugas, Mexico, Cuba, Jamaica, Puerto Rico, Colombia, and Venezuela. DEPTH: 1–37 m (3–121 ft), but generally found at depths greater than 3–4 m (10–13 ft).

BIOLOGY: *A. multifidus* is uncharacteristically animated for a sea cucumber and often crawls slowly across the bottom as it feeds. Glynn (1965) described four distinct patterns of locomotion: crawling, bounding, rolling, and exploratory activity. Individuals do not attach firmly to the substrate. Their forward progression is accomplished by bounding movements, propelling them at velocities up to nearly 2 m/minute (Glynn 1965), and an individual swimming above the seafloor by undulatory movements of the body wall has been photographed (Cutress, personal communication). Stomach contents that have been examined consist almost entirely of calcareous sediment including bivalves, gastropods, foraminiferans, bryozoans, and the remains of other calcified organisms (Miller and Pawson 1984). This species readily eviscerates via the anus when held in stagnant seawater, such as during transport to the laboratory.

Caycedo (1978) reported several symbionts of *A. multifidus;* gastropods (*Balcis* sp.) are concealed among the papillae, and the porcellanid crab *Porcellana sayana* (Leach) lives on the dorsal body surface and in the tentacles and cloaca. Adults and postlarvae of a pearlfish, *Carapus bermudensis* (Jones), have also been reported as internal commensals (Koster and Caycedo 1979). Several individuals from Looe Key were parasitized by the eulimid gastropod *Melanella hypsela* (Verrill & Bush), and one was host to the bumblebee shrimp, *Gnathophylum americanum* Guerin (this report).

SELECTED REFERENCES: Sluiter 1910: 334, figs. a, b [as *Stichopus multifidus*]; Colin 1978:429, 431; Miller and Pawson 1984:51, fig. 41.

Isostichopus badionotus (Selenka)
Figures 156, 187G,H,I

DESCRIPTION: This is a large species that grows to a length of 45 cm (18 in). It is readily distinguished by the presence of several to many dark, conspicuous warts on its dorsal surface and the thick conical papillae protruding from the ventrolateral margin. The upper surface is sharply demarcated from the flattened ventral surface, which carries three bands of crowded, cylindrical tube feet. The body wall is very thick and rigid; it exudes a slimy mucus when the animal is removed from water. The mouth is ventrally placed and surrounded by about 20 large, shield-shaped tentacles with thick stalks.

The body wall ossicles (Figure 187G–I) are tables and C-shaped elements. The tables have a small disk that is perforated with 10–12 marginal holes and four central holes, and a tall spire surmounted by numerous (20–24) small spines.

I. badionotus can vary dramatically in color; hues of orange, yellow, red, brown, or purple are common. The conspicuous dorsal and lateral warts are frequently darker than the surrounding body wall, producing a "chocolate chip" appearance. However, some animals have a dark body coloration and lighter-colored warts, and uniformly pigmented individuals have been observed. Very young specimens may lack pigmentation.

HABITAT: Seagrass beds of *Thalassia* and *Syringodium,* and sandy bottoms with algae. Off eastern Florida, it has been collected on shallow-water coquinoid ledges and deep-water *Oculina* coral reefs (Miller, previously unpublished).

FIGURE 156. *Isostichopus badionotus*.

DISTRIBUTION: Bermuda, South Carolina, Florida, the Bahama Islands, Texas, Mexico, Belize, Panama, Colombia, Venezuela, and many Caribbean Islands, as far south as Brazil. Also in the mid-Atlantic at Ascension Island and in the Gulf of Guinea off western Africa. DEPTH: Low-tide mark to 65 m (213 ft).

BIOLOGY: This is one of the most common shallow-water sea cucumbers in the western Atlantic, and in some areas populations are fairly dense. Adults live fully exposed on mud, sand, or rocks; juveniles attach to the underside of rubble or coral slabs. As noted above, the individuals can secrete mucus. They also may slough off the outer layers of the body wall and may autoeviscerate if roughly handled or stressed by confinement in standing water.

Adults remain fully exposed at all times. They are sluggish, traveling about 0.5 m per day, and completely refill the gut three to four times during their daily peregrinations (Hammond 1982a, 1983). The feeding activity begins in the afternoon and peaks before midnight. A population of *I. badionotus* in Harrington Sound, Bermuda, was estimated to ingest 450–900 metric tons of sediment annually (Crozier 1918). Sloan and von Bodungen (1980) reported that *I. badionotus* is a nonselective feeder, and because of its indiscriminate feeding habits it may take up PCB and petroleum hydrocarbon pollutants (Burns et al. 1990). However, Hammond (1983) reported organic carbon and nitrogen, ATP, and total plant pigment concentrations twice as high in foregut contents as in surrounding sediments, suggesting selective feeding. Much of the assimilated carbon is detrital in origin, and examination of the size of available sand grains, gut contents, and podia morphology suggests that selection apparently is based on something other than grain size (Hammond 1982b). Curiously, meiofaunal species (mostly mollusks and foraminiferans) may be ingested selectively; they are killed but not digested or assimilated (Hammond 1983; Sambrano et al. 1990). This holothuroid appears capable of using extracellular polymer from bacteria as a source of nutrient (Baird and Thistle 1986).

The environment of the foregut is slightly acidic, and there is evidence of some dissolution of carbonate sediments during passage through the gut (Hammond 1981). The feces may have the same organic content as the surrounding sediment (Hammond 1983); they are relatively rapidly dispersed and may enrich the surrounding benthos (Sloan and von Bodungen 1980; Conde et al. 1991). In addition to studies of its feeding behavior, the species has been examined for the presence of neurotransmitters involved in the contractile activity of the holothuroid's intestine (García-Arrarás et al. 1991); numerous studies of cell physiology using tissues from this species are not reviewed here.

I. badionotus is host to a pearlfish, *Carapus bermudensis*. Another commensal, the bumblebee shrimp, *Gnathophyllum americanum*, was found associated with a population of *I. badionotus* in Hobe Sound, Florida (Miller, previously unpublished). Tikasingh (1963) reported that a parasitic gastropod, *Balcis* sp., occurs near the mouth or on the dorsal surface of some individuals found off the Netherlands Antilles. In Bermuda, most individuals are infested by the minute rhabdocoel flatworms *Macrogynium ovalis* Meserve and *Wahlia macrostylifera* Westblad; large numbers may occur in the host's digestive tract and body cavity, but appear to cause the holothuroid no harm (Snyder 1980).

SELECTED REFERENCES: Selenka 1867: 316, pl. 18, fig. 26 [as *Stichopus badionotus*]; Zeiller 1974:115; Miller and Pawson 1984:54, fig. 44, 45; Pawson 1986:540, pls. 178, 14.10.

FAMILY HOLOTHURIIDAE

Actinopyga agassizi (Selenka)
Figures 157, 180G,H,I

DESCRIPTION: Among the largest species of sea cucumbers in the Florida Keys, adult specimens reach 35 cm (14 in) in length. The rounded upper surface of the body is rigid and covered with numerous wartlike papillae of various sizes. It differs markedly from the flattened ventral surface, which is soft and carries broad rows of tube feet tipped with suckers. The body wall is thick, leathery, and usually partially contracted when the species is encountered in its natural habitat. The ventrally positioned mouth is surrounded by 20–30 large, peltate tentacles, which often are extended and prominently visible. Perhaps the most distinctive characteristic of *A. agassizi* is the presence of five conspicuous, white, calcareous teeth surrounding the anus.

The body wall ossicles (Figure 180G–I) are rosettelike elements, which vary from simple "dog biscuit" shapes to complex rods with dichotomously branched ends.

The coloration of *A. agassizi* can be quite variable, but individuals are most commonly mottled or variegated with brown, orange, and yellow. The tube feet and tentacles are orange and yellow or white.

HABITAT: Coral reef, rocky areas, and seagrass beds. In the Bahama Islands it is common in lagoons (Deichmann 1957), living on sand bottoms with turtle grass.

DISTRIBUTION: Bermuda, Florida, the Florida Keys and the Dry Tortugas, the Bahama Islands, Cuba, Belize, Hispaniola, Jamaica, and Barbados. DEPTH: 0–54 m (0–177 ft).

BIOLOGY: This species is active at night, emerging from its daytime shelters, such as coral heads, to forage over the reef. Some individuals return to the same shelter day after day, but others are only lightly concealed, wrapping themselves with blades of seagrass or coral rubble (Hammond 1982a). Tagged individuals may move as much as 100 m in a week (Smith et al. 1981). During daylight hours, *A. agassizi* can usually be located by the presence of fecal casts piled near hiding places in reef structures and rubble. Feeding usually occurs on algal turf and pavement or rubble bottoms, and appears not to involve the selection of food particles based on size (Hammond 1982a,b). Individuals in the Bahama Islands spawn in July and August (Edwards 1889).

When *A. agassizi* is placed in a bucket of seawater, the collector is often amazed to find, some time later, one or two slender, silver and blue, eel-like fish swimming about in the container. A 10- to 15-cm-long pearlfish, *Carapus bermudensis,* is a commensal that lives in the posterior portion of the intestinal tract or respiratory tree. Up to 10 have been reported from a single holothuroid (Smith et al. 1981). They emerge from the host at night to feed, primarily on shrimps (Smith et al. 1981; there is a photograph of an emerging fish in Markle and Olney [1990: pl. 2]). Trott (1981:627) described the fish's means of ingress as follows: "before entry, the host will contract its anus on tactile stimulus by the fish. It is easy to anthropomorphically relate to the holothurian—it must respire—consequently, on expulsion of water from the anus a persistent fish will immediately enter the host." The symbiotic fish has no known effect on the host, but a parasitic interaction remains a possibility (Trott 1981).

The presence of toxins in the tissue and mucus of *A. agassizi* may protect the host and its commensals from predators (Mosher 1956; Nigrelli and Jakowska 1960). Its evisceration behavior is triggered by elevated

FIGURE 157. *Actinopyga agassizi.*

temperatures and certain metallic ions (Mosher 1956). Hyman (1955:163) noted that the Cuvierian tubules, which contain toxins, are ejected "on proper stimulation." However, the tubules of several *Actinopyga* species, differing structurally from those of other holothuroids, cannot elongate or become adhesive and are not expelled. Therefore, it has been questioned whether they serve as defensive structures in this genus (VandenSpiegel and Jangoux 1992, 1993).

SELECTED REFERENCES: Selenka 1867: 311, pl. 17, figs. 10–12 [as *Mülleria Agassizii*]; Voss 1976:145; Pawson 1986:540, pl. 178.

Holothuria (Cystipus) cubana Ludwig
Figures 158, 183A,B,C,D

DESCRIPTION: This is a relatively small species; mature individuals reach a maximum length of approximately 15 cm (6 in); most are 8–10 cm (3–4 in) long. The body is somewhat flattened and bluntly rounded at the ends. The mouth is ventral, surrounded by 20 small tentacles. Because of the presence of numerous skeletal ossicles, the parchmentlike skin is rigid and rough, like fine sandpaper. The tube feet are small and scattered, and most numerous on the ventral surface, where they are often retracted into pits.

The body wall ossicles (Figure 183A–D) are tables with knobbed disks and complex spires, and heavily knobbed buttons with perforations obscured by knobs.

A thin layer of sand grains usually coats the body wall, obscuring the distinctive underlying coloration. When the sediment particles are brushed away, the upper surface is found to be brownish or grayish white, with two longitudinal rows of light to dark brown blotches. There are 6–12 blotches in each row, and often in the center of each blotch lies a small white wart supporting a tiny pa-

FIGURE 158. *Holothuria (Cystipus) cubana.*

pilla. However, in some individuals the blotches are inconspicuous. The ventral surface of the body is dirty white or pale olive-yellow. Large individuals have a decidedly yellow tinge to the body wall.

HABITAT: A wide range of sediments, in the intertidal and shallow subtidal zones.

DISTRIBUTION: Bermuda, Cuba, Puerto Rico, Antigua, Barbados, Curaçao, and Venezuela (Martínez de Rodríguez 1973). Previously unreported localities include the Bahama Islands, the Dry Tortugas, and Florida (Biscayne Bay, the Indian River near Fort Pierce Inlet, and south of St. Lucie Inlet, Stuart) (Miller, previously unpublished). DEPTH: To 7 m (23 ft).

BIOLOGY: *H. cubana* is a cryptic species that clings to the underside of rocks, sponges, or coral rubble or burrows in the sand beneath rocks. It is frequently overlooked because individuals are camouflaged by a fine layer of sand coating the body wall. In the vicinity of Carrie Bow Cay, Belize, this species and two congeners, *H. arenicola* and *H. impatiens,* occurred together under the same piece of calcareous rock (Miller, previously unpublished).

SELECTED REFERENCES: Ludwig 1875: 104, pl. 7, fig. 34 [as *Holothuria cubana*]; Deichmann 1954:394, fig. 66:24–28 [as *Holothuria cubana*]; Rowe 1969:156; Pawson 1986:538, pl. 178 [as *Holothuria cubana*].

Holothuria (Halodeima) floridana
Pourtalès
Figures 159, 183E,F,G

DESCRIPTION: A species of moderate size, *H. floridana* reaches a maximum length of about 25 cm (10 in). The slender and cylindrical body gently tapers near both ends. The mouth is almost ventral, surrounded by 20 peltate tentacles with small side branches. Scattered on the ventral surface are numerous cylindrical tube feet that cannot be completely retracted. The body wall is thin and smooth, but the short cylindrical or pointed feet of the dorsal side are set on low conical warts; they are far less numerous than the ventral tube feet. Warts are most prominent at the anterior and posterior ends and along the lateral margins of the body; they are most pronounced on small individuals.

The body wall ossicles (Figure 183E–G) are tables and rosettes. The tables, far less numerous than the rosettes, have a tall spire composed of four pillars, each pillar ending in three pointed teeth (one vertical, two horizontal). The tiny rosettes are concentrated in heaps that are visible to the unaided eye as small light-colored dots surrounding the bases of the tube feet.

The coloration is highly variable. Some

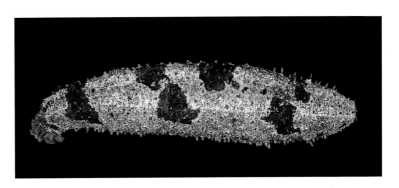

FIGURE 159. *Holothuria (Halodeima) floridana.*

individuals are a fairly uniform light to dark yellow, brown, gray, or red, but others are mottled, with small to large blotches of yellow, white, brown, or black pigment on the upper surface. Generally, small individuals are mottled. On the ventral surface, the tips of the tube feet are yellow, and there may be three parallel stripes of dark pigment running the entire length of the animal between the bands of feet. The tentacles, which are often exposed in living and preserved specimens, vary from translucent to black with yellow fronds; frequently their stalks are flecked with brown or black.

HABITAT: Turtle grass beds, tide pools, or mangrove swamps; near the shoreline.

DISTRIBUTION: Yucatán Peninsula southward along the coast of Central America to Colombia, also the northern Bahama Islands, Cuba, Jamaica, and Aruba. Around Florida, from Stuart on the east coast, to Florida Bay (Miller, previously unpublished); especially common in Biscayne Bay, the Dry Tortugas, and along the Gulf side of the Florida Keys. DEPTH: Less than 1–1.5 m (less than 3–5 ft).

BIOLOGY: In suitable habitats this species may be very abundant. Individuals sometimes occur in the open, but are frequently "hidden" under stationary clumps of drift algae (e.g., *Laurencia* spp.). Small specimens up to 10–15 cm long may be found in large numbers beneath coral rubble or clinging to branches of the coralline alga *Halimeda opuntia*.

The reproductive cycle is seasonal, with a spawning period in late summer in southern Florida and the Bahama Islands (Edwards 1909; Engstrom 1980c). According to Edwards (1909), the species lacks an auricularia larva. The advanced larval stage that emerges from the fertilization membrane after 6 days of development has five tentacles around the mouth and a posterior tube foot. In the larva, the cloaca begins to contract rhythmically before the respiratory tree has developed; in the adult, cloacal contractions fill the respiratory trees with oxygenated water. The juveniles move about differently than adults, relying on their tentacles for locomotion (Edwards 1909).

A symbiotic pearlfish, *Carapus bermudensis*, was found in an individual of *H. floridana* collected from a seagrass bed south of Cape Florida, the first record of an association between the two species (Miller, previously unpublished).

REMARKS: In the Florida Keys, where the distributional ranges of *H. floridana* and *H. mexicana* overlap, these two species produce hybrids with intermediate characteristics (Engstrom 1980b,c). The existence of hybrids, and intraspecific variability, make the two species difficult to identify. However, the body of *H. floridana* usually is slender and gently tapered toward the ends. The body wall is relatively thin and flexible, and the upper surface carries several warts and numerous tube feet, but there is no marked distinction in coloration between the upper and lower surfaces. In contrast, *H. mexicana* has a robust body that tapers little and terminates in bluntly rounded ends. The body wall is thick and extremely rigid, and dorsal warts and tube feet are few and often inconspicuous. In addition, the dark-colored upper surface contrasts conspicuously with the lower surface, which is frequently bright red, orange, yellow, tan, or white. *H. mexicana* is generally found at somewhat greater depths than *H. floridana*, most individuals of the former occurring between 2 and 10 m (7–33 ft).

SELECTED REFERENCES: Pourtalès 1851: 8 [as *Holothuria floridana*]; Deichmann 1930:72, pl. 5, figs. 5–9 [as *Holothuria flor-*

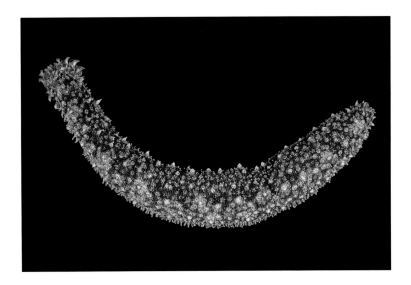

FIGURE 160. *Holothuria (Halodeima) grisea.*

idana]; Caso 1955:505, pl. 2, figs. 1–29 [as *Holothuria floridana*]; Rowe 1969:138.

Holothuria (Halodeima) grisea
Selenka
Figures 160, 184*A,B,C,D,E,F,G,H*

DESCRIPTION: Individuals of this species can reach a length of 25 cm (10 in). The body is subcylindrical, with a distinctly flattened sole covered with numerous cylindrical tube feet. The upper body surface carries six rows (four in young animals) of papillae, borne upon very large warts, each wart surrounded by 5–10 small tube feet. The mouth is directed slightly downward, with 20–25 bushy peltate tentacles.

The striking, harlequin colors of this species are a helpful distinguishing feature. The ground color of living animals, red or yellowish red, contrasts with brown mottling and with white papillae with yellow tips on the upper body. The feet on the sole are yellow-tipped, and the tentacles are yellow.

The body wall ossicles, tables with about 12 marginal spines on the disk, are scattered (Figure 184*A–D*). There is also an inner layer of plates (Figure 184*E–H*), with two or four central holes and some smaller peripheral holes, and the margins of the plates are equipped with blunt teeth.

HABITAT: Generally on seagrass flats, but in Florida it is commonly associated with *Phragmatopoma lapidosa* Kinberg worm reefs, and on sandy bottoms.

DISTRIBUTION: Florida (but not reported from the Florida Keys), the Bahama Islands, Puerto Rico, Jamaica, the Lesser Antilles to Curaçao, Panama, Colombia, southern Brazil, and West Africa. Although reported for coastal Texas, it may not be a year-round inhabitant there (Pomory 1989). DEPTH: Less than 5 m (less than 16 ft).

BIOLOGY: Its numerous tube feet give *H. grisea* a tenacious grip on hard substrates, and it cannot be pulled away from aquarium walls without damaging the feet. On the southeast coast of Florida, large numbers of these sea cucumbers are often associated with *Phragmatopoma lapidosa* worm reefs, at least during the summer months. At one locality (Fort Walton rocks, near Fort Pierce), individuals disappear from the reefs during the winter, possibly migrating some distance offshore. In some localities, individuals are numerous in shallow lagoons, especially where marine grasses are abundant. During extreme low tides, these areas may be almost desiccated and the temperatures can exceed 35°C, but the holothuroids somehow are capable of tolerating the harsh conditions.

A preliminary study of ovarian ultrastructure and oogenesis in this species revealed features typical of several other holothuroids that have been similarly examined (Eckelbarger and Young 1992). However, the oocytes of *H. grisea* have highly branched microvilli and endocytotic vesicles, which may be involved in the uptake of nutrients during yolk production. Its pattern of coelom formation during early development has also been studied, with surprising results (Balser et al. 1993). Although holothuroid larvae were thought to lack an axocoel, the auricularia larva of this and other species was found to possess an axocoel that is lined with mesothelial podocytes and is connected, by a duct, to a pore on the dorsal surface of the larva.

SELECTED REFERENCES: Selenka 1867: 328, pl. 18, figs. 55, 56; Deichmann 1930: 77, pl. 5, figs. 1–4; 1957:11; Caycedo 1978: 168, pl. 6.

Holothuria (Halodeima) mexicana
Ludwig
Figures 161, 162, 183*H,I,J,K*

DESCRIPTION: This is a large species, reaching a length of 30–50 cm (12–20 in). The body is thick-walled and extremely rigid, and bluntly rounded on both ends. In adults, a distinctive folding or wrinkling of the dorsal and lateral integument gives the body a pleated appearance. There are very few warts or tube feet on the smooth upper surface. The tube feet typically hold small pieces of algae or detritus against the upper surface of the body. The lower surface is flat and covered with numerous, nearly equally spaced, cylindrical tube feet. They are completely retractile and are often marked by sunken pits in the skin. The mouth is ventral, surrounded by 20–22 broad shield-shaped tentacles.

The body wall ossicles (Figure 183*H–K*) are tables and small perforated plates. The tables are very similar to those of *H. floridana*, with smooth, circular margins and tall spires composed of four pillars. The numerous tiny, flat plates are subcircular and have approximately 15–20 irregular perforations; they are not collected in distinct heaps at the base of the tube feet, as are the rosettes of *H. floridana*.

The dorsal surface is usually dark gray, brown, or black in adults and brownish yellow in young individuals. The ventral surface is usually reddish pink, yellowish orange, or white (Figures 161, 162); often there is a bright pink or red pigment stripe running along the midline and bordered on either side by white stripes. The tube feet are brown with darker brown or black tips.

HABITAT: Seagrass beds, mangroves, lagoons, reef hard bottoms and sand plains, and rock terraces.

DISTRIBUTION: The Florida Keys, the Bahama Islands, Cuba, Jamaica, Puerto Rico, St. Martin, Antigua, Barbados, Tobago, Curaçao, Aruba, Bonaire, Venezuela, and islands off Colombia, Belize, and the Yucatán. DEPTH: 0.5–20 m (1.5–65 ft), but usually between 2 and 10 m (7–33 ft).

BIOLOGY: This sluggish species shows little or no response to disturbance and little cryptic behavior. It is conspicuous in the open areas around coral reefs, often with its large,

SEA CUCUMBERS

FIGURE 161. *Holothuria (Halodeima) mexicana*: with whitish ventral surface.

FIGURE 162. *Holothuria (Halodeima) mexicana*: with reddish ventral surface.

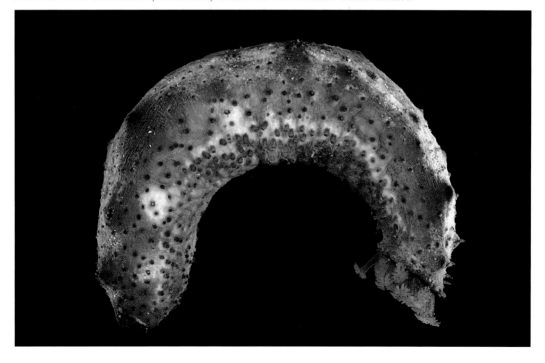

cylindrical fecal strands nearby; the relatively slow breakdown of the feces has been examined (Conde et al. 1991). Individuals generally move less than a meter per day. Their feeding activity is cyclical, beginning in late afternoon and peaking before midnight, sufficient time to refill the gut three to four times (Hammond 1982a). The environment of the foregut is slightly acidic, and there is evidence of some dissolution of carbonate sediment during its passage through the digestive tract (Hammond 1981).

According to Hammond (1983), *H. mexicana* selectively ingests nutrient-rich particles of sediment. Organic carbon and nitrogen, ATP, and total plant pigment are twice as high in its foregut contents as in samples of the surrounding sediment. Examination of size of available sand grains, gut contents, and podia morphology suggests that selection is based on something other than grain size (Hammond 1982b). Much of the assimilated carbon is detrital in origin; ingested meiofauna is killed but not digested or assimilated. Peptides that are similar to those in the mammalian digestive system have been identified from this species, and experiments indicate that the peptides act as neurotransmitters, involved in the contractile activity of the holothuroid's intestine (García-Arrarás et al. 1991).

Unlike *H. floridana*, which can be found in great numbers near the shoreline, *H. mexicana* is usually encountered as solitary individuals on offshore reef habitats. Juveniles of *H. mexicana* appear to be rare, whereas juveniles of *H. floridana* are quite easy to find under coral rocks and rubble. The two species can occur together, both spawn in late summer in the Florida Keys, and hybrid individuals are not uncommon there (Engstrom 1980b,c). Midday spawning of *H. mexicana* has been observed at Curaçao, in September and October (van Veghel 1993). Mosher (1982) found that in both males and females, the release of gametes is preceded by elevation of the anterior one-third to two-thirds of the body off the seafloor, sometimes accompanied by gentle undulatory movements or writhing. Intermittent shedding of gametes, which can be expelled in a stream 10 cm high, can last up to 90 minutes. The species has a pelagic larva that was reared by Lacalli (1988). The auricularia larva has a single, sinuous ciliary band and transforms, in 8 hours, into a doliolaria stage with five circular bands.

Tikasingh (1963) found that some *H. mexicana* from Bonaire harbor commensal pearlfish in their intestine or respiratory tree, the fish later identified as *Carapus bermudensis* (Smith et al. 1981; Trott 1981). He also noted some specimens parasitized by a tiny gastropod, *Balcis* sp.; another parasitic eulimid snail, *Megadenus holothuricola* Rosén, has been reported (Rosén 1910).

REMARKS: Kaplan (1982) referred to *H. mexicana* as the most common sea cucumber in the Caribbean. Although it is found in the Florida Keys, it is not nearly as common in that area as *H. floridana*. Characteristics helpful in distinguishing these closely related species are enumerated above, under *H. floridana*. *Isostichopus badionotus* is sometimes confused with *H. mexicana*, but it is readily distinguished by examining the ventral body surface. In the latter species, the tube feet are scattered on the body wall; in the former, they are arranged in three distinct longitudinal bands, each band with a number of rows in adults.

SELECTED REFERENCES: Ludwig 1875: 101, pl. 7, fig. 47 [as *Holothuria mexicana*]; Caso 1955:501, pl. 1 [as *Holothuria mexicana*]; Rowe 1969:138; Sefton and Webster 1986:99, pl. 170 [as *Holothuria mexicana*].

SEA CUCUMBERS

Holothuria (Platyperona) parvula
(Selenka)
Figures 163, 185A,B,C

DESCRIPTION: The largest specimens of this small species are only 7–10 cm (3–4 in) long. The body is elongate and of uniform diameter throughout its length. On the upper and lateral surfaces there are several prominent, conical warts, each tipped with a papilla. The lower surface is flattened, covered by numerous long, thick tube feet that are incapable of complete contraction. The body wall is thin and leathery, and it is slightly gritty because of the numerous ossicles embedded in the skin. The mouth is ventral, surrounded by 18–20 peltate tentacles with long stalks. Around the whorl of tentacles is a narrow collar of conical warts that resemble those of the upper body surface.

The body wall ossicles (Figure 185A–C) are tables and buttons. The undulating margin of the tables is perforated by eight large holes, and the spire is composed of four pillars that terminate in numerous short, conical spines. The buttons are irregular in outline and perforated by two parallel rows of elongate holes. Often there are more perforations in one row than in the other.

The coloration of this species is fairly consistent throughout its range. The body wall is golden, yellow, or bright greenish brown, and the ventral tube feet and tentacle fronds are yellow. Specimens of *H. parvula* liberate an unidentified green pigment when they are stored in ethanol, but the color of preserved and living specimens is similar.

HABITAT: On shallow reefs, in tide pools, and on reef flats, near the low-tide mark.

FIGURE 163. *Holothuria (Platyperona) parvula.*

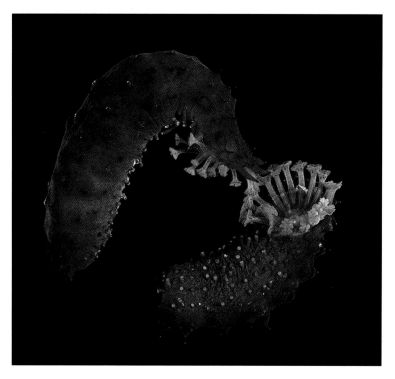

DISTRIBUTION: From Bermuda to Brazil, including the Florida Keys, the Dry Tortugas, the Bahama Islands, Jamaica, Puerto Rico, the Virgin Islands, Anguilla, St. Martin, Antigua, Barbados, Tobago, Aruba, Curaçao, Venezuela, and Colombia. In Florida, it occurs south of St. Lucie Inlet, Stuart (Miller, previously unpublished), and a record exists for the southwestern Gulf of Mexico off Veracruz, Mexico (Henkel 1982). DEPTH: Low-tide mark to 4 m (13 ft).

BIOLOGY: Individuals cling tightly to the undersurface of rocks and rubble, using their powerful tube feet; often, several occur under one shelter. This species is reported to eviscerate but only under extraordinary provocation (Kille 1937). However, it quickly expels copious quantities of white Cuvierian tubules via its anus when disturbed. So readily does it liberate these defensive organs that it has commonly been referred to as the "cotton spinner" sea cucumber. H. L. Clark (1901a:257) reported for this species "Cuvier's organs very noticeable; in one specimen their bulk fully one-tenth of the whole animal." Although tubule discharge, elongation, and adhesion have not been studied for *H. parvula*, the processes have been elucidated for *Holothuria forskali* (Delle Chiaje) (VandenSpiegel and Jangoux 1987).

Crozier (1917), Deichmann (1922 [as *Actinopyga parvula*]), and Kille (1937) were the first to report on the ability of *H. parvula* to reproduce by asexual fission. Emson and Mladenov (1987a) concluded that fission occurs most often in the summer months in Bermuda. They determined that the sea cucumber splits into two nearly equal parts midway between the mouth and anus and found that many asexually reproducing individuals are capable of feeding within 2 months of fission. Complete regeneration of the viscera occurs within 1 year; thus fission could be an annual process for *H. parvula*.

REMARKS: A similar-looking species, *H. surinamensis*, is easily confused with *H. parvula*. Both sea cucumbers have the ability to reproduce by fission, both have an ethanol-soluble green pigment in their body walls, both occur throughout the same geographic range, and both occupy the same habitats (sometimes co-occurring). However, their body wall ossicles are diagnostic. Ossicles in the shape of buttons are abundant in *H. parvula* but lacking in *H. surinamensis*, and there are clear differences in the shape of the tables from these species (compare Figure 185*A* and *E*). In addition, Cuvierian tubules are copious in *H. parvula* but lacking or poorly developed in *H. surinamensis*.

Another similar species, *H. rowei*, is known from a single specimen (see below).

SELECTED REFERENCES: Selenka 1867: 314, pl. 17, figs. 17, 18 [as *Mülleria parvula*]; Deichmann 1930:70, pl. 4, figs. 14–22 [as *Holothuria parvula*]; Rowe, 1969:145; Pawson 1986:538, pl. 178 [as *Holothuria parvula*].

Holothuria (Platyperona) rowei
Pawson & Gust
Figure 164

DESCRIPTION: This appears to be a small species (3 cm [1 in]), based on the dimensions of the single known specimen. It resembles *H. parvula* in its general body plan but has distinctively shaped body wall ossicles (Figure 164). The tables are complex, with approximately 20 peripheral holes. The buttons have thickened rims with a distinctive scalloped inner margin, and they usually carry small knobs along the midline.

According to Pawson and Gust (1981), a preserved specimen of *H. rowei* is a variegated light brown.

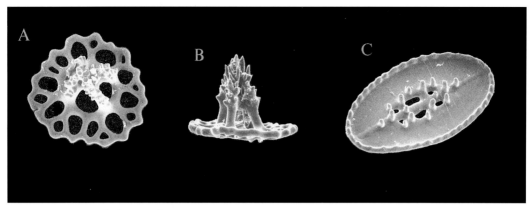

FIGURE 164. *Holothuria (Platyperona) rowei*, body wall ossicles: A, table, dorsal view; B, same, lateral view; C, button. Photos by D. Pawson.

HABITAT: A single specimen was collected from "outer reefs" off Key Largo, Florida.

DISTRIBUTION: Presently known only from Key Largo, Florida. DEPTH: Probably taken at a depth between 3 and 20 m (10–66 ft) because it was found in association with reefs at that locality.

SELECTED REFERENCES: Pawson and Gust 1981:873, fig. 1.

Holothuria (Selenkothuria) glaberrima Selenka
Figures 165, 184I,J,K

DESCRIPTION: This is a small *Holothuria*, reaching a maximum length of about 15 cm (6 in). The body is cylindrical and soft-skinned, and the ventral surface has a distinct sole covered with numerous cylindrical tube feet. The dorsal papillae are small and inconspicuous. The tentacles are large and dendritic.

There are no tables in the body wall; all the ossicles are straight or curved rods, and branches at the ends (or sometimes along the shaft) may anastomose to form a meshwork (Figure 184I–K). The tube feet contain well-developed endplates.

Color of the body in life ranges from almost black to faded brown, with contrastingly darker tentacles.

HABITAT: Typically in the wave-washed surf zone, but individuals may burrow in sand (Caso 1968a).

DISTRIBUTION: The Bahama Islands southward, throughout the West Indian islands to Trinidad, and from Panama and Mexico. DEPTH: Low-tide mark to 42 m (138 ft).

BIOLOGY: Large numbers of individuals may be encountered clinging firmly to the outer surfaces of rocks and lodged in crevices (Deichmann 1930, 1957; Caso 1968a). Their richly branched, dendritic tentacles, a characteristic atypical in the genus *Holothuria*, are periodically extended into swirling waters to capture plankton (Deichmann 1957). The species breeds during June and July in Barbados. The bright pink eggs give rise to an auricularia larva in 3 days; metamorphosis occurs approximately 2 weeks after fertilization (Lewis 1960). The species shows little ability

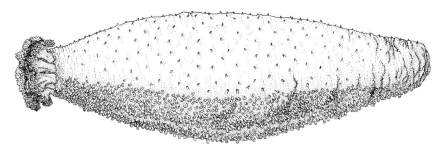

FIGURE 165. *Holothuria (Selenkothuria) glaberrima*. Illustration by C. Messing.

to osmoregulate, and its rate of respiration varies directly with temperature and the concentration of dissolved oxygen in the surrounding water (Sisak and Sander 1985). Neuropeptides from *H. glaberrima* have been characterized; they may be involved in its digestive physiology (García-Arrarás et al. 1991; Díaz-Miranda et al. 1992).

At Veracruz, Mexico, the ectoparasitic snail *Balcis intermedia* (Cantraine) attaches itself to the body of this holothurian (Caso 1968b). The pearlfish *Carapus bermudensis* lives in the respiratory trees of *H. glaberrima* in the Bahama Islands (Smith et al. 1981).

SELECTED REFERENCES: Selenka 1867: 328, pl. 18, figs. 57, 58 [as *Holothuria glaberrima*]; Deichmann 1930:69, pl. 4, figs. 10–13 [as *Holothuria glaberrima*]; Caycedo 1978:163, pl. 3a–c.

Holothuria (Semperothuria) surinamensis Ludwig
Figures 166, 185D,E,F

DESCRIPTION: This is a medium-sized species, reaching a length of approximately 20 cm (8 in). The body tapers gradually toward the anterior end but not toward the bluntly rounded posterior end. The body wall is thin and flexible, and it has a slightly roughened texture. Tube feet are not particularly abundant in this species. Dorsally there are some small, pointed papillae; ventrally they are more numerous, large, cylindrical, and distributed along the radii. The mouth is at the extreme anterior end of the body and is surrounded by 10–20 tentacles that are short and equal in size and resemble stalks of cauliflower. They in turn are encircled by a collar of tube feet. The variable number of tentacles in this species results from its asexual fission.

The body wall ossicles (Figure 185D–F) are tables, in many of which the disk is greatly reduced to a simple X. In others, there is a more typical disk having one large central perforation, zero to eight peripheral holes, and 9–15 marginal teeth. The table spires are composed of four long pillars, each carrying several short teeth at the tip or from the midpoint to the tip. Unlike other *Holothuria* species in the Florida Keys, *H. surinamensis* lacks a layer of buttons, small perforated plates, or rosettes. However, their absence in this species is not unique; for example, they are also missing in *H. glaberrima*.

H. surinamensis tends to cover itself with a thin layer of fine sediment that conceals its color pattern. The species has been described as yellowish brown, reddish brown, purplish brown, chocolate brown, olive, or black. Usually there is a double series of darker spots running along the length of the dorsal surface, each spot centered on a papilla. The ventral surface is uniformly brown and paler than the dorsal surface. The feet are inconspicuous, or, in very dark individuals, the ventral feet may be conspicuously white. The light brownish green tentacles are flecked with numerous, tiny, dark brown spots.

HABITAT: Usually near the shore, and on shallow flats in lagoons or protected bays. It occurs in finger coral (*Porites* sp.) and beneath rocks and rubble on coral reefs, in seagrass beds, and associated with mangrove roots or coralline algae, especially *Penicillus* sp. In deeper environments, it occurs in *Oculina varicosa* Lesueur coral at a depth of 42 m (138 ft) off the central east coast of Florida; on the west coast off Tampa Bay, it has been taken by trawl from a sand and shell bottom at 18 m (59 ft) (Miller and Pawson 1984).

DISTRIBUTION: Bermuda, Jamaica, Cuba, Puerto Rico, some of the Lesser Antilles, the eastern and western Gulf of Mexico, Colombia, Venezuela, Surinam, and southward to Brazil. In addition to the Floridian records above (see *Habitat*), it is reported from Biscayne Bay, off Miami (Deichmann 1939). One specimen has been found off Grand Bahama Island (Miller, previously unpublished). DEPTH: Low-tide mark to 42 m (138 ft).

BIOLOGY: During the day, individuals remain concealed among the branches of corals, under rocks and rubble, in seagrass mats, or below the surface of soft sediment. Crozier (1915) found that *H. surinamensis* is nocturnal, lying exposed on the bottom at night and slowly disappearing into the silt after sunrise. He investigated its reactions to mechanical, chemical, and photic stimuli and reported on its feeding, respiratory movements, locomotion, evisceration, and regeneration. Crozier (1915) found that this species shows no signs of stress in response to diurnal temperature fluctuations between 20 and 31°C. Fisher stated that *H. surinamensis* found off Antigua remained healthy and active, though sluggish, in extremely shallow water where daytime temperatures were "very warm" (Deichmann 1926). The stomach contents of the species consist of amorphous material and sediment (Miller and Pawson 1984).

Crozier (1915, 1917) studied asexual reproduction by fission in *H. surinamensis* and determined that 11% of the individuals sampled were regenerating after fission. They could be recognized by the light color of the anterior or posterior end of the body. He found that regenerating individuals were half as long as those not regenerating, and that they had as few as nine tentacles instead of the complete series of 20.

FIGURE 166. *Holothuria (Semperothuria) surinamensis.*

It has been reported that some *H. surinamensis* from Antigua harbored two species of pearlfish (Fisher, in Deichmann [1926]), and the same association has been noted for a specimen from Fort Pierce, Florida (Miller, previously unpublished). Like many other *Holothuria* species, *H. surinamensis* may eviscerate when left in stagnant water. The internal organs are expelled through the cloaca, the mouth, or through a weakened section of the body wall; afterward, regeneration of the viscera must draw on reserves of nutrient in the body wall (Crozier 1915).

REMARKS: *H. surinamensis* is strikingly similar to *H. parvula;* see *Remarks* under the latter species.

SELECTED REFERENCES: Ludwig 1875: 111, pl. 7, fig. 27 [as *Holothuria surinamensis*]; Miller and Pawson 1984:57, figs. 46, 47; Pawson 1986:540, pl. 178 [as *Holothuria surinamensis*].

Holothuria (Theelothuria) princeps
Selenka
Figures 167, 185G,H,I,J

DESCRIPTION: Adults of this species are medium-sized, reaching a length of 30 cm (12 in). The body is subcylindrical, arched above, flattened below, and distinctly tapered, more so at the posterior end than toward the mouth. The body wall is relatively thin (except in strongly contracted, preserved specimens), yet rigid because of the quantity of embedded skeletal ossicles. On the upper surface, there are conspicuous warts of various sizes, each carrying a small apical papilla. On the lower surface, the tube feet are regularly spaced, and they are usually fully contracted, forming pits in the body wall. The mouth is subterminal, situated on the ventral surface and surrounded by 20 very small tentacles.

In this species, the ossicles (Figure 185G–J) are buttons and two types of tables. The buttons (Figure 185J) are quite complex and irregular. Many are twisted, incomplete, and knobbed along the midline or margin; most are perforated with two parallel rows of three to five holes. The tables (Figure 185H,I) in the body wall have an undulating or dentate margin and a very short spire terminating in few to several short spines. Unusual tacklike tables (Figure 185G) in the tube feet have a thick, perforated disk and an enormous solid spire tapering to an acute point. They are arranged in whorls around the feet with the tips penetrating the integument and are large enough to be seen with a low-power magnifying glass.

A dramatic color pattern distinguishes *H. princeps*. The ground coloration is mottled brown, tan, and white, and clearcut concentric rings of lighter and darker pigment surround the base of the dorsal warts. In addition, on the upper surface, there are two longitudinal rows of dark brown or black blotches of varying size, about 12 blotches per row. The tentacles are white.

HABITAT: Sandy mud and shell substrates, and seagrass *(Thalassia, Syringodium)* beds.

DISTRIBUTION: Florida (Jacksonville to Pensacola), the Florida Keys, the Bahama Islands, the Dominican Republic, Puerto Rico, Mexico, Colombia, and Venezuela. *H. princeps* has the widest depth range of any sea cucumber species covered in this handbook—from the low-tide mark to 229 m (751 ft).

BIOLOGY: This burrowing species is sometimes found in abundance. Following a severe storm in the northern Gulf of Mexico off Florida, Wells and Wells (1961:267) reported "thousands of sea cucumbers in windrows on the outer beach near Fort Walton." It appears to ingest sand, fragments of calcareous or-

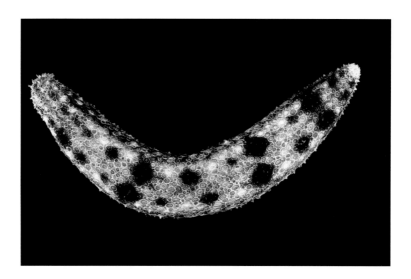

FIGURE 167. *Holothuria (Theelothuria) princeps.*

ganisms, and some amorphous material (Miller and Pawson 1984). Dissections of several hundred specimens revealed the presence of a previously undescribed species of commensal crab, *Pinnaxodes floridensis* Wells & Wells, in the sea cucumber's cloaca and respiratory trees. Infestation rates as high as 61% were noted. The species may also harbor *Carapus bermudensis,* a pearlfish (Haburay et al. 1974).

SELECTED REFERENCES: Selenka 1867: 332, pl. 18, figs. 67–69 [as *Holothuria princeps*]; Deichmann 1930:58, pl. 2, figs 1–8 [as *Holothuria princeps*]; Martínez de Rodríguez and Mago Herminson 1975:193, pl. 4, figs. 1–3 [as *Theelothuria princeps*]; Miller and Pawson 1984:60, figs. 49, 50.

Holothuria (Thymiosycia) arenicola
Semper
Figures 168, 186A,B,C,D,E

DESCRIPTION: This is a medium-sized species that reaches a maximum length of 25 cm (10 in). The body is slender, elongate, swollen at the middle, and distinctly tapered toward both ends; the upper and lower surfaces are similar in appearance. The skin is thin and smooth in the midbody region, but stiff and rough near the ends. Small, cylindrical tube feet are scattered over the entire body. They are somewhat more numerous on the lateral margins and the ventral surface, and tend to be borne on irregular warts toward the anterior and posterior ends. The mouth, at the extreme anterior end of the body, is surrounded by 20 very small tentacles.

The body wall ossicles (Figure 186A–E) are tables and buttons. The tables (Figure 186A,B) have a circular to square disk perforated by four to eight marginal and four central holes; the spire has four short columns terminating in approximately 12 tiny spines. The buttons (Figure 186C–E) are of two types, a button with two parallel rows of three elongate holes, and an irregular, longer button with few to many holes, several of which are nearly closed.

The drab coloration of this species is usually obscured by a thin coating of silt, but the animal is tan, light brown, or pale gray, with contrasting flecks of darker brown, black, or rust color. There quite often are two dorsal longitudinal rows of dusky blotches with up to 20 blotches per row. The stalks

FIGURE 168. *Holothuria* (*Thymiosycia*) *arenicola*.

of the yellowish, transparent tentacles have a sprinkling of brown dots. Juveniles (less than 1 cm [less than 0.4 in]) bear little resemblance to the adults, but are readily distinguished by their terminally placed mouth with yellow tentacles and by their tube feet with conspicuous yellow tips.

HABITAT: Seagrass beds of reef flats, and sand plains, from the fore reef to the back reef zone.

DISTRIBUTION: Reportedly circumtropical. In the western Atlantic, known from Bermuda to Brazil, including Enmedio Reef in the Gulf of Mexico (Henkel 1982). Around Florida it occurs off Fort Pierce, the Florida Keys, and the Dry Tortugas. DEPTH: In the Florida-Caribbean region, the shoreline to at least 13 m (42 ft).

BIOLOGY: This holothuroid generally burrows several centimeters below the sediment surface, producing prominent conical mounds at the sediment surface. It can be abundant, but is seldom observed by divers. Individuals are difficult to excavate from beneath a mound, because the sea cucumber contracts its body, wedging itself tightly in the burrow. Large individuals may reside in dead conch shells or beneath coral rubble. Near Carrie Bow Cay, Belize, they occurred beneath the same rock with *H. cubana* and *H. impatiens* (Miller, previously unpublished).

H. arenicola ingests surface and subsurface sediments via a feeding funnel that terminates 15–20 cm below the sediment surface (Mosher 1980; Hammond 1982a). It selectively feeds on nutrient-rich particles (Hammond 1983). In a month to several months, the position of the feeding funnel changes very little, indicating that considerable time is needed to exhaust the resources within an individual's reach (Hammond 1982a). The intensity of feeding activity does not vary during a 24-hour cycle. The gut is cleared in about 6 hours and is completely emptied and refilled several times a day. The foregut is slightly acidic, and there is evidence of some dissolution of carbonate sediments during passage through the gut (Hammond 1981). This chemical process does not significantly change the composition of the sediment, though the mound-building activities of this species considerably alter the environment.

In areas of high population density (1.2 individuals per square meter), Mosher (1980) estimated that the volume of sediment passing through the intestine of a single animal was 36 liters/m² per year. The transfer of such large quantities of sediment from several centimeters below the seafloor to the sediment surface suggests that this species plays an important role in reworking sediment and intensifying siltation.

Randall (1967) reported that the trunk fish, *Lactophrys trigonus* (Linnaeus), may eat *H. arenicola*. In Bermuda, Snyder (1980) found minute rhabdocoel flatworms, *Anoplodiera* sp., in the digestive tract of the majority of individuals examined.

SELECTED REFERENCES: Semper 1868: 81, pl. 20, pl. 30, fig. 13, pl. 35, fig. 4 [as *Holothuria (Sporadipus) arenicola*]; Caso 1955:513, pl. 5, figs. 1–21 [as *Holothuria arenicola*]; Rowe 1969:147; Pawson 1986:540, pl. 178 [as *Holothuria arenicola*].

Holothuria (Thymiosycia) impatiens
(Forskål)
Figures 169, 186F,G,H,I,J,K

DESCRIPTION: This medium-sized species grows as large as 15–20 cm (6–8 in) in length. The body is slender, elongate, and narrower anteriorly than posteriorly. Often the difference in diameter between the two ends is so lopsided that the sea cucumber takes on the shape of a flask. The skin is thin, rough, and very wrinkled. Tube feet are scattered; many are on distinct conical warts that tend to form longitudinal rows along the lateral margins. There are relatively few cylindrical tube feet on the ventral surface and more abundant papillae on the dorsal surface. The mouth, at the extreme anterior end of the body, is surrounded by 20 medium-sized to large tentacles. A narrow yet distinct collar of papillae surrounds the tentacles.

The body wall ossicles (Figure 186F–K) are tables and buttons. The tables (Figure 186F–I) have a characteristic square disk with eight large subequal, circular to oblong, peripheral holes and one central hole. Often there are a few smaller holes near the margin. The short spire bears numerous (up to 30 or more) tiny spines. The buttons (Figure 186J,K) are all long, slender, and smooth, with three or four pairs of large, elongate holes.

The body wall is mottled gray, brown, and purple. Some specimens have contrasting blotches of darker pigment arranged in two series running the length of the dorsal surface. Occasionally, near the anterior end of the body the blotches merge, forming distinct transverse bands. Individuals may be darker near the anterior end, and some have a wide black collar.

HABITAT: Near the low-tide mark in tide pools and seagrass flats, as well as along the deeper reaches of the reef complex; among rocks and rubble, and in sandy sediments.

DISTRIBUTION: Reported from most tropical regions of the world; in the western Atlantic, throughout the Greater and Lesser Antilles, and from Florida, the Florida Keys (Key Largo, Key West), the Dry Tortugas, the Bahama Islands, Mexico (off Veracruz), Belize, Panama, Colombia, and Venezuela. DEPTH: 0.5–27 m (1.5–89 ft).

BIOLOGY: This is possibly the most active *Holothuria* species in the tropical western Atlantic; animals removed from their hiding place twist and contort their bodies continuously. When placed in an aquarium, they frequently raise their anterior end off the bottom, much like a snake. Individuals held in running seawater systems have been noted to escape from confinement by crawling through the drain hole (Miller, previously unpublished). When disturbed, individuals eject quantities of Cuvierian tubules.

Although this species has a widespread

FIGURE 169. *Holothuria (Thymiosycia) impatiens.*

distribution in the West Indies region, it does not appear to be very abundant at any particular locality. Near Carrie Bow Cay, Belize, *H. impatiens* and two congeners, *H. cubana* and *H. arenicola,* were found together under the same piece of rubble (Miller, previously unpublished).

Harriott (1985) reported that the diameters of fully developed eggs of *H. impatiens* from the Great Barrier Reef range from 155 to 209 μm and that this species spawns there during the fall (December to February).

SELECTED REFERENCES: Forskål 1775: 121, pl. 39, fig. B [as *Fistularia impatiens*]; Deichmann 1930:64, pl. 3, figs. 17, 18 [as *Holothuria impatiens*]; Rowe, 1969:145, fig. 13; Martínez de Rodríguez and Mago Herminson 1975:190, pl. 2, figs. 1–3 [as *Brandothuria impatiens*].

Holothuria (Thymiosycia) thomasi
Pawson & Caycedo
Figures 170, 187A,B,C

DESCRIPTION: This is by far the largest sea cucumber species in the western Atlantic, with adults reaching a length of 2 m (6.5 ft). However, entire animals are rarely seen because individuals, anchored in coral reef crevices, usually extend only the anterior portion of the body to feed. The body is cylindrical, elongate, and expanded at the oral end when the animal is feeding. The skin is soft. The upper surface carries distinct rows of conical papillae, placed on warts that are most prominent along the lateral margins and around the mouth. On the lower surface, cylindrical tube feet are scattered and not particularly abundant. Twenty conspicuous, shield-shaped tentacles extend from the ventrally positioned mouth.

The body wall ossicles (Figure 187A–C) are tables and buttons. The table margins are irregular to square in outline and generally perforated by a ring of 12 marginal holes and four central holes (Figure 187A,B). The spires are short, composed of four pillars, and terminate in 18–20 short spines surrounding a large perforation. The buttons (Figure 187C) are usually swollen in the middle and perforated by two longitudinal rows of elongate holes, three holes per row.

The color is generally a mottled yellowish to golden brown; the dorsal papillae have white tips. Darker hues of brown and maroon also occur, and the lower surface is usually lighter than the upper. Some individuals have distinctive irregular bands of white and golden brown. Tentacles are light pink to yellow or chocolate brown.

HABITAT: Associated with reef-building corals, often on fore reef escarpments between the outer fore reef ridge and the steep reef slope.

DISTRIBUTION: The Bahama Islands, the Florida Keys, Cuba, Puerto Rico, St. Croix, St. Vincent, Mexico, Isla Cozumel, Panama, Colombia, and Carrie Bow Cay, Belize (Miller and Hendler, previously unpublished). DEPTH: 3–30 m (10–98 ft).

BIOLOGY: It seems remarkable that this large sea cucumber was not described until 1980, although before that it was frequently encountered by sport divers in the Bahama Islands and undoubtedly at other localities (e.g., Corvea 1990). Bahamian divers referred to it by the common name "tiger's tail"

in reference to the appearance of the sea cucumber's anterior end.

H. thomasi is cryptic and generally is seen only when feeding with the anterior of the body extended from the coral reef structure. Individuals ingest sand or algae-covered substrates (Hammond 1982a). As described by Pawson and Caycedo (1980), feeding usually commences at dusk. In Jamaica, the majority of a monitored population was active at night (Hammond 1982a). However, Caycedo observed some individuals off Colombia to be active from midday to 1700 hours. They appear to "vacuum" the sediment while feeding; the extended anterior end (one-fourth to one-third of the total body length) gently sweeps across the substrate as the tentacles rapidly grasp sediment particles, pushing them into the mouth. Individuals are capable of ingesting pieces of rubble up to 2 cm long.

H. thomasi is very difficult to dislodge from its niche. When the anterior end is grabbed by a diver, the posterior end of the body swells and anchors the animal. The animal contracts, withdrawing into the reef, and the diver may find himself being pulled toward the reef as the sea cucumber retreats. Individuals occasionally are observed on open sand, moving between shelters at a speed of 0.25 m/minute, and up to a third of the population may shift shelters each night (Hammond 1982a).

An intact specimen collected at 18 m during a night dive off San Salvador, the Bahama Islands, ejected a large specimen of a commensal pearlfish, *Carapus bermudensis*, when brought to the surface (Cameron, personal communication).

SELECTED REFERENCES: Pawson and Caycedo 1980:454, figs. 1, 2; Miller and Pawson 1984:63, figs. 51, 52; Sefton and Webster 1986:100, pl. 171 [as *Holothuria thomasae*].

FIGURE 170. *Holothuria* (*Thymiosycia*) *thomasi*.

FAMILY CAUDINIDAE

Paracaudina chilensis obesacauda
(H. L. Clark)
Figures 171, 187*J*,*K*

DESCRIPTION: This species is medium-sized, reaching a length of 15 cm (6 in). Its lack of tube feet, unusual shape, and size make this species easy to identify. The body is cylindrical, slightly tapered anteriorly toward a bluntly rounded mouth, and markedly tapered posteriorly to form a tail that may represent one-third of the total body length. The tail is frequently curled toward the anterior end. The body wall is thin, flexible, relatively smooth, but slightly gritty. Near the mouth and along the tail, the skin is usually contracted into pleated folds. The mouth is surrounded by 15 small, digitate tentacles, each ending in four sharp points.

The body wall ossicles (Figure 187*J*,*K*) are crossed baskets. The cross on one side of the basket has a hole in the center; on the opposite side it is solid.

In life, *P. c. obesacauda* is uniform white to light gray.

HABITAT: Soft, sandy sediments.

DISTRIBUTION: Florida (Jupiter Inlet, Everglades City, Sanibel Island, Sarasota), the Florida Keys (Cedar Key, Key West), the Dry Tortugas, Texas, and Brazil. DEPTH: Low-tide mark to 10 m (33 ft).

BIOLOGY: *P. c. obesacauda* is a deposit feeder. Miller and Pawson (1984) found that the intestines of Florida west coast individuals contain 70% quartz sand and 30% calcareous remains, including small ostracod crustaceans, mollusk fragments, echinoid spines, and foraminiferans.

Individuals collected near Jupiter Inlet had been deposited on the shore by a beach renourishment dredge working 400–500 m offshore. Thirty specimens were found in a rather isolated area of dredge spoil, and many other dead or moribund individuals were noted nearby, suggesting that the offshore population was fairly dense.

The Pacific subspecies *Paracaudina chilensis chilensis* (J. Müller), like some other

FIGURE 171. *Paracaudina chilensis obesacauda*.

sea cucumbers that live in low-oxygen habitats, has hemoglobin-containing cells that circulate through its body cavity and its water vascular and hemal systems. These red hemocytes have a high affinity for oxygen and may assist in respiration when oxygen levels are reduced (Baker and Terwilliger 1993).

REMARKS: The closely related subspecies *P.*

c. chilensis is widely distributed in the Pacific Ocean (Miller and Pawson 1984).

SELECTED REFERENCES: H. L. Clark 1907:38, pl. 9, figs. 1–5 [as *Caudina obesacauda*]; Miller and Pawson 1984:66, figs. 53, 54.

FAMILY SYNAPTIDAE

Epitomapta roseola (Verrill)
Figures 172, 188E,F,G,H,I,J

DESCRIPTION: This tiny, worm-shaped holothurian can reach a maximum length of 10 cm (4 in), but most individuals are only half that size. Those from Florida do not exceed 2.5 cm (1 in). The body is slender and cylindrical, and the body wall is thin and transparent, revealing internal structures such as the intestine, gonad, and longitudinal muscle bands (Figure 172). A conspicuous diagnostic feature of this species is the presence of small, circular to oval bumps on the surface of the body wall. The dark reddish brown structures are so numerous that the sea cucumber appears pink. The mouth is surrounded by 12 digitate tentacles, each with seven fingerlike projections. The oral (inner) side of the tentacles bears two to five sensory cups.

The body wall ossicles (Figure 188E–J) are anchors, anchor plates, and miliary granules. Anchors from the anterior portion of the body have smooth flukes; those near the posterior end are longer and have serrate

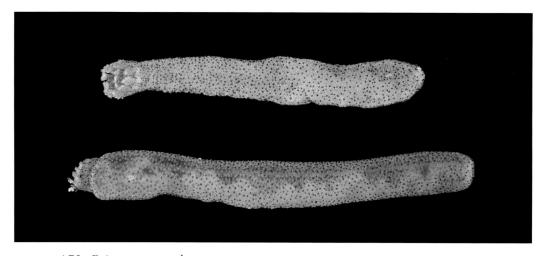

FIGURE 172. *Epitomapta roseola*.

flukes. The presence of tiny C- or O-shaped miliary granules (Figure 188*I,J*) distinguishes this species from other synaptids of the Florida Keys.

HABITAT: Beneath rocks in fine gravel and coarse sand, and in sand. An individual was found in an oyster bar in the Indian River near Fort Pierce, Florida (Miller, previously unpublished).

DISTRIBUTION: Connecticut, Massachusetts, Bermuda, and Florida. H. L. Clark (1924, 1933) tentatively identified a synaptid from off Jamaica as *E. roseola*. DEPTH: Low-tide mark to 11 m (35 ft).

BIOLOGY: This species is delicate, and it readily fragments while being sieved from sediment and during preservation, if it is not narcotized. The intestine generally is filled with sediment, suggesting that *E. roseola* is a deposit feeder.

SELECTED REFERENCES: Verrill 1873: 422 [as *Leptosynapta roseola*]; Heding 1928:234, fig. 40 (12–22).

Euapta lappa (Müller)
Figures 173, 188*K,L,M,N*

DESCRIPTION: This species is unique in appearance among sea cucumbers of the Florida Keys and the Bahama Islands. It is a moderately large, skinny animal, reaching a maximum length of 1 m (3 ft) or more. The body wall is extremely thin and flexible, and active individuals can rapidly alter their length. The general shape is cylindrical and vermiform, and a segmented appearance is produced by alternate widened (pouchlike) and constricted rings at regular intervals along the body. The mouth is surrounded by 15 large, pinnate tentacles, each with 20–35 pairs of lateral digits. The tentacles of living animals are usually extended and very active; preserved specimens become contracted and flaccid.

The body wall ossicles (Figure 188*K–N*) are anchors, anchor plates, and miliary granules. The bases of the anchors (Figure 188*M*) in *E. lappa* are branched, unlike those of other synaptids from the Florida Keys and the Bahama Islands. The miliary granules (Figure 188*N*) are in the shape of tiny rosettes.

The color is variable, and its intensity is related to the degree of expansion or contraction of the body wall; fully contracted specimens appear most colorful. The dorsal and lateral surfaces are usually striped, with longitudinal bands of alternating brown and golden yellow. Irregular patches of white may be scattered along the brown color bands, and the entire body wall may be flecked with minute black or white dots. The ventral surface is less pigmented and often white or silver.

HABITAT: Shallow reef zones such as reef flats.

DISTRIBUTION: Widespread throughout the Caribbean region, including the southwestern Gulf of Mexico (Henkel 1982). Off Florida, it occurs in the Keys and the Dry Tortugas.

DEPTH: Low-tide mark to 24 m (79 ft).

BIOLOGY: During the day, *E. lappa* remains concealed in coral rubble, corals, conch shells, or beneath rock slabs. At dusk, individuals rapidly emerge from shelters and crawl about on sand, rubble, or seagrass until dawn (Hammond 1982a). This is a very active species; when disturbed, individuals may writhe about. In normal locomotion, one end of the body becomes long and slender, as the opposite end is compressed into tight folds. This process, called "direct overlapping peristalsis," requires the coordinated activities of the circular and longitudinal muscles, the tentacles, and the anchor plates of the body wall (Heffernan and Wainwright 1974). As it feeds in sandy and algae-covered habitats, the tentacles continuously convey sediment

FIGURE 173. *Euapta lappa.*

to the mouth (Hammond 1982a). Individuals feed only at night and process food rapidly, totally voiding the intestines within an hour. Animals collected during the day have empty digestive tracts (Hammond 1982a).

Anyone who has handled a live specimen of this species can vouch for the sticky nature of the body wall, which will readily adhere to flesh or fabric; an individual placed in a small container will generally get snagged upon itself! This adhesion is caused by the numerous anchor ossicles in the skin. The flukes of the anchors slightly elevate the overlying skin and tend to project through the body wall.

SELECTED REFERENCES: Müller 1850: 134 [as *Synapta lappa*]; Pawson 1976:374, fig. 2B; Colin 1978:431, 432 (pl.).

Genus *Leptosynapta*

Leptosynapta, one of the most perplexing genera of sea cucumbers in the tropical western Atlantic, is critically in need of taxonomic revision. One-half of the described species may be synonyms. At least 11 nominal species from the Florida and Caribbean region have been referred to *Leptosynapta* since H. L. Clark's (1924) review of the genus. Several descriptions of species are based on only one or a few specimens, and some of the species have not been collected since first described.

Members of this genus are all superficially similar in appearance (Figure 174). It is impossible to distinguish between most of the nominal species without careful examination of the ossicles, especially the minute miliary granules. Even then, the remarkable variability in ossicle shapes and sizes, within and between populations of the same species, is confusing. Moreover, the anchors and anchor plates characteristic of *Leptosynapta* species vary significantly in size from the anterior to the posterior end of the body.

In this study, two species treated in detail, *Leptosynapta multigranula* and *L. tenuis*, were studied at first hand from Florida localities. The descriptions of two other species reported to occur off the Florida Keys, *L. crassipatina* and *L. parvipatina*, are based on the account of H. L. Clark (1924).

Leptosynapta crassipatina
H. L. Clark

DESCRIPTION: This small species reaches a length of 4 cm (1.6 in) and is very similar in overall appearance to *L. tenuis* (Figure 174). The mouth is surrounded by 12 tentacles, each with four or five pairs of lateral digits. Four to 10 sensory cups arise near the base of the tentacles on their inner margin.

The body wall ossicles are anchors, anchor plates, and miliary granules. The posterior anchors and anchor plates are larger than those found near the anterior end. The anterior anchor flukes are smooth or armed with a single, small tooth; posterior anchor flukes carry three to seven distinct teeth. The miliary granules are not abundant; they are slender curved rods, with branched tips and one or two small knobs on the convex margin.

HABITAT: This species burrows in sandy mud near mangroves. Additional information on its biology is lacking.

DISTRIBUTION: *L. crassipatina* was first described from two specimens collected near Key West, Florida, probably at a depth of less than 1 m (less than 3 ft) (H. L. Clark 1924). There are later records, which are suspect, for the west and north coasts of Florida and west to Mississippi (Deichmann 1954; Richmond 1962; Smith 1971a,b) (see *Remarks* under *L. tenuis*).

SELECTED REFERENCES: H. L. Clark

1924:479, pl. 6, figs. 1–4; Deichmann 1930: 208.

Leptosynapta multigranula
H. L. Clark

Figure 189G,H,I,J,K,L,M

DESCRIPTION: This slender, vermiform species reaches a maximum length of 10 cm (4 in) and a diameter of 0.8 cm (0.3 in). It is very similar in appearance to *L. tenuis* (Figure 174). The mouth is surrounded by 12 tentacles, each ending in five or six pairs of digits that project at right angles to the stem. There are two to nine sensory cups on the inner (oral) surface of each tentacle. In comparison with other *Leptosynapta* species discussed in this book, the body wall is relatively thick and opaque; the large ossicles elevate the body wall and give the skin a bumpy appearance.

The body wall ossicles (Figure 189G–M) are anchors, anchor plates, and miliary granules. The anchors near the mouth are approximately one-half the length of those near the anus. The flukes of the anchors (Figure 189G–I) are serrate, with up to 10 conspicuous teeth. The miliary granules (Figure 189L,M) are in the shape of tiny, irregular C-shapes, most of which have swollen ends.

The body wall of this species has a pink tinge.

HABITAT: *L. multigranula* burrows among the roots of manatee grass *(Syringodium)* and turtle grass *(Thalassia)* in calcareous sediments. Additional information on the biology of this sea cucumber is lacking.

DISTRIBUTION: Florida (Biscayne Bay) and the Dry Tortugas (Deichmann 1939); also Looe Key in the Florida Keys (Miller, previously unpublished). DEPTH: 0.3–2 m (1–6 ft).

SELECTED REFERENCES: H. L. Clark 1924:486, pl. 8, figs. 3–7; Deichmann 1930: 207 [as *Leptosynapta multigranulata*].

Leptosynapta parvipatina
H. L. Clark

DESCRIPTION: This species is fragile and fragments readily, but intact individuals are as long as 15 cm (6 in). It is very similar in appearance to *L. tenuis* (Figure 174). The mouth is surrounded by 12 tentacles, each with five lateral pairs of digits and one longer terminal digit.

The numerous miliary granules in the body wall and tentacles are variable in shape, ranging from simple disks, some perforated at the center, to curved rods that are enlarged and notched at the tips. The anchors and anchor plates at the posterior part of the body are larger than those toward the anterior end. Posterior anchor flukes carry four to five teeth; anterior anchor flukes may lack teeth or have one small tooth near their tips.

L. parvipatina, in life, is reported to be pinkish white.

HABITAT: Muddy sand or seagrass beds.

DISTRIBUTION: *L. parvipatina* was first found at Tobago, where it was collected with shovel and sieve in shallow water, and "numerous specimens" have been reported from Biscayne Bay, south of Cape Florida (Deichmann 1939); however, see *Remarks* under *L. tenuis*.

BIOLOGY: This species burrows in fine sediment.

SELECTED REFERENCES: H. L. Clark 1924:490, pl. 4, figs. 8, 9, pl. 6, figs. 5–8.

Leptosynapta tenuis (Ayres)
Figures 174, 189A,B,C,D,E,F

DESCRIPTION: This slender, worm-shaped species reaches approximately 15 cm (6 in) in length and 0.6 cm (0.2 in) in diameter. The thin, transparent body wall reveals within five white, longitudinal muscle bands. Scattered about the body, between the sites of anchor plates, lie numerous, minute "warts," which give the skin a papillose appearance. Surrounding the mouth are 12 tentacles, each with 9–11 digits of which the terminal one is the longest, and 15–25 sensory cups on the inner surface.

The ossicles (Figure 189A–F) consist of anchors, anchor plates, and miliary granules. The posterior anchors and plates are larger than those in the anterior body wall. The number of teeth on the anchor flukes vary from none to approximately seven (Figure 189A,B). The miliary granules (Figure 189E,F) are numerous, especially in the tentacle stalks. They are in the shape of tiny, straight to curved rods, C-shapes, or dumbbells. In the tentacle branches, there are longer curved rods with enlarged ends; the ends are scalloped and sometimes perforate. The tentacle rods of some individuals have scalloped margins throughout their length. Ciliated funnels occur in two forms, small conical ones and much larger ones shaped like mint flowers (Heding 1928).

L. tenuis is translucent white when alive.

HABITAT: Sandy and muddy sediments, sometimes associated with seagrass rhizomes.

DISTRIBUTION: Previously thought to extend from New England to South Carolina; now known from the east and west coasts of Florida (see *Remarks* below). DEPTH: Less than 1 m (less than 3 ft).

BIOLOGY: *L. tenuis* is fragile, and it readily autotomizes when handled or when held in unfavorable water conditions. Smith (1971a,b) gave details of the regeneration process in *L. tenuis* (referred to as *L. crassipatina*). Unlike H. L. Clark (1899), who suspected that the oral complex was necessary for regeneration of the posterior body region, Smith found that complete regeneration was

FIGURE 174. *Leptosynapta tenuis*.

possible even in the absence of anterior structures (i.e., mouth, tentacles, calcareous ring, nerve ring, and water ring).

The species is hermaphroditic, its gonadal tubules containing a mixture of testicular and ovarian tissue. In North Carolina, individuals can spawn twice a year, in the fall and spring, releasing sperm before shedding mature oocytes 0.2 mm in diameter (Green 1978).

The body wall of *L. tenuis* is thrown into antiperistaltic waves during burrowing activity; the tentacles are used to displace sediment (H. L. Clark 1899; Hunter and Elder 1967). Ruppert and Fox (1988) reported that it constructs a U-shaped burrow and ventilates it by conducting peristaltic waves along the body from the anterior to the posterior end. These sea cucumbers ingest sediment as they burrow, and the tail end of the burrow is often marked by a fecal mound (Powell 1977). Individuals may periodically reverse position in their burrow to feed on material in the fecal mound (Ruppert and Fox 1988). Their activities tend to stabilize the upper few centimeters of sediment, but destabilize the sediment below (Myers 1977a,b). Ruppert and Fox (1988) indicated that a large population of *Leptosynapta* may turn over 4–25 metric tons of sediment per hectare in a year.

This species, and other members of the family Synaptidae, feel sticky when they are alive because of the numerous anchor ossicles in their skin. The anchors, which pivot from associated anchor-plate ossicles, have sharp flukes that project from the integument and snag objects that contact the sea cucumber (Stricker 1985, 1986).

The tiny bivalves *Montacuta percompressa* Dall and *Mysella* sp., seemingly host-specific, are sometimes found attached to *L. tenuis* (Ruppert and Fox 1988).

REMARKS: *L. tenuis* was thought to be restricted to temperate waters from Massachusetts to South Carolina. However, it has been collected in Florida, from the Indian River (near Stuart), Biscayne Bay (off Miami), and from the west and northwest Gulf coast off Levy and Franklin counties (Miller, previously unpublished). The size and shape of the ossicles from the Floridian specimens differ somewhat from the dimensions of ossicles of their northern counterparts, but this variability does not appear to be significant.

Confounding this interpretation are reports of *L. parvipatina* from Biscayne Bay (Deichmann 1939) and of *L. crassipatina* in the northeastern Gulf of Mexico (Deichmann 1954; Richmond 1962; Smith 1971a,b). If Deichmann's (1939) identification of *L. parvipatina* is correct, then Biscayne Bay is the only known locality, other than Tobago (the type locality), from which this species has been reported. However, as mentioned above, synaptids taken in Biscayne Bay, very near the site where Deichmann collected, are referable to *L. tenuis* (Miller, previously unpublished). Furthermore, it seems likely that the specimens Deichmann (1954), Richmond (1962), and Smith (1971a,b) identified as *L. crassipatina* actually represent *L. tenuis*. Synaptids examined from near Tallahassee (adjacent to Smith's collecting site) and from Cedar Keys appear to be *L. tenuis*, making this species the most common *Leptosynapta* of shallow intertidal sand flats along the coast of Florida.

SELECTED REFERENCES: Ayres 1851a:11 [as *Synapta tenuis*]; H. L. Clark 1924:483, pl. 7, figs. 12–16 [as *Leptosynapta inhaerens* (western Atlantic material, in part)]; Heding 1928:208, figs. 28/1–12, 29/2–6; Ruppert and Fox 1988:78, figs. 118, 159.

Protankyra ramiurna Heding
Figures 175, 190A,B,C,D,E

DESCRIPTION: This medium-sized species can reach a length of 15 cm (6 in). The body is worm-shaped and circular in cross section. Five distinctively wide bands of muscle run the length of the body. The body wall, between the longitudinal muscles, is so thin and transparent that the internal organs are visible within the body cavity. The mouth is surrounded by 12 equal, stout, digitate tentacles, each of which carries two pairs of pointed, elongate projections and a much smaller, rounded digit near the tip.

The body wall ossicles (Figure 190A–E) are anchors, anchor plates, and miliary granules. The anterior anchors and plates are significantly smaller than those at the posterior end of the body. The anchor arms have four to seven prominent teeth, and the anchor bases are covered with minute spines (Figure 190C). The anchor plates (Figure 190A,B) have more numerous large holes than those of other synaptids from the Florida Keys, and they have a distinct bridge to which the anchor articulates. The miliary granules (Figure 190D,E) are small, plump rods, slightly constricted at the middle.

The body wall of *P. ramiurna* is light pink, but so thin and transparent that the ossicles are visible to the naked eye as small white spots; the longitudinal muscle bands show through as white bands. In some individuals, miniscule flecks of brown pigment are scattered over the body and on the stalks of the otherwise transparent tentacles.

HABITAT: In addition to the specimens reported from sand and shell sediments, several have been sieved from coarse, "dirty" sediments adjacent to a mangrove island off Puerto Rico, and one has been found off Miami, among the roots of the seagrass *Thalassia* (Miller, previously unpublished).

DISTRIBUTION: Known only from the southwestern coast of Puerto Rico near La Parguera and from the mouth of Biscayne Bay, off Miami, Florida (Miller, previously unpublished). The incomplete specimen from which Heding (1928) described the species was from an unspecified site in the West Indies. DEPTH: Less than 1 m (less than 3 ft).

BIOLOGY: The body wall of *P. ramiurna* is sticky, like that of other synaptids, because of the anchor ossicles lying in the thin epidermis. Specimens of *P. ramiurna* collected off Puerto Rico in July had well-developed gonads, suggesting that this species may breed during summer months.

The name of this species acknowledges the presence of vibratile urnae (also called ciliated funnels) on the inner surface of its body wall. These microscopic organs, typical of most synaptid holothuroids, are vase-shaped structures that act, in concert with mucus produced in the coelom, to clear impurities from the sea cucumber's coelomic fluid (Jans and Jangoux 1989, 1992). The species-specific shapes of urnae are of importance in synaptid taxonomy (Heding 1928; Jans and Jangoux 1989).

REMARKS: This is the first report of *P. ramiurna* since Heding's description of the species in 1928. Although Heding's specimen was lacking an anterior end, his description of the ossicles and internal anatomy leaves little doubt that our material is identical. However, it is possible that Heding's specimen and the present material represent another species, *P. benedeni* (Ludwig), known only from the coast of Brazil (Ludwig 1881). To resolve the issue, specimens of *P. benedeni* are needed for comparison.

SELECTED REFERENCE: Heding 1928: 275, fig. 54.

SEA CUCUMBERS

FIGURE 175. *Protankyra ramiurna*.

Synaptula hydriformis (Lesueur)
Figures 176, 190*F,G,H,I*

DESCRIPTION: The vermiform body of this holothuroid is reported to reach a maximum length of 10 cm (4 in). Most specimens are less than half that size, and juveniles less than 1 cm (less than 0.4 in) are not uncommon. The body wall is thin, semitransparent, and very elastic. Around the mouth, full-grown specimens have 12 pinnate tentacles with a webbed base. Each tentacle stalk carries up to 20 pairs of lateral digits; each has a pair of dark pigment spots on the inner surface, near the mouth.

The body wall ossicles (Figure 190*F–I*) are anchors, anchor plates, and miliary granules. The anchor flukes (Figure 190*H*) are smooth, with a series of prominent knobs at their apex. Near the base of the anchor plates (Figure 190*F,G*) there is a distinct bridge, the point of articulation for the anchor. The minuscule miliary granules (Figure 190*I*) look like flower-shaped rosettes.

The color of living *S. hydriformis* is useful for identification. It varies from grayish green to reddish brown, with contrasting white patches of miliary granules clumped just beneath the epidermis. Young individuals are usually more pale and transparent than adults.

HABITAT: Coral reefs, seagrass beds, and around mangroves; usually associated with marine plants. Pawson (1986) found this species to be common in land-locked saltwater ponds at Bermuda.

DISTRIBUTION: Bermuda, Florida (Indian River at Titusville and Fort Pierce, Biscayne Bay), the Florida Keys, the Dry Tortugas, many islands of the Greater and Lesser Antilles, Belize, and Brazil. Museum specimens (USNM) confirm the presence of this species in the Bahama Islands, Texas, and Panama (Miller, previously unpublished). DEPTH: Less than 1–7 m (less than 3–24 ft).

BIOLOGY: *S. hydriformis* is an epiphytic species that clings to fleshy and calcareous algae (e.g., *Batophora*, *Ulva*, *Penicillus*, and *Halimeda* species), seagrass blades *(Thalassia, Halodule)*, and the submerged prop roots of mangrove trees *(Rhizophora)*. In Bermuda, the predominant color form is reddish brown, and the animal's color contrasts with the green host plants. However, the body wall coloration frequently matches the color of its

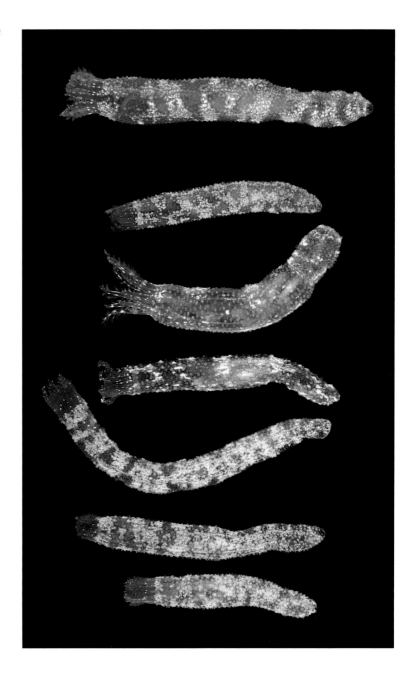

FIGURE 176. *Synaptula hydriformis*.

host plants. So, it is only when one carefully examines a clump of algae or seagrass that this active little synaptid can be seen, crawling with its featherlike tentacles and rapidly expanding or contracting in reaction to disturbance. Its stomach contents consist largely of diatoms and lesser amounts of green and brown algae (Martínez 1989). Although the species lacks a creeping sole, it always moves with the dorsal side up (Olmsted 1917). As in other synaptids, protruding anchor ossicles in the body wall cause this species to feel sticky; it is not uncommon to find a number of individuals clinging to one's wetsuit after a walk through a seagrass meadow.

S. hydriformis, one of a small number of

viviparous sea cucumber species, broods its young internally. Observation of live adults with a microscope often reveals embryos "floating" in the coelomic cavity fluid. H. L. Clark (1907) noted that brooding continues throughout the year, with up to 176 embryos developing in a single adult. Juveniles 2–3 mm (0.1 in) long are released through a rupture in the body wall near the anus (H. L. Clark 1907). H. L. Clark (1898) provided a detailed account of the brooding behavior in this species (as *Synapta vivipara* Ludwig). As a consequence of brooding, the young emerge close to their parents; thus large numbers of *S. hydriformis* are often found together.

REMARKS: *S. hydriformis* is distinguished from other Florida and Bahama Islands synaptids by its body wall coloration. Its ossicles are similar to those of *E. lappa*, but the two species are easily separated by habitat, coloration, body form, and size. However, if ossicle preparations are required to confirm an identification, note that the anchor stocks of *S. hydriformis* (Figure 190H) are straight, but those of *E. lappa* are branched (Figure 188M). *Leptosynapta* species and *Epitomapta roseola* lack the dark red and green pigmentation characteristic of *S. hydriformis*. Furthermore, the anchor flukes of *E. roseola* and the *Leptosynapta* spp. are serrate, but those of *S. hydriformis* are smooth.

SELECTED REFERENCES: Lesueur 1824: 162 [as *Holothuria hydriformis*]; H. L. Clark 1907:82, pl. 6; Pawson 1976:375, fig. 2C; 1986:541, pls. 178, 14.8.

FAMILY CHIRIDOTIDAE

Chiridota rotifera (Pourtalès)
Figures 177, 188A,B,C,D

DESCRIPTION: This small to moderate-sized holothurian is reported to reach a length of 10 cm (4 in). Most individuals are less than 5 cm long, with a diameter of approximately 5 mm (0.2 in). The body wall is quite smooth and slimy. However, it appears rough and bumpy because of numerous hemispherical, wartlike protuberances that contain aggregations of wheel ossicles. The skin between the warts is semitransparent, and the longitudinal muscle bands and other internal structures are visible in living individuals. Twelve digitate tentacles surround the mouth; each has five pairs of digits, four lateral pairs and one terminal pair.

The body wall ossicles (Figure 188A–C) are wheels and tiny, irregular C-shapes. The wheels have a serrate inner margin and six spokes that converge around a small hole in the central hub. Aggregations of these wheels form warts or "wheel papillae." The wheels of young individuals have more spokes than those found in adults (H. L. Clark 1910).

Though this species is small, individuals are easily recognized in the field by their striking coloration. The pink or red ground color of the body wall contrasts sharply with the conspicuous, white wheel warts.

HABITAT: In tide pools and along rocky shores, on the sand beneath rubble, and in seagrass beds among *Thalassia* roots or among the branches of small corals *(Porites)* and coralline algae *(Halimeda, Penicillus)*. On the east coast of Florida, near Fort Pierce, it is associated with *Phragmatopoma lapidosa* worm reefs.

DISTRIBUTION: Bermuda, Florida (Fort

Pierce), the Dry Tortugas, the Berry Islands of the Bahamas (Miller, previously unpublished), Mexico, Belize, Jamaica, Puerto Rico, the Virgin Islands, Antigua, Barbados, Tobago, Trinidad, Aruba, Bonaire, and Brazil. In the Florida Keys it occurs at Conch Key, Looe Key (Miller, previously unpublished), Key Biscayne, and Key West. Based on museum specimens (USNM), it is also found at the Florida Keys (Soldier, Ragged, and Grassy keys), Tortola, Virgin Gorda, Guadeloupe, Mexico (Quintana Roo), Panama (San Blas Islands), and Venezuela (Isla de Margarita). DEPTH: Low-tide mark to 10 m (33 ft), though most individuals are found at depths less than 1 m (less than 3 ft).

BIOLOGY: *C. rotifera*, like *Synaptula hydriformis*, broods its young internally, and it is interesting that both species have nearly identical geographic ranges. It seems astounding that these tiny brooders, which release slow-moving young in the vicinity of the adults, are distributed over thousands of kilometers. However, juveniles of both species can disperse by attaching to marine plants or flotsam that can drift for great distances in ocean currents.

H. L. Clark (1910) described the brooding behavior in *C. rotifera*, noting its similarity to that in *S. hydriformis*, but *C. rotifera* females may each brood more than 500 embryos that are all at the same developmental stage. Engstrom (1980a) provided more details of the late stages of brooding and of the behavior of newly released juveniles.

REMARKS: The most obvious feature distinguishing this sea cucumber from other apodids found in the Florida Keys is the smooth, slimy nature of its body wall. Other apodous genera in the Florida Keys (i.e., *Epitomapta, Leptosynapta, Euapta, Synaptula*) are sticky, because of the presence of anchor-shaped ossicles in the body wall.

SELECTED REFERENCES: Pourtalès 1851: 15 (as *Synapta rotifera*); Ludwig 1881:41, pl. 3, figs. 1–15; Heding 1928:293, fig. 60.

FIGURE 177. *Chiridota rotifera.*

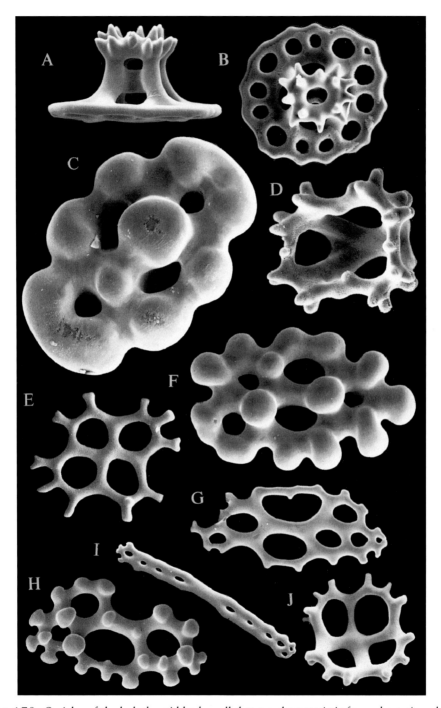

FIGURE 178. Ossicles of the holothuroid body wall that are characteristic for each species, photographed at high magnification using scanning electron microscopy. *Duasmodactyla seguroensis:* A, body wall table, lateral view; B, same, dorsal view; *Ocnus surinamensis:* C, button; D, basket; *Ocnus suspectus:* E, developing button; F, button; G, H, plates; I, tube foot supporting rod; J, basket.

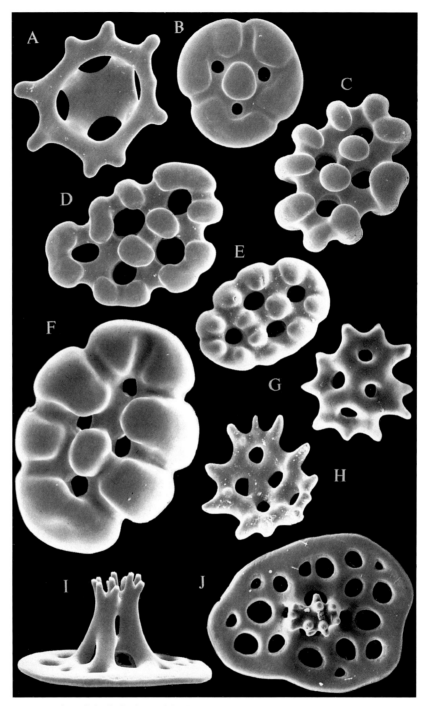

FIGURE 179. Ossicles of the holothuroid body wall that are characteristic for each species, photographed at high magnification using scanning electron microscopy. *Thyonella gemmata:* A, basket; B–D, buttons; *Thyonella pervicax:* E, F, buttons; G, H, baskets; *Neothyonidium parvum:* I, body wall table, lateral view; J, same, dorsal view.

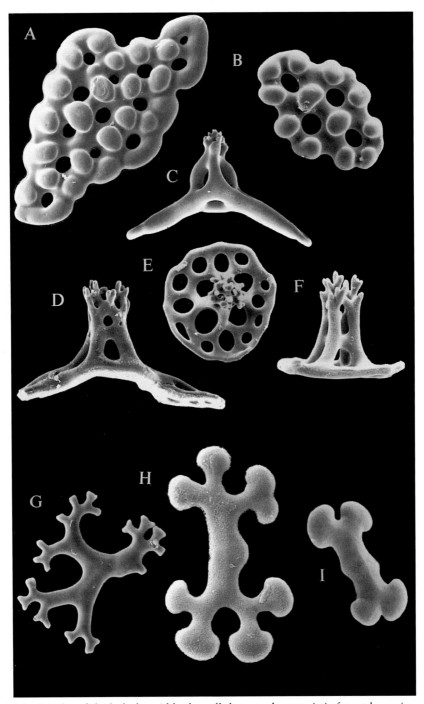

FIGURE 180. Ossicles of the holothuroid body wall that are characteristic for each species, photographed at high magnification using scanning electron microscopy. *Pseudothyone belli:* A, perforated plate; B, button; C, tube foot supporting table, lateral view; *Sclerodactyla briareus:* D, tube foot supporting table, lateral view; E, body wall table, dorsal view; F, same, lateral view; *Actinopyga agassizi:* G–I, rosettes from body wall.

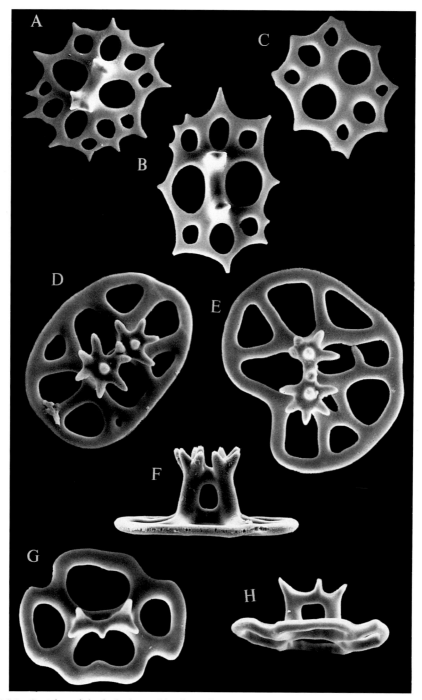

FIGURE 181. Ossicles of the holothuroid body wall that are characteristic for each species, photographed at high magnification using scanning electron microscopy. *Phyllophorus (Urodemella) occidentalis:* A, B, body wall tables, dorsal view; C, same, ventral view; *Euthyonidiella destichada:* D, E, body wall tables, dorsal view; F, same, lateral view; *Euthyonidiella trita:* G, body wall table, dorsal view; H, same, lateral view.

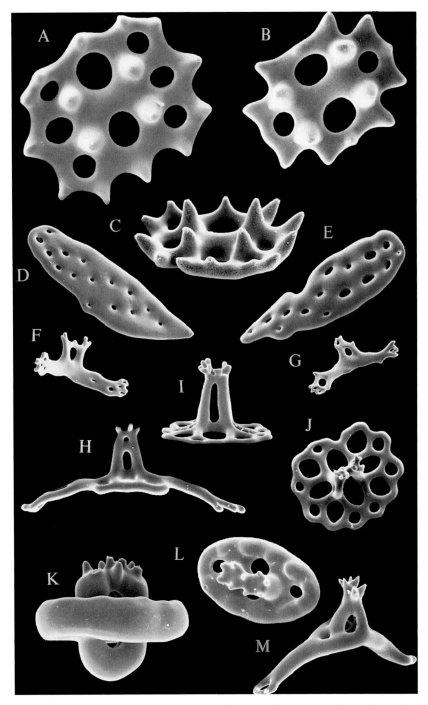

FIGURE 182. Ossicles of the holothuroid body wall that are characteristic for each species, photographed at high magnification using scanning electron microscopy. *Phyllophorus (Urodemella) arenicola:* A, B, body wall tables, dorsal view; C, same, oblique view; *Stolus cognatus:* D, E, perforated plates; F, G, tube feet supporting rods; *Thyone deichmannae:* H, tube foot supporting table, lateral view; I, body wall table, lateral view; J, same, dorsal view; *Thyone pseudofusus:* K, body wall table, lateral view; L, same, dorsal view; M, tube foot supporting table, lateral view.

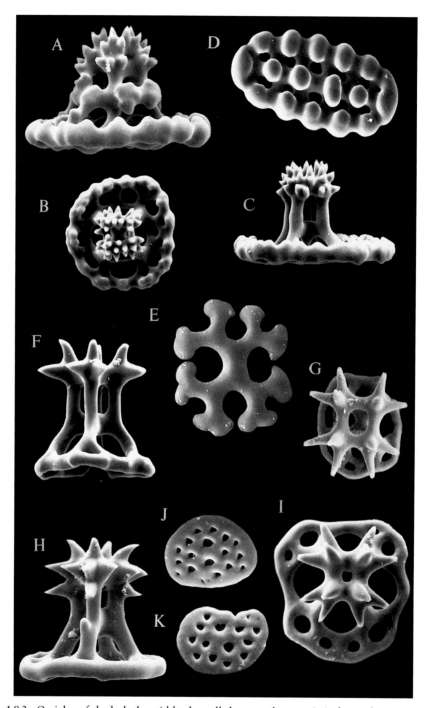

FIGURE 183. Ossicles of the holothuroid body wall that are characteristic for each species, photographed at high magnification using scanning electron microscopy. *Holothuria (Cystipus) cubana*: A, C, body wall tables, lateral view; B, same, dorsal view; D, button; *Holothuria (Halodeima) floridana*: E, rosette; F, body wall table, lateral view; G, same, dorsal view; *Holothuria (Halodeima) mexicana*: H, body wall table, lateral view; I, same, dorsal view; J, K, perforated plates.

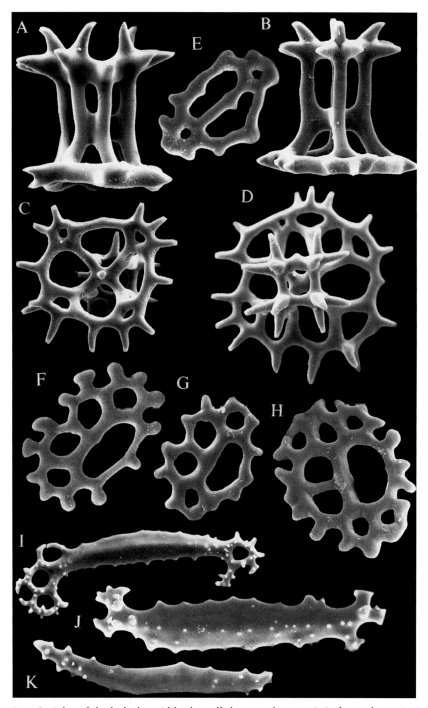

FIGURE 184. Ossicles of the holothuroid body wall that are characteristic for each species, photographed at high magnification using scanning electron microscopy. *Holothuria (Halodeima) grisea: A, B,* body wall tables, lateral view; *C,* same, ventral view; *D,* same, dorsal view; *E–H,* perforated plates; *Holothuria (Selenkothuria) glaberrima: I–K,* rods from body wall.

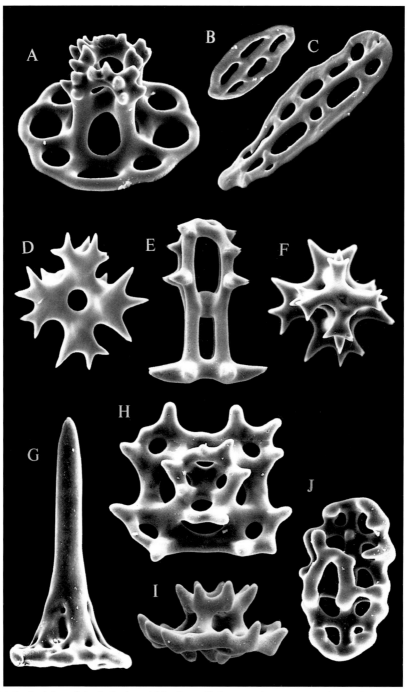

FIGURE 185. Ossicles of the holothuroid body wall that are characteristic for each species, photographed at high magnification using scanning electron microscopy. *Holothuria (Platyperona) parvula*: A, body wall table, oblique view; B, C, buttons; *Holothuria (Semperothuria) surinamensis*: D, body wall table, ventral view; E, same, lateral view; F, same, dorsal view; *Holothuria (Theelothuria) princeps*: G, tube foot "tacklike" table, lateral view; H, body wall table, dorsal view; I, same, lateral view; J, button.

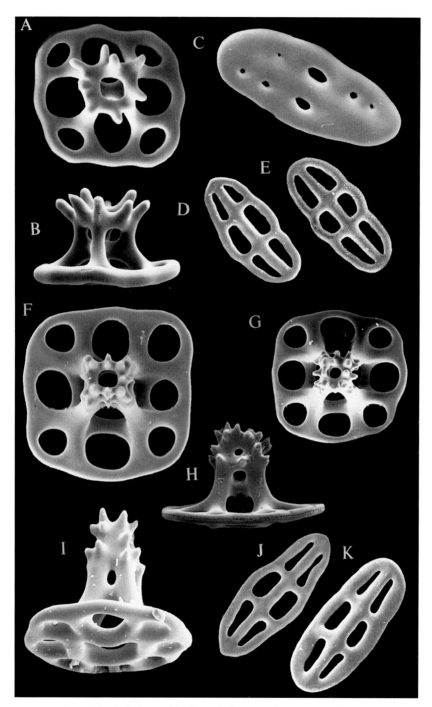

FIGURE 186. Ossicles of the holothuroid body wall that are characteristic for each species, photographed at high magnification using scanning electron microscopy. *Holothuria (Thymiosycia) arenicola*: A, body wall table, dorsal view; B, same, lateral view; C–E, buttons; *Holothuria (Thymiosycia) impatiens*: F, G, body wall tables, dorsal view; H, same, lateral view; I, same, oblique view; J, K, buttons.

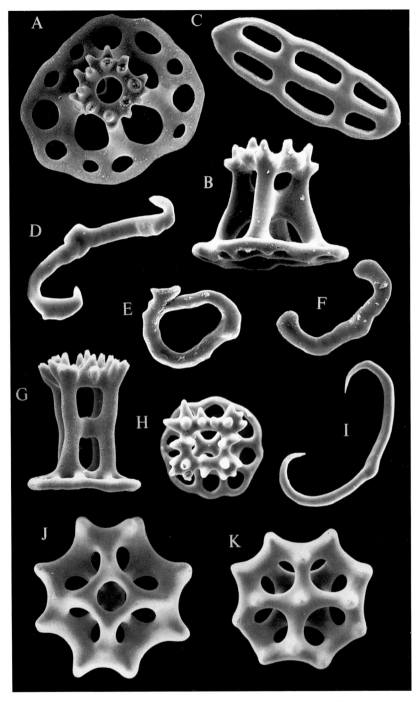

FIGURE 187. Ossicles of the holothuroid body wall that are characteristic for each species, photographed at high magnification using scanning electron microscopy. *Holothuria (Thymiosycia) thomasi:* A, body wall table, dorsal view; B, same, lateral view; C, button; *Astichopus multifidus:* D–F, "C-, O-, S-shaped" elements; *Isostichopus badionotus:* G, body wall table, lateral view; H, same, dorsal view; I, "C-shaped" element; *Paracaudina chilensis obesacauda:* J, body wall table, dorsal view; K, same, ventral view.

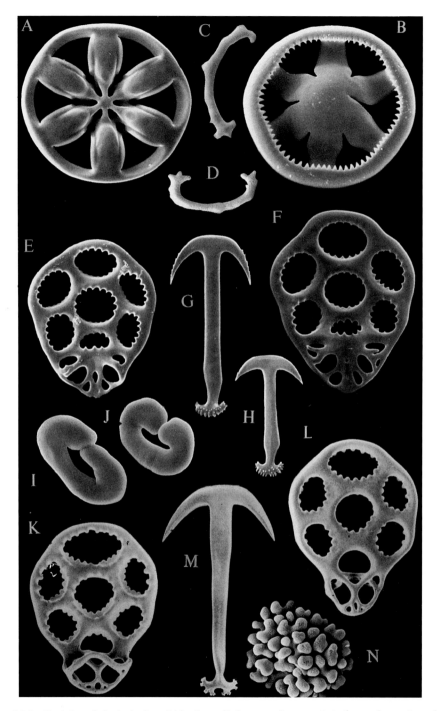

FIGURE 188. Ossicles of the holothuroid body wall that are characteristic for each species, photographed at high magnification using scanning electron microscopy. *Chiridota rotifera*: A, B, wheel elements; C, D, "C-shaped" elements; *Epitomapta roseola*: E, anchor plate, outer surface; F, same, inner surface; G, H, anchors; I, J, miliary granules; *Euapta lappa*: K, anchor plate, outer surface; L, same, inner surface; M, anchor; N, miliary granule.

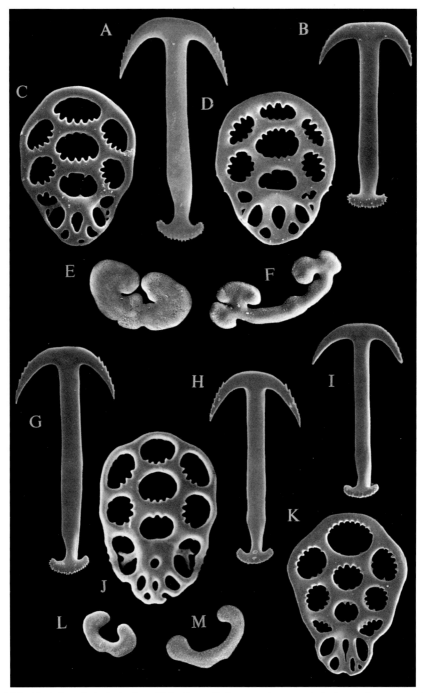

FIGURE 189. Ossicles of the holothuroid body wall that are characteristic for each species, photographed at high magnification using scanning electron microscopy. *Leptosynapta tenuis:* A, B, anchors; C, anchor plate, inner surface; D, same, outer surface; E–F, miliary granules; *Leptosynapta multigranula:* G–I, anchors; J, anchor plate, inner surface; K, same, outer surface; L, M, miliary granules.

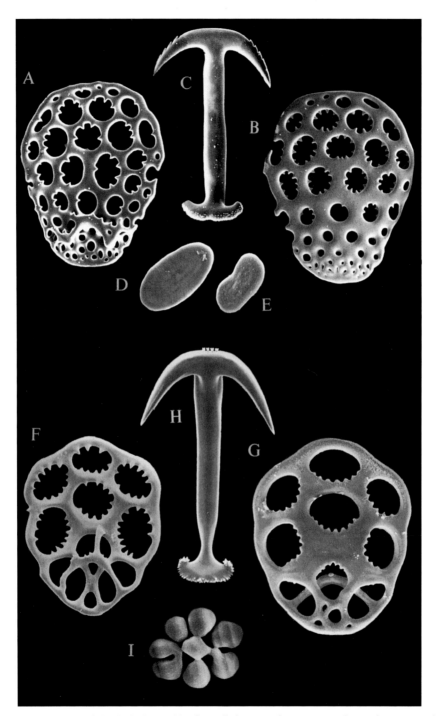

FIGURE 190. Ossicles of the holothuroid body wall that are characteristic for each species, photographed at high magnification using scanning electron microscopy. *Protankyra ramiurna:* A, anchor plate, outer surface; B, same, inner surface; C, anchor; D, E, miliary granules; *Synaptula hydriformis:* F, anchor plate, outer surface; G, same, inner surface; H, anchor; I, miliary granule.

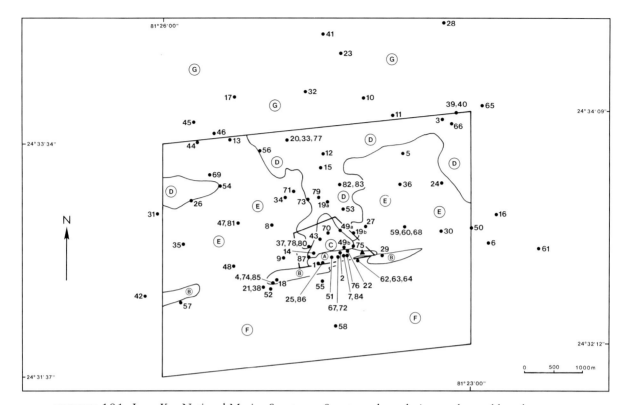

FIGURE 191. Looe Key National Marine Sanctuary. Sanctuary boundaries are denoted by a large quadrangle, and the sanctuary core area is indicated by a small quadrangle. Sampling stations are indicated by numbered solid dots. The solid triangle marks the location of Sanctuary Navigational Marker Buoy no. 24. Habitats (circled) are as follows: A, Fore Reef; B, Intermediate and Deep Reef; C, Reef Flat; D, Back Reef; E, Sand Plain; F, Deep Sand Bed; G, Hawk Channel. Illustration by P. Mikkelsen.

APPENDIX

Data on Looe Key National Marine Sanctuary Station and the Distribution of Echinoderm Species by Station

TABLE 1. Descriptions of stations sampled within and adjacent to Looe Key National Marine Sanctuary. Stations located outside sanctuary boundaries are in **boldface** type and marked with a diamond symbol (◆). Station positions indicated are approximate. Fieldwork was carried out from August 1984 to August 1986. Depending on the station, skin or scuba diving was used to make collections. Observations were made at night and during the day, and a variety of substrates were sampled, including sand, seagrass, rock and rubble, living corals, sponges, and algae. Limited use was made of fish toxicant at selected sites, to obtain species living in the interstices of the coral reef.

LK-1 Fore Reef, at Sanctuary Marker Float no. 8; spur-and-groove reef, to shallow areas of reef with coral rubble; depth: 7.6 m (25 ft); 28 August 1984.

LK-2 Fore Reef, at edge of Reef Flat, N of Sanctuary Marker Float no. 19; coral rubble on sand; depth: 0.6–3.0 m (2–10 ft); 28 August 1984.

LK-3 Back Reef, near NE Sanctuary corner; sandy patches among seagrass *(Thalassia, Syringodium)* and algae *(Penicillus, Udotea)*; depth: 8.5 m (28 ft); 28 August 1984.

LK-4 Intermediate and Deep Reef, at Sanctuary Marker Float no. 28; ancient reef rock topped with sparse coral growth; depth: 13.7 m (45 ft); 29 August 1984.

LK-5 NE Sand Plain; rippled coarse sand, seagrass *(Thalassia)*, isolated hard patches of calcareous algae and sponges; depth: 6.1 m (20 ft); 30 August 1984.

◆**LK-6** E of E Sanctuary boundary; low-profile spur-and-groove reef, abundant soft corals, small mounds of solitary corals and abundant sponges; depth: 6.1–9.1 m (20–30 ft); 30 August 1984.

LK-7 Fore Reef, at Sanctuary Marker Float no. 19; high-relief reef with sand patches, coral rubble in open areas and beneath overhangs; depth: 3.0–6.1 m (10–20 ft); 30 August 1984.

LK-8 W Sand Plain, coarse sand with algae and sponges; depth: 7.6 m (25 ft); 30 August 1984.

LK-9 W Sand Plain, just W of Sanctuary core area; patch reef; depth: 9.8 m (32 ft); 30 August 1984.

◆ **LK-10**   Hawk Channel, N of Marker Buoy no. 24; mud; depth: 15.2 m (50 ft); 31 August 1984.

◆ **LK-11**   N of E Sand Plain, just N of Sanctuary boundary; seagrasses *(Syringodium, Thalassia)*, algae *(Penicillus, Udotea)*, and scattered sponges; depth: 9.8 m (32 ft); 31 August 1984.

LK-12   Back Reef, near N Sanctuary boundary; sand, coral rubble, seagrasses, and sponges; abundant hogfish and gray angelfish; depth: 6.1–6.7 m (20–22 ft); 31 August 1984.

LK-13   W Sand Plain, on N Sanctuary boundary; sand on clay and heavily epiphytized seagrass *(Thalassia);* depth: 7.6 m (25 ft); 31 August 1984.

LK-14   Reef Flat, just N of W Fore Reef; slab rubble zone, abundant coral rubble covered with brown algae, isolated coral heads, sea fans; depth: 1.5–2.4 m (5–8 ft); 31 August 1984.

LK-15   Back Reef, N of Sanctuary core area; seagrasses *(Thalassia, Syringodium)* with sand patches and scattered loggerhead sponges; depth: 6.1–7.6 m (20–25 ft); 6 May 1985.

◆ **LK-16**   E of E Sanctuary boundary; seagrasses *(Thalassia, Syringodium)*, sand, algae *(Avrainvillea, Halimeda)*, and scattered sponges; depth: 6.7 m (22 ft); 6 May 1985.

◆ **LK-17**   Hawk Channel, N of W Sand Plain; thin veneer of sand on mud, and very sparse calcareous algae *(Halimeda);* depth: 11.6 m (38 ft); 7 May 1985.

LK-18   Intermediate and Deep Reef, at Sanctuary Marker Float no. 28; spur-and-groove reef, coral rubble, silty, hard pavement, heavily fouled corals *(Acropora, Porites);* depth: 11.6–13.7 m (38–45 ft); 7 May 1985.

LK-19   Back Reef, N (19a) and E (19b) of Sanctuary core area marker; seagrasses *(Thalassia, Syringodium)* and coral rubble (denser at 19b); depth: 2.4 m (8 ft); 7 May 1985.

LK-20   Back Reef, near N Sanctuary boundary; extensive sand plains with silt-covered seagrass *(Thalassia)*, widely scattered sponges and algae *(Penicillus, Codium, Halimeda);* depth: 5.5–6.1 m (18–20 ft); 7 May 1985.

LK-21   Intermediate and Deep Reef, SW of Fore Reef; spur-and-groove reef, coral rubble, large sponges, corals *(Montastrea)*, and abundant gorgonians; very silty; depth: 19.8–21.3 m (65–70 ft); 8 May 1985.

LK-22   Fore Reef, adjacent to Marker Buoy no. 24; spur-and-groove reef; depth: 2.4–6.7 m (8–22 ft); 8 May 1985.

◆ **LK-23**   Hawk Channel, due N of Marker Bouy no. 24; mud; depth: 12.5 m (41 ft); 9 May 1985.

LK-24   E Sand Plain, near E Sanctuary boundary; shallow sand on hard platform, seagrasses *(Thalassia, Syringodium)*, loggerhead sponges, and coral rubble with algae; depth: 7.0 m (23 ft); 9 May 1985.

LK-25   Fore Reef, at Sanctuary Marker Float no. 9; spur-and-groove reef; depth: 3.7–4.6 m (12–15 ft); 9 May 1985.

LK-26   W Back Reef, near W Sanctuary boundary; back reef slope with brain corals, gorgonians, some algae and large basket sponges, gradually merging into sand; depth: 6.7–12.8 m (22–42 ft); 9 May 1985.

LK-27   E Sand Plain, just E of Sanctuary core area; clean rippled sand patches, thick seagrass *(Thalassia)* with algae-covered coral rubble; depth: 4.6 m (15 ft); 9 May 1985.

◆ **LK-28**   Hawk Channel, N of NE corner of Sanctuary; mud and clay; depth: 12.8 m (42 ft); 10 May 1985.

LK-29   Intermediate and Deep Reef, E of Fore Reef; coarse sand, gorgonians, small coral rubble with algae, calcareous algae *(Halimeda)*, and filamentous algae; strong current; depth: 15.8 m (52 ft); 10 May 1985.

LK-30   E Sand Plain, near E Sanctuary boundary; bare sand; depth: 9.1–9.8 m (30–32 ft); 10 May 1985.

◆ **LK-31**   W Back Reef, W of W Sanctuary boundary; gorgonians and coral rubble; very silty; depth: 6.1–7.6 m (20–25 ft); 10 May 1985.

◆ **LK-32**   Hawk Channel, N of N Sanctuary boundary midpoint, mud and broken shell fragments; depth: 12.8 m (42 ft); 12 May 1985.

LK-33   Back Reef, near N Sanctuary boundary; sand, seagrass *(Thalassia),* and scattered sponges; depth: 4.6 m (15 ft): 12 May 1985.

LK-34   W Sand Plain, NW of Reef Flat; clean sand between seagrass *(Thalassia)* patches; depth: 5.5 m (18 ft); 12 May 1985.

LK-35   W Sand Plain, near W Sanctuary boundary; hard bottom, gorgonians, sponges, and small coral heads; very silty; depth: 11.0 m (36 ft); 12 May 1985.

LK-36   Center of E Sand Plain, NE of Reef Flat; sand, seagrass *(Thalassia),* sponges, rocks and coral rubble with algae, scattered gorgonians, and abundant loggerhead sponges; depth: 6.7 m (22 ft); 12 May 1985.

LK-37   Back Reef, adjacent to Reef Flat, at Sanctuary Marker Float no. 3; large pieces of coral rubble, live corals, and gorgonians; depth: 3.0 m (10 ft); 12 May 1985.

LK-38   Intermediate and Deep Reef, SW of Fore Reef; 30–35° slope with sponges and corals; silty; depth: 24.4–30.5 m (80–100 ft); 13 May 1985.

LK-39   Back Reef, NE corner of Sanctuary; high-relief reef with large corals, gorgonians, and sand patches; depth: 6.7 m (22 ft); 13 May 1985.

LK-40   Back Reef, NE corner of Sanctuary; high-relief reef; same locality as LK-39, night dive; depth: 6.7 m (22 ft); 13 May 1985.

◆ LK-41   Hawk Channel, N of N Sanctuary boundary midpoint; thin layer of sand and mud covering shell debris; depth: 12.5 m (41 ft); 14 May 1985.

◆ LK-42   Intermediate and Deep Reef, W of W Sanctuary boundary; silt-covered corals and gorgonians; depth: 16.8–18.3 m (55–60 ft); 14 May 1985.

LK-43   Reef Flat; seagrass *(Thalassia)* and sand; depth: 2.7 m (9 ft); 14 May 1985.

LK-44   W Sand Plain, near NW Sanctuary corner; clay; depth: 9.1 m (30 ft); 14 May 1985.

◆ LK-45   N of N Sanctuary boundary, near NW Sanctuary corner; small coral rubble, sponges, and coral; depth: 10.7 m (35 ft); 15 May 1985.

◆ LK-46   N of N Sanctuary boundary, N of W Sand Plain; sand with sparse seagrass *(Thalassia);* depth: 9.1 m (30 ft); 15 May 1985.

LK-47   Center of W Sand Plain; bare sand; depth: 9.8 m (32 ft); 15 May 1985.

LK-48   W Sand Plain, shallow edge of Intermediate and Deep Reef; patch reefs with abundant coral *(Acropora)* rubble; depth: 13.7 m (45 ft); 16 May 1985.

LK-49   Reef Flat, near NE Sanctuary core boundary (49a) to just N of Fore Reef (49b), adjacent to Marker Buoy no. 24; seagrass *(Thalassia)* and coral rubble (49b); depth: 0.3–1.8 m (1–6 ft); 16 May 1985.

LK-50   E Sand Plain, on E Sanctuary boundary; seagrass *(Thalassia)* and scattered coral heads; depth: 7.3 m (24 ft); 16 May 1985.

LK-51   Fore Reef, at Sanctuary Marker Float no. 12; spur-and-groove reef; depth: 7.6–9.1 m (25–30 ft); 17 May 1985.

LK-52   Intermediate and Deep Reef, SW of Fore Reef; coral ridges and sand grooves; depth: 13.7–15.2 m (45–50 ft); 16 August 1985.

LK-53   Back Reef, adjacent to NE Sanctuary core area marker; seagrass *(Thalassia),* coral rubble and sand patches; depth: 1.8–3.0 m (6–10 ft); 16 August 1985.

LK-54   W Back Reef; gorgonians, large corals, and sponges; night dive; depth: 8.5–9.1 m (28–30 ft); 16 August 1985.

LK-55   Intermediate and Deep Reef slope, S of Fore Reef; patch reef surrounded by clean, unconsolidated sand with some burrows, 50% hard bottom; depth: 24.4–25.9 m (80–85 ft); 17 August 1985.

LK-56   Back Reef, near N Sanctuary boundary, N of W Sand Plain; seagrasses *(Syringodium, Thalassia),* sand with thin flocculent crust and scattered coral rubble; depth: 7.6 m (25 ft); 17 August 1985.

LK-57   Intermediate and Deep Reef, near W Sanctuary boundary; sand, more than 50% hard bottom, gorgonians and large corals; depth: 24.4 m (80 ft); 18 August 1985.

APPENDIX

LK-58   Deep Sand Bed, S of Fore Reef; 100% cover of blue-green algae on sand; depth: 36.6–38.1 m (120–125 ft); 18 August 1985.

LK-59   Center of E Sand Plain; rippled sand with algae, short, epiphytized seagrass *(Thalassia)*, patches of delicate coral rubble and calcareous algae *(Halimeda)*; depth: 9.1 m (30 ft); 18 August 1985.

LK-60   Center of E Sand Plain; same locality as LK-59, night dive; rippled sand and sparse seagrass *(Thalassia)*; depth: 9.1 m (30 ft); 18 August 1985.

♦ **LK-61**   E of E Sanctuary boundary; high-relief reef with more dead than living coral, and undercut rocky ledges surrounded by silty sand; depth: 22.9–27.4 m (75–90 ft); 19 August 1985.

LK-62   S edge of Fore Reef, adjacent to Marker Bouy no. 24; spur-and-groove reef; depth: 3.0–9.1 m (10–30 ft); 19 August 1985.

LK-63   S edge of Fore Reef, adjacent to Marker Bouy no. 24; bare sand; depth: 6.1 m (20 ft); 19 August 1985.

LK-64   S edge of Fore Reef, adjacent to Marker Bouy no. 24; spur-and-groove reef; night dive; depth: 3.0–9.1 m (10–30 ft); 19 August 1985.

♦ **LK-65**   E of NE Sanctuary corner; silty sand, patch reef, abundant coral rubble, sparse seagrass *(Thalassia)* and calcareous algae; depth: 6.1–9.1 m (20–30 ft); 20 August 1985.

LK-66   Back Reef, near NE Sanctuary corner; hard bottom, sand and gorgonians; night dive; depth: 8.2 m (27 ft); 21 May 1986.

LK-67   Fore Reef, at Sanctuary Marker Float no. 16; spur-and-groove reef; depth: 1.8–6.7 m (6–22 ft); 22 May 1986.

LK-68   Center of E Sand Plain; sand and patch reefs; depth: 6.7 m (22 ft); 22 May 1986.

LK-69   W Sand Plain, adjacent to W Back Reef; seagrass *(Thalassia)*; depth: 9.1 m (30 ft); 22 May 1986.

LK-70   Reef Flat, N of Sanctuary Marker Float no. 12; seagrass *(Thalassia)* and sand; depth: 2.4 m (8 ft); 23 May 1986.

LK-71   W Sand Plain, NW of Reef Flat; bare sand with seagrass patches; depth: 4.6 m (15 ft); 23 May 1986.

LK-72   Fore Reef, at Sanctuary Marker Float no. 15; spur-and-groove reef; depth: 0.9–6.7 m (3–22 ft); 23 May 1986.

LK-73   Back Reef, NW of Reef Flat; dense seagrass *(Thalassia)* surrounded by bare sand; depth: 6.1 m (20 ft); 24 May 1986.

LK-74   Intermediate and Deep Reef, at Sanctuary Marker Float no. 28; spur-and-groove reef and coral rubble; depth: 9.1–14.6 m (30–48 ft); 24 May 1986.

LK-75   Reef Flat, just N of E Fore Reef; coral rubble; depth: 0.9–1.8 m (3–6 ft); 24 May 1986.

LK-76   Fore Reef, at Sanctuary Marker Float no. 20; spur-and-groove reef; depth: 5.5–6.1 m (18–20 ft); 24 May 1986.

LK-77   Back Reef, near N Sanctuary boundary, at Sanctuary Marker Float no. 26; patch reef surrounded by shell rubble and sand; depth: 4.9–6.1 m (16–20 ft); 12 August 1986.

LK-78   Back Reef, adjacent to Reef Flat, at Sanctuary Marker Float no. 3; spur-and-groove reef and coral rubble; depth: 7.6 m (25 ft); 12 August 1986.

LK-79   Back Reef, just N of Reef Flat; sand, coral rubble, and seagrass; depth: 1.5–3.0 m (5–10 ft); 12 August 1986.

LK-80   Back Reef, adjacent to Reef Flat, at Sanctuary Marker Float no. 3; spur-and-groove reef and coral rubble; depth: 7.6 m (25 ft); 13 August 1986.

LK-81   Center of W Sand Plain; sand, seagrass, and patch reef; depth: 9.8 m (32 ft); 13 August 1986.

LK-82   Back Reef, N of Reef Flat; sand and seagrass; depth: 5.5 m (18 ft); 13 August 1986.

LK-83   Back Reef, N of Reef Flat; seagrasses *(Thalassia, Halodule)* and sand; depth: 3.7 m (12 ft); 13 August 1986.

LK-84   Fore Reef, at Sanctuary Marker Float no. 19; spur-and-groove reef; depth: 6.1 m (20 ft); 13 August 1986.

LK-85 Intermediate and Deep Reef, at Sanctuary Marker Float no. 28; patch reef and coral rubble; depth: 15.2 m (50 ft); 14 August 1986.

LK-86 Fore Reef, at Sanctuary Marker Float no. 9; spur-and-groove reef and coral rubble; depth: 7.6 m (25 ft); 14 August 1986.

LK-87 Reef Flat, just N of W Fore Reef; coral rubble field; depth: 1.5–2.4 m (5–8 ft); 14 August 1986.

TABLE 2. Stations at which echinoderms were collected and observed, within and adjacent to Looe Key National Marine Sanctuary. Species are listed alphabetically within classes. LK-numbers refer to stations listed in Table 1. Stations located outside sanctuary boundaries are in **boldface** type. Refer to Sanctuary Map (Figure 191) and Table 1 for station positions.

Asteroidea
  Asterina folium LK-2
  Astropecten duplicatus LK-20, 30, 47, 59
  Echinaster sentus LK-21, **45**
  Ophidiaster guildingi LK-2, 14, 37, 48, 78, 84
  Oreaster reticulatus LK-34, 56, 59
  Poraniella echinulata LK-4, 21
Crinoidea
  Nemaster discoideus LK-38
Echinoidea
  Arbacia punctulata LK-21, 22, **31**, 35, 37, **45**, 49, 56, **65**, 67, 68, 78, 84, 87
  Astropyga magnifica LK-**45**, 60
  Brissopsis elongata LK-**17**, **32**
  Clypeaster luetkeni LK-30
  Clypeaster rosaceus LK-1, **11**, 12, 15, 20, 24, 33, 43, **45**, 50, 54, 56, 69, 73, 82
  Clypeaster subdepressus LK-8, **11**, 12, 13, 15, 20, 24, **31**, 33, 36, 39, 40, 43, **45**, 47, 50, 53, 54, 56, 59, 60, 66, 71
  Diadema antillarum LK-2, 4, 6, 8, 9, **11**, 12, 14, 18, 25, 26, **31**, 35, 37, 39, 40, 52, 53, 54, 56, **61**, 64, **65**, 66, 77, 78, 84, 87
  Echinometra l. lucunter LK-2, 14, 19
  Echinometra viridis LK-2, 7, 14, 22, 25, **31**, 48, 49, 53, 62, **65**, 67, 78, 87
  Echinoneus cyclostomus LK-1, 4, 6, 7, 9, 14, 25, **31**, 51
  Encope michelini LK-13, 44, **46**, 56, **65**, 81

Eucidaris tribuloides LK-1, 2, 4, 5, 6, 7, 9, **11**, 12, 14, 15, 18, 19, 20, 21, 22, 24, 25, 26, 27, **31**, 33, 35, 36, 37, 38, 39, 40, **42**, 49, 50, 51, 53, 54, 55, 56, 57, 58, **61**, 62, 63, 64, **65**, 66, 68, 69, 73, 77, 78, 87
Leodia sexiesperforata LK-3, 5, 8, 12, 13, 15, **16**, 20, 27, 30, 43, 44, 47, 59, 60, 62, **65**, 68, 71, 77, 81
Lytechinus variegatus LK-20, **31**, **45**, 56, **65**, 87
Meoma v. ventricosa LK-3, 8, **11**, 12, 13, 15, 19, 20, 27, 30, **31**, 33, 36, 40, 43, **45**, 46, 47, 50, 53, 54, 56, 58, 59, 60, **65**, 66, 69, 82
Moira atropos LK-13, **23**, **32**, **46**, 59
Paraster doederleini LK-**10**, **28**, **32**, **41**
Paraster floridiensis LK-7, 59
Plagiobrissus grandis LK-3, 6, 13, 15, **16**, 27, 29, 30, 47, 59, 60, 62, 71
Tripneustes ventricosus LK-2, 14, 24, 87
Holothuroidea
  Actinopyga agassizi LK-14, 19, 49, 64, 78
  Astichopus multifidus LK-12, 20, 24, 27, 33, 44, **45**, **46**, 56, 60, 69, 82
  Chiridota rotifera LK-2, 3
  Epitomapta roseola LK-24, 27, 33, **45**, 59
  Euapta lappa LK-7, 78, 87
  Holothuria arenicola LK-5, 7, 14, 15, 20, 27, 37, 53, 56, 69, 71, 87
  Holothuria cubana LK-24, **65**
  Holothuria mexicana LK-19, 52, 64
  Holothuria parvula LK-25
  Holothuria princeps LK-59
  Holothuria thomasi LK-64
  Isostichopus badionotus LK-4, 19, 27
  Leptosynapta multigranula LK-53
  Pseudothyone belli LK-53
  Synaptula hydriformis LK-27, 37, 50, 53, 62, 63
Ophiuroidea
  Amphiodia planispina LK-12, 33, 49
  Amphiodia pulchella LK-43, 50
  Amphioplus sepultus LK-**45**
  Amphipholis gracillima LK-56
  Amphipholis januarii LK-20, 24, 33, **45**, 50
  Amphipholis squamata LK-12, 20, 22, 25, 27, 33, 37, 48, 50, 53
  Amphiura fibulata LK-**11**
  Amphiura palmeri LK-**31**
  Amphiura stimpsonii LK-7, **42**, 48, 63

Astrophyton muricatum LK-6, 9, 12, 14, 19, 22, 26, **31**, 36, 38, 39, 40, 50, 54, 57, 60, **61**, 63, 64, **65**, 66

Ophiactis algicola LK-9, 18, 37, 48, 52, 55, **61**

Ophiactis quinqueradia LK-4, **6**, 7, 9, **16**, 18, 20, 21, 22, 24, 26, **31**, 35, 37, **42**, 48, 51, 55, 57, **61**, 64, **65**, 81

Ophiactis savignyi LK-5, 7, 15, 22, 25, 26, 27, 33, 37, 53, 63

Ophiocnida scabriuscula LK-27, 43

Ophiocoma echinata LK-1, 2, **6**, 7, 9, 14, 18, 19, 22, 25, 26, **31**, 35, 37, 50, 53, 54, 62, 67, 78, 87

Ophiocoma paucigranulata LK-7

Ophiocoma pumila LK-1, 2, 4, 5, **6**, 7, 9, 12, 14, 18, 19, 22, 24, 25, **31**, 35, 37, **42**, 48, 51, 53, 57, 62, 63, **65**, 67, 75, 78, 87

Ophiocoma wendtii LK-1, 2, 4, 5, **6**, 7, 9, 14, 18, 19, 21, 22, 24, 25, 26, 35, 37, 39, 48, 49, 52, 55, 57, 64, **65**, 67, 75, 77, 78, 87

Ophiocomella ophiactoides LK-24, 25

Ophioderma appressum LK-2, 4, 67, 9, 14, 18, 19, 21, 22, 25, 37, **42**, 62, 67, 78

Ophioderma brevispinum LK-11, 14, 15, 20, 40, **45**, 56, 65

Ophioderma cinereum LK-7, 9, 14, 18, 22, 24, 25, 26, **31**, 35, 37, 48, 62, 67, 78

Ophioderma guttatum LK-2, 4, **6**, 7, 9, 22, **31**, 35, 37, 39, **65**, 77, 78, 84

Ophioderma phoenium LK-1, 4, 7, 22

Ophioderma rubicundum LK-1, 4, **6**, 7, 9, 18, 21, 22, 37, 39, 40, **42**, 48, 52, 57, 62, **65**, 74, 78

Ophiolepis impressa LK-1, 4, **6**, 7, 9, 14, 18, 21, 22, 25, 26, **31**, 35, 37, 39, **42**, 48, 62, **65**, 67, 78

Ophiolepis paucispina LK-14, 37

Ophiomyxa flaccida LK-2, **6**, 7, 9, 14, 18, 21, 22, 25, **31**, 35, 37, **42**, 57, 62, 67, 78

Ophionereis reticulata LK-1, 7, 9, 12, 14, 19, 21, 24, 25, **31**, 33, 37, 48, 53, **65**, 75, 78

Ophionereis squamulosa LK-**6**, 12, 20, 22, 27, 35, 48, 50, 53, 63

Ophionereis vittata LK-55

Ophiophragmus pulcher LK-**11**, 15, 24, 33, 37, **45**, 50, 53, 56, 77

Ophiopsila riisei LK-**6**, 12, 22, 35, 36, 40, **42**, 48, 64

Ophiopsila vittata LK-33

Ophiostigma isocanthum LK-4, 12, 20, 22, 37, **42**, **45**, 48, 50, 52, 55, 57

Ophiostigma siva LK-4, 12, 20, 22, 37, **42**, 48, 50, 52, 55, 57

Ophiothrix angulata LK-1, 2, 3, 4, 5, **6**, 7, 9, 12, 14, 15, **16**, 18, 19, 20, 22, 24, 26, 27, **31**, 33, 35, 36, 37, 38, 39, 40, **42**, **45**, 48, 51, 52, 53, 54, 55, 57, 58, 60, **61**, 62, 63, 64, **65**, 75, 77, 78, 85

Ophiothrix lineata LK-1, 4, 5, **6**, 7, 9, **11**, 12, 18, 21, 22, 24, 26, **31**, 33, 35, 36, 37, 38, 39, **42**, 48, 54, 55, 57, 64, 65

Ophiothrix orstedii LK-1, 2, 4, 5, **6**, 7, 9, 12, 14, 15, 18, 19, 20, 22, 24, 25, 26, **31**, 33, 35, 36, 37, 39, **45**, 48, 49, 53, 62, 63, **65**, 75, 77, 78, 87

Ophiothrix suensonii LK-1, 4, **6**, 9, 14, 18, 21, 22, 26, **31**, 35, 38, 40, **42**, 48, 52, 54, 63, 64, **65**, 85

GLOSSARY

ABORAL. In a direction away from the mouth; the part of the body opposite the mouth.

ACCESSORY DORSAL ARM PLATE. In some ophiuroids, one or several small, symmetrically arranged plates that are inserted between the dorsal arm plate and the lateral arm plate.

ADAPICAL. In echinoids, toward the highest part of the test.

ADORAL SHIELDS. In ophiuroids, a pair of plates, one of which is found at each side of the oral shield.

ADPRESSED. Squeezed against. The adpressed arm spines of ophiuroids are flattened against the sides of the arm.

AMBULACRAL GROOVE. In asteroids, the arm furrow on the oral (ventral) surface of the arm, in which the tube feet are carried. Its sides are formed by the adambulacral plates, and it is roofed by the ambulacral plates. In crinoids, a furrow on the oral (dorsal) surface of the pinnules, arms, and central body, which is lined with cilia and bordered by the tube feet.

AMBULACRUM. A zone of the body that carries tube feet (pl. ambulacra). Echinoderms generally have five ambulacra. The midline of an ambulacrum is a radius.

ANAL CONE (OR ANAL TUBE). In crinoids and echinoids, a fleshy projection bearing the anus at its apex.

ANASTOMOSING. Said of branching structures that are joined to form a network pattern.

ANCHOR. See Ossicle types.

ANCHOR PLATE. See Ossicle types.

ANNULUS. A ring-shaped structure (pl. annuli).

APICAL SYSTEM. In echinoids, a ring of specialized skeletal plates, including the genital plates and ocular plates. Usually located on the highest point of the test.

APODOUS (APODAN). Lacking tube feet, in reference to holothurians.

APPENDAGE. A tube foot, spine, pedicellaria, or arm of an adult, or a projection from the larval body. In holothuroids, the tube feet may be modified to form a papilla lacking terminal suckers.

ARBORESCENT. Branching, and thus treelike in form.

ARISTOTLE'S LANTERN. The dental complex of most echinoids, which consists of up to 40 movable ossicles and 60 controlling muscles arranged in units of five. Five teeth, which are part of the lantern, are protruded through the mouth for feeding. The lantern is reduced or lacking in irregular echinoids, except the clypeasteroids.

ARM. In asteroids, crinoids, and ophiuroids, a movable, jointed ambulacral projection, distal to the disk or calyx, that carries a radial branch of the water vascular system and the nervous system. Sometimes called a ray.

ARM JOINT. In ophiuroids, one of a series of articulating units composing the arm, consisting of an internal vertebral ossicle, surrounding dorsal, lateral, and ventral arm plates, and associated structures. Also referred to as an arm segment.

ARM LENGTH. A body dimension of an ophiuroid measured from the edge of the disk to the tip of an arm.

ARM SPINES. Spines attached to the lateral arm plate in ophiuroids and to the various arm plates in asteroids.

ASEXUAL REPRODUCTION. Reproduction that occurs without the fusion of male and female gametes, usually by splitting of the body into two parts that regenerate. The genetically identical offspring of an asexual parent is a clone.

AUTOEVISCERATION. In holothuroids, expulsion of the digestive tract and associated organs through the anus; in some species, the anterior end of the body ruptures and the calcareous ring and associated organs are expelled. Believed to be a defensive mechanism.

AUTOTOMY. A process of self-mutilation involving loss of portions of the body. Sometimes defensive, as in ophiuroids, which detach the disk or parts of the arms. Also, part of the normal processes of asexual reproduction and the production of crinoid arm branches.

BASAL. In crinoids, one of a circlet of five plates that form part of the calyx.

BASKET. See Ossicle types.

BILATERAL SYMMETRY. A pattern of symmetry, based upon an anterior-posterior axis, in which the left side of the body is a mirror image of the right side.

BINARY FISSION. See Fission.

BIPINNARIA. A free-swimming larval stage of asteroids. Bipinnaria larvae have blunt larval appendages that support a ciliated band. The bipinnaria either may develop into a brachiolaria larva or may undergo metamorphosis, during which the juvenile arises on the left side of the larval body.

BROODING. Reproductive mode in which the embryos are protected on, in, or beneath the parent and emerge as tiny, crawl-away juveniles.

BURSA. In ophiuroids, an organ within the disk formed by an inpouching of epidermis (see Bursal slit). Bursae function as respiratory structures and are associated with the gonoducts. They house the developing embryos of brooding species.

BURSAL SLIT. The opening of a bursa, located on the ventral interradius of the disk at the base of the arm. There generally is one bursal slit on each side of an arm.

BUTTON. See Ossicle types.

CALCAREOUS RING. A ring of large ossicles surrounding the holothurian esophagus. It forms a point of insertion for longitudinal muscles and, when present, retractor muscles.

CALCITE. The mineral form of calcium carbonate that makes up the echinoderm skeleton.

CALYX. The cuplike central portion of the crinoid body, which supports the arms and visceral mass.

CENTRODORSAL. The middle ossicle attached to the aboral surface of the crinoid calyx; commonly carries cirri.

CIDAROIDS. A subclass of echinoids that arose in the Triassic; typified by species that have few, large, solid spines.

CIRRI. The unbranched, jointed appendages arising from the crinoid centrodorsal (sing. cirrus); they are used for attachment to the substratum.

CLOACA. In holothuroids, the posterior part of the intestine; it carries the openings to the respiratory trees and Cuvierian tubules, when present.

COELOMIC CAVITY. A fluid-filled space within an animal's body, which is lined with peritoneal tissue. The largest coelom of echinoderms is the body cavity separating the digestive tract and the body wall.

COMMENSAL. An organism that lives in association with another organism and which usually benefits from the partnership without harming its host.

CONGENERS. The species belonging to a single genus.

CONTRA. In opposition.

COQUINOID. Refers to a hard, cemented deposit of shell fragments. Named for the small and abundant coquina clam.

CRENULATE. Grooved; refers to the ribbed edge of certain echinoid tubercles.

CRYPSIS. Concealment, or the habit of concealment, which protects an organism by rendering it inconspicuous to predators and safe from physical damage.

CUVIERIAN TUBULES. Defensive structures of some holothurians, discharged through the anus as sticky threads that entangle and discourage predators.

DENDRITIC. Branching in a treelike manner, as in certain holothuroid tentacles.

DENTAL PAPILLAE. See Papillae.

DERMIS. The stratum of cells beneath the epidermal covering of the body wall. In echinoderms, the skeleton develops in the dermis and is filled with dermal tissue.

DIGITATE. Fingerlike, or carrying fingerlike structures; applied to certain holothuroid tentacles.

DISK. The round or pentagonal central body region of ophiuroids and asteroids. Also, in many echinoderms, the tube foot is equipped with a terminal disk, usually employed for attachment to substrates.

DISK DIAMETER. A body dimension of an ophiuroid measured from the distal edge of a pair of radial shields to the disk edge in the opposite interradius.

DISTAL. In a direction away from the center of the body; for example, toward the tip of the arm in asteroids or the tip of a spine in echinoids.

DORSAL. In echinoderms, this term is variously applied. In asteroids, ophiuroids, and echinoids, it usually refers to the surface of the body that is opposite the mouth, the surface that is uppermost. In holothuroids, with mouth and anus at opposite ends of a cylindrical body, the uppermost surface is considered dorsal. In crinoids, by convention, the surface opposite the mouth is considered dorsal even though it is functionally the ventral (lower) side.

DORSAL ARM PLATE. A plate on the aboral surface of an ophiuroid arm joint; one of the plates on the aboral surface of an asteroid arm.

ECHINOPLUTEUS. The free-swimming larval stage of an echinoid. Echinoplutei have appendages that are supported by skeleton and that bear ciliary bands. The process of metamorphosis in which a juvenile echinoid arises from a rudiment on the left side of the echinopluteus body is completed in only a few minutes.

ECHINULATE. Something spiny or prickly, usually referring to the microscopic texture of a skeletal element such as a spine.

EMBRYO. An early developmental stage that is enclosed in a fertilization membrane or protected by the body of the parent. It transforms into a juvenile through metamorphosis.

EPIFAUNA. Animals living on the surface of the ocean floor, as opposed to the burrowing infauna.

EPT (=EXPANDED PERIPHERAL TRABECULAE). Microscopic, transparent nodules on the surface of skeletal plates. In ophiocomid ophiuroids they are a component of a photoreceptor system.

EVISCERATION. See Autoevisceration.

EXTANT. In existence, as opposed to extinct.

EXTINCT. No longer in existence, as opposed to extant.

FASCIOLES. In many irregular echinoids, narrow bands of small, specialized spines; visible on the denuded test as bands of densely packed, tiny tubercles. Fascioles provide a flow of water to aid in respiration.

FENCE PAPILLAE. See Papillae.

FISSION. Asexual reproduction by splitting of the body into two parts, each of which regenerates into a complete animal. Fissiparity refers to the process of asexual reproduction by means of fission.

FISSIPAROUS. Reproducing asexually by fission (see Fission).

FOOD GROOVE. In crinoids, furrows lined with cilia, which conduct particles of food from the pinnules to the arms and then on to the mouth (see Ambulacral groove).

GAMETOGENESIS. The process of formation of reproductive cells, the eggs and sperm.

GENITAL PAPILLAE. In ophiuroids, granules or spinules attached to the edge of the bursal slit. The term is also used (perhaps more appropriately) for the fleshy outlets of the gonoducts, in ophiuroids that lack bursae and in some echinoids. In holothuroids, a single, fleshy genital papilla opens to the exterior on the dorsal surface of the body immediately posterior to the tentacles.

GENITAL PLATE. In ophiuroids, a barlike ossicle connecting the radial shield to the arm and supporting the radial edge of the bursal slit.

GENITAL PINNULE. In crinoids, a pinnule distal to the oral pinnules that contains gonad tissues.

GLOBIFEROUS PEDICELLARIA. A three-valved echinoid pedicellaria that is equipped with venom sacs.

GONOCHORIC. Used of organisms possessing separate sexes, as opposed to hermaphroditic.

GRANULES. See Skeletal elements.

HEART URCHIN. A more or less heart-shaped burrowing echinoid; the term is usually used in reference to the order Spatangoida.

HERMAPHRODITIC. Used of organisms with functional male and female reproductive structures. Hermaphroditic individuals may express both sexes simultaneously, alternately, or sequentially. See Gonochoric.

HOOKS. Minute, movable, curved, and sharply pointed ossicles that articulate with the dorsal arm scales in gorgonocephalid ophiuroids (other ophiuroids may have hook-shaped *arm spines* attached to the lateral arm plates).

INFAUNA. Animals living within sediment on the ocean floor, as opposed to epifauna.

INFEROMARGINALS. In asteroids, a row of plates that define the ventral edge of the body; inferomarginals are overlain by a row of superomarginals.

INFRADENTAL PAPILLAE. See Papillae.

INTERAMBULACRUM. An oral or aboral sector of the body lying between two ambulacra. An interradius.

INTERRADIAL (INTERRADIUS, INTERRADII). Referring to interambulacral sectors of the body.

INTERTIDAL. A zone of the seashore that is exposed by low tides.

INTROVERT. In some holothurians, the anterior of the body including tentacles and associated structures. It can be withdrawn into the body by means of retractor muscles.

IRREGULAR ECHINOID. A heart-shaped or disk-shaped echinoid, usually covered with very short spines, generally living in or on soft sediment. The anus of irregular echinoids is dissociated from the apical system of the test, and the body is bilaterally symmetrical about an axis crossing the anus and the mouth.

JAW. A movable triangular structure that extends into the mouth in ophiuroids and asteroids. In the latter group, it is also referred to as a mouth-angle plate.

JOINT. See Arm joint.

LABRUM. A posterior or lower projection of the border of the mouth in irregular echinoids.

LAPPETS. In crinoids, small movable plates that support the tube feet and form a protective covering over the food grooves.

LARVA. An early developmental stage that is independent of the fertilization membrane and the parent. Larvae (pl.) transform into juveniles through the process of metamorphosis.

LATERAL ARM PLATES. In ophiuroids, paired plates covering the sides of each arm joint and bearing the arm spines.

LECITHOTROPHY. A mode of reproduction in which free-swimming larvae develop using nutrient deposits in the egg. Lecithotrophic larvae do not feed on particulate matter, but they may supplement yolk reserves by the uptake of nutrients dissolved in seawater.

LUNULE. A slit in the echinoid test, as in the five- or six-holed sand dollars.

MADREPORITE. A plate with numerous perforations that is connected to the water vascular ring by a so-called stone canal. In most holothuroids it is internal. In asteroids and echinoids it opens to the exterior on the dorsal surface of the body; in ophiuroids it opens on the ventral surface, near the mouth. Crinoids lack a madreporite, having instead a series of small pores in the tegmen, opening to the body cavity.

MEIOFAUNA. Microscopic animals between 0.1 and 1 mm (0.004–0.04 in) in size, which are smaller than macrofauna and larger than microfauna. Meiofaunal organisms typically live among sand grains.

MESENTERY. A membrane composed of the peritoneal tissue lining the body cavity, which connects the viscera to the body wall.

MILIARY GRANULE. See Ossicle types.

OCULAR PLATE. A plate in the apical system of echinoids. The five ocular plates are radial (ambulacral) in position, and new ambulacral plates develop at their distal edges. In asteroids, the most distal plate of the arm, enclosing the terminal tube foot.

OPHIOPLUTEUS. The free-swimming larval stage of an ophiuroid. Ophioplutei (pl.) have appendages that are supported by skeleton and bear ciliary bands. During metamorphosis, the juvenile ophiuroid develops from a rudiment on the ventral surface of the larva; the process may involve resorption or loss of parts of larval structures.

OPTIC CUSHION. A pigmented light-sensory structure of asteroids, projecting from the base of the ventral surface of the terminal tube foot.

ORAL. In a direction toward the mouth; a part of the body on the same surface as the mouth.

ORAL FRAME. In ophiuroids, a ring of ossicles surrounding the mouth opening. The frame is composed of the jaw structures, oral shields, and adoral shields; the arms radiate from the sides of the frame, and its dorsal surface is capped by the disk.

ORAL PAPILLAE. See Papillae.

ORAL SHIELD. A relatively large plate at the distal end of the ophiuroid jaw. At least one of the oral shields is modified as a madreporite.

ORAL TENTACLES. In ophiuroids, tube feet inside the mouth, arising from the jaws. Also referred to as oral tube feet.

OSMOREGULATE. To control the concentration of dissolved chemicals in the body fluids. Osmoconformer animals have body fluids at the same concentration as the surrounding liquid; osmoregulator animals maintain body fluids at a concentration that is independent of the surrounding medium.

OSSICLE. A small, usually microscopic skeletal element, embedded in integument. Commonly found in the body wall of holothurians, but also known from the body wall and body cavity tissues of other echinoderms (e.g., the ophiuroid stomach) and the tube feet of echinoids and asteroids. The term is sometimes used to refer to all skeletal elements, especially in crinoids.

OSSICLE TYPES. In holothuroids, the various types of microscopic skeletal ossicles can be broadly classified, based on their shapes. The shanks of fluked anchors are associated with shield-shaped anchor plates in the body wall of synaptid holothuroids. Baskets are minute cup-shaped ossicles usually with four perforations. Buttons are also minute, with few perforations; they may be smooth or knobbed. Solid, irregularly shaped miliary granules are found in the body wall and muscles of apodous holothuroids. Sievelike perforated plates are widespread in holothuroids and may be found in other echinoderm classes, especially in juvenile individuals. Rods are commonly found as supporting structures in tentacles or tube feet. Rosettes are tiny ossicles formed by dichotomous branching of rods in a single plane. Tables consist of a basal perforated disk and a central spire composed of three or more rods joined by crossbars. Wheels are common in some apodous holothuroids; they usually have six spokes and a serrated rim.

PAPILLAE. In holothuroids, specialized dorsal tube feet that lack a suckered tip. In ophiuroids, certain skeletal elements of the jaws or disk: dental papillae—a cluster of small, blunt, spinelike structures on the dental plate, near the ventral tip of the jaw; fence papillae—peg-shaped spinelets fringing the edge of the disk of *Ophiophragmus* species; infradental papillae—in amphiurid ophiuroids, a pair of small, blocklike plates attached at the ventral tip of the jaw, below the teeth; oral papillae—small plates at the edge of the mouth, attached to the edges of the jaw plate and/or to the adoral shield. Oral papillae can be variously shaped, from spinelike to scalelike. Also see Genital papillae.

PAPILLOSE. Covered with papillae.

PAPULAE. Small, soft, retractable extensions of the body cavity that project through pores in the body wall of asteroids; used for respiration. Papulae may be finger- or glove-shaped and are sometimes aggregated in dense patches.

PAXILLAE. In some asteroids, columnar plates that bear an apical cluster of spinelets or granules.

PEDICELLARIAE. Small stalked or unstalked pincerlike organs on the body of asteroids and echinoids, used for defense and grooming.

PELTATE. Shield-shaped; used to describe the tentacles of some holothuroids.

PENICILLATE TUBE FOOT. In some echinoids, a long, extensible tube foot bearing a brushlike array of glandular projections at the tip.

PENTACTULA. The developmental stage following metamorphosis; occurring in holothuroids with either planktotrophic or lecithotrophic larvae. It has an anterior mouth and buccal podia and one or two tube feet.

PERFORATED PLATE. See Ossicle types.

PERFORATE TUBERCLE. In echinoids, a primary or secondary tubercle with an apical perforation for the insertion of a ligament.

PERIPROCT. In echinoids, a flexible region surrounding the anus, which consists of a membrane containing embedded plates and often bearing spines and pedicellariae.

PERISTALTIC. Used to describe a rhythmic, wavelike motion generated by alternating muscular contractions and relaxation.

PETALS. In irregular echinoids, five expanded, aboral parts of the ambulacra, which have a flowerlike shape (also known as petaloids). Within the petals are tube feet that generally are modified for respiration.

PINNATE. Featherlike.

PINNULAR COMB. A group of modified pinnulars (see Pinnule) of an oral pinnule, which has a comblike profile. Present in Comasteridae and some Antedonidae.

PINNULE. In crinoids, an unbranched appendage arising from a brachial ossicle and composed of a series of pinnular ossicles.

PLANKTOTROPHY. The mode of development of free-swimming larvae that feed on particulate matter. Planktotrophic larvae grow using nutrients obtained from the plankton.

PLATES. See Skeletal elements.

POLIAN VESICLE. Fluid-filled sacs attached to the water vascular ring. They act as reservoirs, holding fluid that is displaced when tube feet contract.

PRIMARY PLATES. The first-formed plates on the dorsal side of the disk. In ophiuroids, for example, these are the central and five radial plates. In adults, they may form a rosette of scales near the center of the disk, or they may be separated by numerous secondarily developed scales.

PRIMARY SPINES. In echinoids, the first-formed, larger spines; carried on primary tubercles in the ambulacra and interambulacra.

PROXIMAL. Toward the center of the body.

R (=MAJOR RADIUS). A body dimension of an asteroid, measured from the center of the disk to the tip of the longest arm.

r (=MINOR RADIUS). A body dimension of an asteroid, measured from the center of the disk to the edge of the disk in the middle of an interradius.

RADIAL. In a direction toward the central axis of an arm or ambulacrum; a part of the body near an arm or ambulacrum.

RADIAL SHIELDS. Pairs of plates on the dorsal surface of the ophiuroid disk, which lie near the base of each arm. They are usually relatively large and conspicuous, but may be hidden by granules or superficial scales.

RADIAL SYMMETRY. A pattern of symmetry in which identical segments of the body are arranged around a central axis. Echinoderms generally have a five-part (pentamerous) radial symmetry.

REGULAR ECHINOID. A more or less spherical echinoid, with long spines, and with the anus situated at the center of the aboral surface. Typically lives on hard bottoms or among marine plants.

RESPIRATORY TREES. Paired respiratory organs of some holothuroids. They are attached to the cloaca, just inside the anus, and project anteriorly in the body cavity.

RHEOPHILIC. Literally, "current-loving." Rheophilic organisms populate habitats with significant water movement. Rheophobic organisms avoid currents and occupy sheltered positions or are restricted to low-energy habitats.

ROD. See Ossicle types.

ROSETTE. Modified basal ossicles of some crinoids; a microscopic ossicle of some holothuroids (see Ossicle types).

SCALES. See Skeletal elements.

SECONDARY SPINES. In echinoids, the smaller spines carried on secondary tubercles in the ambulacra and interambulacra (see Primary spines).

SKELETAL ELEMENTS. Supporting and protective dermal structures consisting of a calcite meshwork (stereom) and invested with a thin layer of tissue (stroma). Various skeletal elements are distinguished as follows: plates are tabular structures with a characteristic shape and a fixed position; scales are flat, thin structures that are overlapping, tesselate, or haphazardly arrayed; spines are *movable*, articulating structures that are long, slender, and attenuated. Small structures *fixed* to the surface of scales or plates include the following: granules that are minute and nearly equidimensional; spinelets that are enlarged, elongate cylindrical, or angular granules (this term is sometimes applied to small spines); spinules that have various numbers of pointed apical projections (e.g., bifid, trifid, multifid). Also fixed, elongate, blunt-tipped, and larger than granules are the small structures referred to as stumps, which are usually prickly, and tubercles, which are smooth and more massive. In addition, tubercles (see below) can refer to outgrowths of plates, rather than to articulated elements. Also see Papillae.

SENSORY CUP. A stalked cuplike sensory organ on the tentacle stems of some holothuroids.

SOLE. In some holothuroids, the flattened ventral part of the body, either covered with or surrounded by tube feet.

SPHAERIDIA. Organs of orientation generally situated near the mouth of echinoids, consisting of a tiny spine with a swollen tip, set in a pit or a closed chamber.

SPINES. See Skeletal elements.

SPINELETS. See Skeletal elements.

STATOCYST. In some holothuroids, a sensory organ of balance, consisting of a hollow sphere lined by sensory receptor cells, which contains movable granules.

STONE CANAL. A tube, usually reinforced with ossicles, leading from the madreporite to the water vascular ring canal.

SUBTIDAL. Refers to the region of the sea, below the extreme low-tide level, that is perpetually covered with water.

SUPEROMARGINALS. In asteroids, a row of plates defining the dorsal edge of the body. They overlie a row of inferomarginals.

SUPRAGENERIC. Above the rank of a genus.

TABLE. See Ossicle types.

TEETH. In ophiuroids, small plates or spines attached to the dental plate on the inner edge of the jaw, a series of them extending into the mouth. In echinoids, the five hard, sharp, and movable ossicles incorporated in Aristotle's lantern. The term also refers to five movable ossicles that surround the anus of some holothuroids.

TEGMEN. The surface of the crinoid visceral mass that bears the mouth and confluent food grooves leading to the arms, as well as the anus. It may be naked or reinforced with ossicles or plates.

TENTACLE PORE. In ophiuroids, an opening between the ventral arm plate and the lateral arm plate, through which a tube foot projects. Each arm joint has two tentacle pores.

TENTACLES. In holothuroids, feeding structures in the form of highly modified tube feet arranged in a ring around the mouth.

TENTACLE SCALES. Small, movable spines or scales, associated with ophiuroid tube feet, which are attached to the ventral arm plate and/or lateral arm plate. Tentacle scales may cover the tentacle pores and protect the retracted tube feet.

TERRIGENOUS. Refers to sediments derived from the land, as opposed to sediments of a marine origin.

TEST. The "shell" of an echinoid, made up of many small skeletal plates. A "naked" test is one from which soft tissue and projecting structures such as spines have been removed. This process occurs naturally after the death of a sea urchin. To identify some urchins, it is necessary to clean a portion of the test with bleach, to see the underlying plates.

TRIFID. Divided into three parts. Regarding "trifid spinule," see Skeletal elements.

TUBE FEET. Fluid-filled, fingerlike extensions of the water vascular system that protrude through openings in the skeleton or between skeletal elements. Muscles and nerves in the shaft of the tube feet control their movements; glands (and sometimes a muscular sucker) at the tip function in adhesion. Specialized tube feet are used for locomotion, feeding, burrowing, respiration, and a combination of functions (see Papillae, Tentacles).

TUBERCLE. A rounded prominence on the skeleton. In echinoids and some asteroids, a spine articulates with a tubercle (see also Skeletal elements).

VENTRAL. In echinoderms, this term is variously applied. In asteroids, echinoids, and ophiuroids, it is the surface of the body that carries the mouth; this surface is in contact with the substrate. In the holothuroids, with mouth and anus at opposite ends of a cylindrical body, the ventral surface is lowermost—in contact with the substrate. In crinoids, the ventral surface carries the mouth and is functionally the uppermost surface.

VENTRAL ARM PLATE. A plate on the oral surface of each ophiuroid arm joint; one of the plates on the oral surface of an asteroid arm.

VERMIFORM. Worm-shaped.

VERTEBRA. An internal ossicle within every ophiuroid arm joint, connected by ligament and muscle to the vertebrae of adjacent joints. It is so named because of a resemblance to bones in the human spinal column.

VITELLARIA. A type of free-swimming lecithotrophic echinoderm larva. It is barrel-shaped, has several transverse rings of locomotory cilia, and lacks a mouth.

WHEEL. See Ossicle types.

LITERATURE CITED

Abbott, D. P., J. C. Ogden, and I. A. Abbott. 1974. *Studies on the Activity Pattern, Behavior, and Food of the Echinoid Echinometra lucunter (Linnaeus) on Beachrock and Algal Reefs at St. Croix, U.S. Virgin Islands.* West Indies Laboratory Special Publication No. 4. Fairleigh Dickinson University, Christiansted, St. Croix, U.S. Virgin Islands. iv + 111 pp.

Ablanedo, N., H. González, M. Ramírez, and I. Torres. 1990. Evaluación del erizo de mar *Echinometra lucunter* como indicador de contaminación por metales pesados, Cuba. *Aquatic Living Resources—Ressources vivantes aquatiques* 3:113–120.

Absalao, R. S. 1990. Ophiuroid assemblages off the Lagoa dos Patos outlet, Southern Brazil. *Ophelia* 31:133–143.

Agassiz, A. 1863. List of the echinoderms sent to different institutions in exchange for other specimens, with annotations. *Bulletin of the Museum of Comparative Zoölogy at Harvard College* 1:17–28.

———. 1877. North American starfishes. *Memoirs of the Museum of Comparative Zoölogy at Harvard College* 5: iv + 1–137, 20 pls.

Agassiz, L. 1841. *Monographies d'echinodermes vivants et fossiles. Échinites. Famille des Clypéastroides. Seconde Monographie. Des Scutelles.* Petitpierre, Neuchâtel. vi + 151 pp., 26 pls.

Alexander, D. E., and J. Ghiold. 1980. The functional significance of the lunules in the sand dollar, *Mellita quinquiesperforata*. *Biological Bulletin* (Woods Hole) 159:561–570.

Alexander, S., and K. Haburay. 1977. First record of *Ophiophragmus moorei* (Echinodermata, Ophiuroidea) in Florida coastal waters. *Florida Scientist* 40:254–255.

Allain, J. Y. 1978. Déformations du test chez l'oursin *Lytechinus variegatus* (Lamarck) (Echinoidea) de la Baie de Carthagène. *Caldasia* 12: 363–375.

Allee, W. C. 1927. Studies in animal aggregations: Some physiologial effects of aggregation on the brittle starfish, *Ophioderma brevispina*. *Journal of Experimental Zoology* 48:475–495.

Alvà, V., and M. Jangoux. 1990. Fréquence et causes presumées de la régénération brachiale chez *Amphipholis squamata* (Echinodermata: Ophiuroidea). In *Echinoderm Research.* Proceedings of the Second European Conference on Echinoderms, Brussels, Belgium, 18–21 September 1989, eds. C. De Ridder, P. Dubois, M.-C. Lahaye, and M. Jangoux, 157–153. Balkema, Rotterdam.

———. 1992. Démographie et reproduction de l'ophiure *Amphipholis squamata* à Luc-sur-Mer (Baie-de-Seine, France). In *Echinoderm Research 1991.* Proceedings of the Third European Conference on Echinoderms, Lecce, Italy, 9–12 September 1991, eds. L. Scalera-Liaci and C. Canicattì, 203. Balkema, Rotterdam.

Alvà, V., and C. Vadon. 1989. Ophiuroids from the western coast of Africa (Namibia and Guinea-Bissau). *Scientia Marina* 53:827–845.

Ambrose, W. G., Jr. 1993. Effects of predation and disturbance by ophiuroids on soft-bottom community structure on Oslofjord: Results of a mesocosm study. *Marine Ecology Progress Series* 97:225–236.

Anderson, J. M. 1960. Histological studies on the digestive system of a starfish, *Henricia*, with notes on Tiedemann's pouches in starfishes. *Biological Bulletin* (Woods Hole) 119: 371–398.

———. 1966. Aspects of nutritional physiology. In *Physiology of Echinodermata*, ed. R. A. Boolootian, 329–357. Interscience Publishers, New York.

———. 1978. Studies on functional morphology in the digestive system of *Oreaster reticulatus* (L.) (Asteroidea). *Biological Bulletin* (Woods Hole) 154:1–14.

Aronson, R. B. 1987. Predation on fossil and Recent ophiuroids. *Paleobiology* 13:187–192.

———. 1988. Palatability of five Caribbean ophiuroids. *Bulletin of Marine Science* 43:93–97.

———. 1991. Predation, physical disturbance, and sublethal arm damage in ophiuroids: A Jurassic-Recent comparison. *Marine Ecology Progress Series* 74:91–97.

———. 1992. The effects of geography and hurricane disturbance on a tropical predator–prey interaction. *Journal of Experimental Marine Biology and Ecology* 162:15–33.

———. 1993. Hurricane effects on backreef echinoderms of the Caribbean. *Coral Reefs* 12: 139–142.

Aronson, R. B., and C. A. Harms. 1985. Ophiuroids in a Bahamian saltwater lake: The ecology of a Paleozoic-like community. *Ecology* 66:1472–1483.

Austin, W. C., and M. G. Hadfield. 1980. Ophiuroidea: The brittle stars. In *Intertidal Invertebrates of California*, eds. R. H. Morris, D. P. Abbott, and E. C. Haderlie, 146–159. Stanford University Press, Stanford, California.

Avent, R. M., M. E. King, and R. H. Gore. 1977. Topographic and faunal studies of shelf-edge prominences off the central East Florida coast. *Internationale Revue der Gesamten Hydrobiologie* 62:185–208.

Ayres, W. O. 1851a. A description of a new species of *Synapta* under the name of *Synapta tenuis*. *Proceedings of the Boston Society of Natural History* 4:11–12.

———. 1851b. A description and drawings of a new species belonging to the genus *Ophiolepis*. *Proceedings of the Boston Society of Natural History* 4:133–135.

———. 1852. Descriptions of two new species of *Ophiuridae* with the names *Ophiothrix hispida*, and *Ophiolepis uncinata*. *Proceedings of the Boston Society of Natural History* 4:248–250.

Baird, B. H., and D. Thistle. 1986. Uptake of bacterial ultracellular polymer by a deposit-feeding holothurian (*Isostichopus badionotus*). *Marine Biology* (Berlin) 92:183–187.

Bak, R. P. M., M. J. E. Carpay, and E. D. de Ruyter van Steveninck. 1984. Densities of the sea urchin *Diadema antillarum* before and after mass mortalities on the coral reefs of Curaçao. *Marine Ecology Progress Series* 17:105–108.

Baker, S. M., and N. B. Terwilliger. 1993. Hemoglobin structure and function in the rat-tailed sea cucumber, *Paracaudina chilensis*. *Biological Bulletin* (Woods Hole) 185:86–96.

Balser, E. J., and E. E. Ruppert. 1993. Ultrastructure of axial vascular and coelomic organs in comasterid featherstars (Echinodermata: Crinoidea). *Acta Zoologica* (Stockholm) 74:87–101.

Balser, E. J., E. E. Ruppert, and W. B. Jaeckle. 1993. Ultrastructure of the coeloms of auricularia larvae (Holothuroidea: Echinodermata): Evidence for the presence of an axocoel. *Biological Bulletin* (Woods Hole) 185:86–96.

Barel, C. D. N., and P. G. N. Kramers. 1977. A survey of the echinoderm associates of the north-east Atlantic area. *Zoologische Verhandelingen* (Leiden) 156:1–159.

Bartsch, I. 1974. *Ophioderma tonganum* Lütken und *Ophioderma leonis* Döderlein (Ophiuroidea, Echinodermata). *Mitteilungen aus dem Hamburgischen Zoologischen Museum und Institut* 70:97–104.

Bauer, J. C. 1982. On the growth of a laboratory-reared sea urchin, *Diadema antillarum* (Echinodermata: Echinoidea). *Bulletin of Marine Science* 32:643–645.

Bauer, J. C., and C. J. Agerter. 1987. Isolation of bacteria pathogenic for the sea urchin *Dia-*

dema antillarum (Echinodermata: Echinoidea). *Bulletin of Marine Science* 40:161–165.

Baumiller, T. K., and M. LaBarbera. 1989. Metabolic rates of Caribbean crinoids (Echinodermata), with special reference to deep-water stalked and stalkless taxa. *Comparative Biochemistry and Physiology, Part A, Comparative Physiology* 93A:391–394.

Beddingfield, S. D., and J. B. McClintock. 1993. Feeding behavior of the sea star *Astropecten articulatus* (Echinodermata: Asteroidea): An evaluation of energy-efficient foraging in a soft-bottom predator. *Marine Biology* (Berlin) 115: 669–676.

Bell, B. M., and R. W. Frey. 1969. Observations on ecology and the feeding and burrowing mechanisms of *Mellita quinquiesperforata* (Leske). *Journal of Paleontology* 43:553–560.

Bell, S. S., and J. B. McClintock. 1982. Invertebrates associated with echinoderms from the west coast of Florida with special reference to harpacticoid copepods. In *Echinoderms*: Proceedings of the International Conference, Tampa Bay, 14–17 September 1981, ed. J. M. Lawrence, 229–234. Balkema, Rotterdam.

Bernasconi, I. 1926. Una ofiura vivipara de Necochea. *Anales del Museo nacional de historia natural de Buenos Aires* 34:145–153, pls. 1–4.

Bernasconi, I., and M. M. D'Agostino. 1977. Ofiuroideos del Mar Epicontinental Argentino. *Revista del Museo Argentino de Ciencias Naturales "Bernardino Rivadavia" e Instituto Nacional de Investigacion de las Ciencias Naturales (Hidrobiología)* 5:65–114, pls. 1–11.

Bianconcini, M. S. C., E. G. Mendes, and D. Valente. 1985. The respiratory metabolism of the lantern muscles of the sea urchin *Echinometra lucunter* L.—I. The respiratory intensity. *Comparative Biochemistry and Physiology, Part A, Comparative Physiology* 80A:1–4.

Binaux, R., and C. Bocquet. 1971. Sur le polychromatisme de l'ophiure *Amphipholis squamata* (Delle Chiaje). *Comptes Rendus Hebdomadaires des Séances de l'Academie des Sciences, Série D, Sciences Naturelles* 273: 1618–1619.

Binyon, J. 1966. Salinity tolerance and ionic regulation. In *Physiology of Echinodermata*, ed. R. A. Boolootian, 359–377. Interscience Publishers, New York.

Birkeland, C. 1989. The influence of echinoderms on coral-reef communities. In *Echinoderm Studies*, Volume 3, eds. M. Jangoux and J. M. Lawrence, 1–79. Balkema, Rotterdam.

Birkeland, C., and J. S. Lucas. 1990. *Acanthaster planci: Major Management Problem of Coral Reefs*. CRC Press, Boca Raton, Florida. viii + 257 pp.

Bishop, C. D., and S. A. Watts. 1992. Biochemical and morphometric study of growth in the stomach and intestine of the echinoid *Lytechinus variegatus* (Echinodermata). *Marine Biology* (Berlin) 114:459–467.

———. 1994. Two-stage recovery of gametogenic activity following starvation in *Lytechinus variegatus* Lamarck (Echinodermata: Echinoidea). *Journal of Experimental Marine Biology and Ecology* 177:27–36.

Blake, D. B. 1982. Somasteroidea, Asteroidea, and the affinities of *Luidia (Platasterias) latiradiata*. *Palaeontology* 25:167–191.

———. 1983. Some biological controls on the distribution of shallow water sea stars (Asteroidea: Echinodermata). *Bulletin of Marine Science* 33:703–712.

———. 1987. A classification and phylogeny of post-Palaeozoic sea stars (Asteroidea: Echinodermata). *Journal of Natural History* 21:481–528.

———. 1990. Adaptive zones of the class Asteroidea (Echinodermata). *Bulletin of Marine Science* 46:701–718.

Boffi, E. 1972. Ecological aspects of ophiuroids from the phytal of S. W. Atlantic Ocean warm waters. *Marine Biology* (Berlin) 15:316–328.

Boidron-Metairon, I. F. 1988. Morphological plasticity in laboratory-reared echinoplutei of *Dendraster excentricus* (Eschscholtz) and *Lytechinus variegatus* (Lamarck) in response to food conditions. *Journal of Experimental Marine Biology and Ecology* 119:31–41.

Boolootian, R. A., and M. H. Cantor. 1965. A preliminary report on respiration, nutrition, and behavior of *Arbacia punctulata*. *Life Sciences* 4:1567–1571.

Boone, L. 1928. Scientific results of the first oceanographic expedition of the "Pawnee," 1925. Echinodermata from tropical East American seas. *Bulletin of the Bingham Oceanographic Collection Yale University* 1:1–22.

———. 1933. Scientific results of cruises of the yachts "Eagle" and "Ara," 1921–1928, William K. Vanderbilt, commanding. Coelenterata, Echinodermata, and Mollusca. *Bulletin of the Vanderbilt Marine Museum* 4:1–217, pls. 1–133.

Boxshall, G. A. 1988. A review of the copepod endoparasites of brittle stars (Ophiuroidea). *Bulletin of the British Museum (Natural History) Zoology* 54:261–270.

Brattström, H. 1992. Marine biological investigations in the Bahamas 22. Littoral zonation at three Bahamian beachrock communities. *Sarsia* 77:81–109.

Bray, R. D. 1975. Community structure of shallow-water Ophiuroidea of Barbados, West Indies. M.S. thesis, McGill University, Montreal, Canada. viii + 91 pp.

———. 1981. Size variation of the rubble-dwelling ophiuroid *Ophiocoma echinata* of Barbados, West Indies. In *The Reef and Man*. Proceedings of the Fourth International Coral Reef Symposium, Volume 2, eds. E. D. Gomez, C. E. Birkeland, R. W. Buddemeier, R. E. Johannes, J. A. Marsh, Jr., and R. T. Tsuda, 619–621. Marine Sciences Center, University of the Philippines, Quezon City, Philippines.

———. 1985. Stereom microstructure of vertebral ossicles of the Caribbean ophiuroid *Ophiocoma echinata*. Proceedings of the Fifth International Coral Reef Congress, Tahiti, 27 May–1 June 1985, Volume 5: *Miscellaneous Papers (A)*, eds. V. M. Harmelin and B. Salvat, 279–284. Antenne Museum—E.P.H.E., Moorea, French Polynesia.

Brehm, P., and J. G. Morin. 1977. Localization and characterization of luminescent cells in *Ophiopsila californica* and *Amphipholis squamata* (Echinodermata: Ophiuroidea). *Biological Bulletin* (Woods Hole) 152:12–25.

Brito, I. M. 1960. Os ofiuróides do Rio de Janeiro. Parte I—Ophiothrichidae, Ophiochitonidae e Ophiactidae. *Avulso Centro de estudos zoológicos, Universidade do Brasil* 6:1–4.

Britton, J. C., and B. Morton. 1989. *Shore Ecology of the Gulf of Mexico*. University of Texas Press, Austin. viii + 387 pp.

Brown, W. I., and J. M. Shick. 1979. Bimodal gas exchange and the regulation of oxygen uptake in holothurians. *Biological Bulletin* (Woods Hole) 156:272–288.

Buchanan, J. B. 1963. Mucus secretion within the spines of ophiuroid echinoderms. *Proceedings of the Zoological Society of London* 141:251–259.

Buckland-Nicks, J., C. W. Walker, and F.-S. Chia. 1984. Ultrastructure of the male reproductive system and of spermatogenesis in the viviparous brittle-star, *Amphipholis squamata*. *Journal of Morphology* 179:243–262.

Burke, T. E. 1974. Echinoderms. In *Biota of the West Flower Garden Bank*, eds. T. J. Bright, and L. H. Pequegnat, 311–331. Gulf Publishing Company, Houston, Texas.

Burns, K. A., M. G. Ehrhardt, J. MacPherson, J. A. Tierney, G. Kananen, and D. Connelly. 1990. Organic and trace metal contaminants in sediments, seawater and organisms from two Bermudan harbours. *Journal of Experimental Marine Biology and Ecology* 138:9–34.

Burrage, B. R. 1964. The possibility of paralytic effects of selected sea stars and brittle-stars on *Octopus bimaculatus*. *Transactions of the Kansas Academy of Science* 67:496–498.

Byrne, M. 1988. Evidence for endocytotic incorporation of nutrients from the haemal sinus by the oocytes of the brittlestar *Ophiolepis paucispina*. In *Echinoderm Biology*. Proceedings of the Sixth International Echinoderm Conference, Victoria, 23–28 August 1987, eds. R. D. Burke, P. V. Mladenov, P. Lambert, and R. L. Parsley, 557–563. Balkema, Rotterdam.

———. 1989. Ultrastructure of the ovary and oogenesis in the ovoviviparous ophiuroid *Ophiolepis paucispina* (Echinodermata). *Biological Bulletin* (Woods Hole) 176:79–95.

———. 1991. Reproduction, development and population biology of the Caribbean ophiuroid *Ophionereis olivacea*, a protandric hermaphrodite that broods its young. *Marine Biology* (Berlin) 111:387–399.

Byrne, M., and G. Hendler. 1988. Arm structure of the ophiomyxid brittlestars (Echinodermata: Ophiuroidea: Ophiomyxidae). In *Echinoderm Biology*. Proceedings of the Sixth International Echinoderm Conference, Victoria, 23–28 August 1987, eds. R. D. Burke, P. V. Mladenov, P. Lambert, and R. L. Parsley, 687–695. Balkema, Rotterdam.

Cabanac, A., J.-F. Hamel, and J. H. Himmelman. 1993. Morphological adaptations of the sand dollar *Echinarachnius parma* to righting itself. *Canadian Journal of Zoology* 71:1274–1276.

Cairns, S. D. 1987. A revision of the Northwest

Atlantic Stylasteridae (Coelenterata: Hydrozoa). *Smithsonian Contributions to Zoology* 418:iv + 1–131.

Camargo, T. M. de. 1982. Changes in the echinoderm fauna in a polluted area on the coast of Brazil. In *Papers from the Echinoderm Conference. The Australian Museum Sydney, 1978*, ed. F. W. E. Rowe. *Australian Museum Memoir* 16:165–173.

Cameron, R. A. 1986. Reproduction, larval occurrence and recruitment in Caribbean sea urchins. *Bulletin of Marine Science* 39:332–346.

Camp, D. K., S. Cobb, and J. F. van Breedveld. 1973. Overgrazing of seagrasses by a regular urchin, *Lytechinus variegatus*. *BioScience* 23:37–38.

Carpenter, P. H. 1884. Report on the Crinoidea collected during the voyage of H.M.S. Challenger, during the years 1873–1876. Part I. General morphology, with descriptions of the stalked crinoids. *Report on the Scientific Results of the Voyage of H.M.S. Challenger during the Years 1873–76. Zoology*, Volume 11 (Part 32): xii + 1–442, pls. 1–62.

———. 1888. Report on the Crinoidea collected during the voyage of H.M.S. Challenger, during the years 1873–1876. Part II. The Comatulae. *Report on the Scientific Results of the Voyage of H.M.S. Challenger during the Years 1873–76. Zoology*, Volume 26 (Part 60): ix + 1–399, pls. 1–70.

Carpenter, R. C. 1981. Grazing by *Diadema antillarum* (Philippi) and its effects on the benthic algal community. *Journal of Marine Research* 39:749–765.

———. 1984. Predator and population density control of homing behavior in the Caribbean echinoid *Diadema antillarum*. *Marine Biology* (Berlin) 82:101–108.

———. 1986. Partitioning herbivory and its effects on coral reef algal communities. *Ecological Monographs* 56:345–363.

———. 1988. Mass-mortality of a Caribbean sea urchin: Immediate effects on community metabolism and other herbivores. *Proceedings of the National Academy of Sciences of the U.S.A.* 85:511–514.

———. 1990. Mass mortality of *Diadema antillarum*. *Marine Biology* (Berlin) 104:67–77.

Carrera, C. J. 1974. The shallow water amphiurid brittle stars (Ophiuroidea: Echinodermata) of Puerto Rico. M.S. thesis, University of Puerto Rico, Mayaguez, Puerto Rico. vi + 103 pp.

Caso, M. E. 1955. Contribución al conocimiento de los holoturoideos de Mexico. II. Algunas especies de holoturoideos litorales de la costa atlántica mexicana. *Anales del Instituto de Biología, Universidad de México* 26:501–525.

———. 1968a. Contribucion al estudio de los holoturoideos de Mexico. Ecologia y morphologia de *Holothuria glaberrima* Selenka. *Anales del Instituto de Biología, Universidad Nacional Autónoma de México* 39:21–30.

———. 1968b. Contribucion al estudio de los holoturoideos de Mexico. Un caso de parasitismo de *Balcis intermedia* (Cantraine) sobre *Holothuria glaberrima* Selenka. *Anales del Instituto de Biología, Universidad Nacional Autónoma de México* 39:31–40.

Caycedo, I. E. 1978. Holothuroidea (Echinodermata) de aguas someras en la costa norte de Colombia. *Anales del Instituto de Investigaciones Marinas de Punta de Betín* 10:149–198.

———. 1979. Observaciones de los equinodermos en las Islas del Rosario. *Anales del Instituto Investigaciones Marinas de Punta de Betín* 11:39–47.

Chesher, R. H. 1963. The morphology and function of the frontal ambulacrum of *Moira atropos* (Echinoidea: Spatangoida). *Bulletin of Marine Science of the Gulf and Caribbean* 13:549–573.

———. 1966a. The R/V *Pillsbury* Deep-Sea Biological Expedition to the Gulf of Guinea, 1964–65. 10. Report on the Echinoidea collected by R/V *Pillsbury*. *Studies in Tropical Oceanography* (Miami) 4:209–223.

———. 1966b. Redescription of the echinoid species *Paraster floridiensis* (Spatangoida: Schizasteridae). *Bulletin of Marine Science* 16:1–19.

———. 1968a. The systematics of sympatric species in West Indian spatangoids: A revision of the genera *Brissopsis*, *Plethotaenia*, *Palaeopneustes*, and *Saviniaster*. *Studies in Tropical Oceanography* (Miami) 7:1–165.

———. 1968b. *Lytechinus williamsi*, a new sea urchin from Panama. *Breviora* 305:1–13.

———. 1969. Contributions to the biology of *Meoma ventricosa* (Echinoidea: Spatangoida). *Bulletin of Marine Science* 19:72–110.

———. 1970. Evolution in the genus *Meoma*

(Echinoidea: Spatangoida) and a description of a new species from Panama. *Bulletin of Marine Science* 20:731–761.

———. 1972. Biological results of the University of Miami Deep-Sea Expeditions. 86. A new *Paraster* (Echinoidea: Spatangoida) from the Caribbean. *Bulletin of Marine Science* 22:10–25.

Chia, F.-S., and C. W. Walker. 1991. Echinodermata: Asteroidea. In *Reproduction of Marine Invertebrates,* Volume VI, *Echinoderms and Lophophorates,* eds. A. C. Giese, J. S. Pearse, and V. B. Pearse, 301–353. The Boxwood Press, Pacific Grove, California.

Clark, A. H. 1915. A monograph of the existing crinoids. Volume 1, The comatulids. Part 1. *United States National Museum Bulletin* 82:vi + 1–406, 17 pls.

———. 1917. Four new echinoderms from the West Indies. *Proceedings of the Biological Society of Washington* 30:63–70.

———. 1921a. A monograph of the existing crinoids. Volume 1, The comatulids. Part 2. *United States National Museum Bulletin* 82: xxv + 1–795, pls. 1–57.

———. 1921b. Report on the ophiurans collected by the Barbados-Antigua Expedition from the University of Iowa in 1918. *University of Iowa Studies in Natural History* 9:29–63.

———. 1931. A monograph of the existing crinoids. Volume 1, The comatulids. Part 3. Superfamily Comasterida. *United States National Museum Bulletin* 82:vii + 1–816, pls. 1–82.

———. 1934. A new sea urchin from Florida. *Journal of the Washington Academy of Science* 24:52–53.

———. 1939a. Echinoderms of the Smithsonian-Hartford Expedition, 1937 with other West Indian records. *Proceedings of the United States National Museum* 86:441–456, pls. 53–54.

———. 1939b. Echinoderms (other than holothurians) collected on the Presidential Cruise of 1938. *Smithsonian Miscellaneous Collections* 98:1–18, pls. 1–5.

———. 1941. A monograph of the existing crinoids. Volume 1, The comatulids. Part 4a. Superfamily Mariametrida (except the family Colobometridae). *United States National Museum Bulletin* 82:vii + 1–603, pls. 1–61.

———. 1947. A monograph of the existing crinoids. Volume 1, The comatulids. Part 4b. Superfamily Mariametrida (concluded—the family Colobometridae) and Superfamily Tropiometrida (except the families Thalassometridae and Charitometridae). *United States National Museum Bulletin* 82:vii + 1–473, pls. 1–43.

———. 1948. Two new starfishes and a new brittle-star from Florida and Alabama. *Proceedings of the Biological Society of Washington* 61:55–66.

———. 1950. A monograph of the existing crinoids. Volume 1, The comatulids. Part 4c. Superfamily Tropiometrida (the families Thalassometridae and Charitometridae). *United States National Museum Bulletin* 82:vii + 1–382, pls. 1–32.

———. 1952. *Schizostella,* a new genus of brittle-star (Gorgonocephalidae). *Proceedings of the United States National Museum* 102: 451–454, pl. 40.

Clark, A. H., and A. M. Clark. 1967. A monograph of the existing crinoids. Volume 1, The comatulids. Part 5. Suborders Oligophreata (concluded) and Macrophreata. *United States National Museum Bulletin* 82:xiv + 1–860.

Clark, A. M. 1953. A revision of the genus *Ophionereis* (Echinodermata, Ophiuroidea). *Proceedings of the Zoological Society of London* 123:65–94, pls. 1–3.

———. 1955. Echinodermata of the Gold Coast. *Journal of the West African Science Association* 1:16–56, pl. 2.

———. 1964. On the identity of *Clypeaster rosaceus* (Linnaeus) and some other irregular echinoids. Z. N. (S.) 1616. *Bulletin of Zoological Nomenclature* 21:297–302.

———. 1967a. Variable symmetry in fissiparous Asterozoa. *Symposia of the Zoological Society of London* 20:143–157.

———. 1967b. Notes on the family Ophiotrichidae (Ophiuroidea). *Annals and Magazine of Natural History* Ser. 13, 9:637–655, pls. 10–11.

———. 1970. Notes on the family Amphiuridae (Ophiuroidea). *Bulletin of the British Museum (Natural History) Zoology* 19:1–81.

———. 1974. Notes on some echinoderms from southern Africa. *Bulletin of the British Mu-

seum (Natural History) Zoology 26:423–487, pls. 1–3.

———. 1976a. Tropical epizoic echinoderms and their distribution. *Micronesica* 12:111–116, 1 pl.

———. 1976b. Echinoderms of coral reefs. In *Biology and Geology of Coral Reefs*, Volume 3 (Biology 2), eds. O. A. Jones and R. Endean, 95–123. Academic Press, New York.

———. 1977. *Starfishes and Related Echinoderms*. T.F.H. Publications, Neptune City, New Jersey. 160 pp.

———. 1987. Notes on Atlantic and other Asteroidea. 5. Echinasteridae. *Bulletin of the British Museum (Natural History) Zoology* 53:65–78.

Clark, A. M., and J. Courtman-Stock. 1976. *The Echinoderms of Southern Africa*. British Museum (Natural History), London. iv + 277 pp.

Clark, A. M., and M. E. Downey. 1992. *Starfishes of the Atlantic*. Chapman & Hall, London. xxvi + 794 pp.

Clark, H. L. 1898. *Synapta vivipara*: A contribution to the morphology of echinoderms. *Memoirs of the Boston Society of Natural History* 5:53–88, pls. 11–15.

———. 1899. The synaptas of the New England coast. *Bulletin of the United States Fish Commission* 19:21–31.

———. 1901a. The echinoderms of Porto Rico. *Bulletin of the United States Fish Commission* 20 (Pt. 2):231–263, pls. 14–17.

———. 1901b. Bermudan echinoderms. A report on observations and collections made in 1899. *Proceedings of the Boston Society of Natural History* 29:339–344.

———. 1907. The apodous holothurians; a monograph of the Synaptidae and Molpadiidae, including a report on the representatives of these families in the collections of the United States National Museum. *Smithsonian Contributions to Knowledge* 35:1–231, pls. 1–13.

———. 1910. The development of an apodous holothurian *(Chirodota rotifera)*. *Journal of Experimental Zoology* 9:497–516, pls. 1–2.

———. 1914. Growth changes in brittle stars. Carnegie Institution of Washington Publication No. 182, *Papers from the Tortugas Laboratory of the Carnegie Institution of Washington* 5:91–126, pls. 1–3.

———. 1915. Catalog of recent ophiurans: Based on the collection of the Museum of Comparative Zoölogy. *Memoirs of the Museum of Comparative Zoölogy of Harvard College* 25:165–376, pls. 1–20.

———. 1918. Brittle-stars, new and old. *Bulletin of the Museum of Comparative Zoölogy at Harvard College* 62:265–338, pls. 1–8.

———. 1919. The distribution of the littoral echinoderms of the West Indies. Carnegie Institution of Washington Publication No. 281, *Papers from the Department of Marine Biology of the Carnegie Institution of Washington* 13:49–74, pls. 1–3.

———. 1921. The echinoderm fauna of Torres Strait: Its composition and its origin. Carnegie Institution of Washington Publication No. 214, *Papers from the Department of Marine Biology of the Carnegie Institution of Washington* 10:vii+1–224, 40 pls.

———. 1922. The echinoderms of the Challenger Bank, Bermuda. *Proceedings of the American Academy of Arts and Sciences, Boston* 57:354–361, 1 pl.

———. 1924. The holothurians of the Museum of Comparative Zoology. The Synaptidae. *Bulletin of the Museum of Comparative Zoölogy at Harvard College* 65:459–501, pls. 1–12.

———. 1933. A handbook of the littoral echinoderms of Porto Rico and the other West Indian Islands. *Scientific Survey of Porto Rico and the Virgin Islands* 16:1–147, pls. 1–7.

———. 1938. Echinoderms from Australia. *Memoirs of the Museum of Comparative Zoölogy of Harvard College* 55:vii+1–596, pls. 1–28.

———. 1939. Two new ophiurans from the Smithsonian-Hartford Expedition, 1937. *Proceedings of the United States National Museum* 86:415–418, pl. 52.

———. 1941. Reports on the scientific results of the Atlantis Expedition to the West Indies, under the joint auspices of the University of Havana and Harvard University. The echinoderms (other than holothurians). *Memorias de la Sociedad Cubana de Historia Natural "Felipe Poey"* 15:1–154, pls. 1–10.

———. 1942. The echinoderm fauna of Bermuda. *Bulletin of the Museum of Comparative Zoölogy at Harvard College* 89:367–391 + 1 pl.

———. 1946. *The Echinoderm Fauna of Austra-*

lia: Its Composition and Its Origin. Carnegie Institution of Washington Publication No. 566:iv+1–567.

Clarke, F. W., and W. C. Wheeler. 1915. The inorganic constituents of echinoderms. *Professional Papers. United States Geological Survey 90-L, Shorter Contributions to General Geology* 1914-L:191–196.

Clements, L. A. J. 1985. Post-autotomy feeding behavior of *Micropholis gracillima* (Stimpson): Implications for regeneration. In *Echinodermata*. Proceedings of the Fifth International Echinoderm Conference, Galway, 24–29 September 1984, eds. B. F. Keegan and B. D. S. O'Connor, 609–615. Balkema, Rotterdam.

———. 1986. Amino acid uptake by a regenerating brittlestar. *American Zoologist* 26:42A.

Clements, L. A. J., and S. E. Stancyk. 1984. Particle selection by the burrowing brittlestar *Micropholis gracillima* (Stimpson) (Echinodermata: Ophiuroidea). *Journal of Experimental Marine Biology and Ecology* 84:1–13.

Clements, L. A. J., K. T. Fielman, and S. E. Stancyk. 1988. Regeneration by an amphiurid brittlestar exposed to different concentrations of dissolved organic material. *Journal of Experimental Marine Biology and Ecology* 122:47–61.

Cobb, J. L. S., and G. Hendler. 1990. Neurophysiological characterization of the photoreceptor system in a brittlestar, *Ophiocoma wendtii* (Echinodermata: Ophiuroidea). *Comparative Biochemistry and Physiology, Part A, Comparative Physiology* 97A:329–333.

Coe, W. R. 1912. Echinoderms of Connecticut. *Connecticut Geological and Natural History Survey Bulletin* 19:1–152, pls. 1–32.

Colin, P. I. 1978. *Caribbean Reef Invertebrates and Plants: A Field Guide to the Invertebrates and Plants Occurring on Coral Reefs of the Caribbean, the Bahamas and Florida*. T.F.H. Publications, Neptune City, New Jersey. 512 pp.

Colwin, L. H. 1948. Note on the spawning of the holothurian *Thyone briareus* (Lesueur). *Biological Bulletin* (Woods Hole) 95:296–306.

Conde, J. E., H. Díaz, and A. Sambrano. 1991. Disintegration of holothurian fecal pellets in beds of the seagrass *Thalassia testudinum*. *Journal of Coastal Research* 7:853–862.

Cooke, C. W. 1961. Cenozoic and Cretaceous echinoids from Trinidad and Venezuela. *Smithsonian Miscellaneous Collections* 142:1–35, pls. 1–14.

Corry, B. G. 1974. An ultrastructural study of muscle in tube feet of selected ophiuroids. M.S. thesis, University of Victoria, British Columbia, Canada. xii + 117 pp.

Corvea, A. 1990. Nuevo registro holoturoideo (Echinodermata) para Cuba. *Poeyana* 392:1–6.

Corvea, A., M. Abreau, and P. Alcolado. 1985. Distribución de la abundancia y composición específica de los equinodermos en el Golfo de Batabanó. In *Simposio de Ciencias del Mar y VII Jornado Científica del Instituto Oceanologia XX Aniversario*, 78–82. Academia de Ciencias de Cuba, Havana.

Cowles, R. P. 1910. Stimuli produced by light and by contact with solid walls as factors in the behavior of ophiuroids. *Journal of Experimental Zoölogy* 9:387–416.

Criales, M. M. 1984. Shrimps associated with coelenterates, echinoderms, and molluscs in the Santa Marta Region, Colombia. *Journal of Crustacean Biology* 4:307–317.

Crozier, W. J. 1915. The sensory reactions of *Holothuria surinamensis* Ludwig. *Zoologische Jahrbücher Abteilung für Allgemeine Zoologie und Physiologie der Tiere* 5:233–298.

———. 1917. Multiplication by fission in holothurians. *American Naturalist* 51:560–566.

———. 1918. The amount of bottom material ingested by holothurians *(Stichopus)*. *Journal of Experimental Zoölogy* 26:379–389.

———. 1920. Notes on the bionomics of *Mellita*. *American Naturalist* 54:435–442.

Cutress, B. M. 1965. Observation on growth in *Eucidaris tribuloides* (Lamarck), with special reference to the origin of the oral primary spines. *Bulletin of Marine Science* 15:797–834.

———. In press (1995). Changes in dermal ossicles during somatic growth in Caribbean littoral sea cucumbers (Echinodermata: Holothuroidea: Aspidochirotida). *Bulletin of Marine Science* 57.

Davis, W. P. 1966. Observations on the biology of the ophiuroid *Astrophyton muricatum*. *Bulletin of Marine Science* 16:435–444.

Deichmann, E. 1922. Papers from Dr. Th. Mortensen's Pacific Expedition 1914–16. IX. On some cases of multiplication by fission and of coalescence in holothurians; with notes on the synonymy of *Actinopyga parvula* (Sel.). *Videnskabelige Meddelelser fra den naturhistorisk Forening i Kjøbenhavn* 73:199–214.

———. 1926. Echinoderms and insects from the Antilles. Report on the holothurians collected by the Barbados-Antigua Expedition from the University of Iowa. *University of Iowa Studies in Natural History* 11:9–31, pls. 1–3.

———. 1930. The holothurians of the western part of the Atlantic Ocean. *Bulletin of the Museum of Comparative Zoölogy at Harvard College* 71:43–226, pls. 1–24.

———. 1939. Holothurians from Biscayne Bay, Florida. *Proceedings of the Florida Academy of Sciences* 3:128–137.

———. 1954. The holothurians of the Gulf of Mexico. *U.S. Fish and Wildlife Service Fishery Bulletin* 55:381–410.

———. 1957. The littoral holothurians of the Bahama Islands. *American Museum Novitates* No. 1821:1–20.

Delle Chiaje, S. 1828. *Memorie sulla storia e notami degli animali senza vertebre del Regno di Napoli*. Volume 3. xx + 232 pp.; Atlas, pls. 32–49. Naples.

De Ridder, C., and J. M. Lawrence. 1982. Food and feeding mechanisms: Echinoidea. In *Echinoderm Nutrition*, eds. M. Jangoux and J. M. Lawrence, 57–115. Balkema, Rotterdam.

de Roa, E. Z. 1967. Contribucion al estudio de los equinodermos de Venezuela. *Acta Biologica Venezuelica* 5:267–333.

de Ruyter van Steveninck, E. D., and R. P. M. Bak. 1986. Changes in abundance of coral-reef bottom components related to mass mortality of the sea urchin *Diadema antillarum*. *Marine Ecology Progress Series* 34:87–94.

de Ruyter van Steveninck, E. D., and A. M. Breeman. 1987. Deep water vegetations of *Lobophora variegata* (Phaeophyceae) in the coral reef of Curaçao: Population dynamics in relation to mass mortality of the sea urchin *Diadema antillarum*. *Marine Ecology Progress Series* 36:81–90.

Devaney, D. M. 1970. Studies on ophiocomid brittlestars. I. A new genus *(Clarkcoma)* of Ophiocominae with a reevaluation of the genus *Ophiocoma*. *Smithsonian Contributions to Zoology* No. 51:1–41.

———. 1974a. Shallow-water asterozoans of southeastern Polynesia. II. Ophiuroidea. *Micronesica* 10:105–204.

———. 1974b. Shallow-water echinoderms from British Honduras, with a description of a new species of *Ophiocoma* (Ophiuroidea). *Bulletin of Marine Science* 24:122–164.

de Vore, D., and E. D. Brodie, Jr. 1982. Palatability of the tissues of the holothurian *Thyone briareus* (Lesueur) to fish. *Journal of Experimental Marine Biology and Ecology* 61:279–285.

De Voss, L. 1985. Occurrence of specialized papillae on the tube-feet of the ophiuroid *Amphipholis squamata*. In *Echinodermata*. Proceedings of the Fifth International Echinoderm Conference, Galway, 24–29 September 1984, eds. B. F. Keegan and B. D. S. O'Connor, 655. Balkema, Rotterdam.

Díaz-Miranda, L., D. A. Price, M. J. Greenberg, T. D. Lee, K. E. Doble, and J. E. García-Arrarás. 1992. Characterization of two novel neuropeptides from the sea cucumber *Holothuria glaberrima*. *Biological Bulletin* (Woods Hole) 182:241–247.

Dobson, W. E. 1985. A pharmacological study of neural mediation of disc autotomy in *Ophiophragmus filograneus* (Lyman) (Echinodermata: Ophiuroidea). *Journal of Experimental Marine Biology and Ecology* 94:223–232.

———. 1986. Early disc regeneration after autotomy in the brittlestar *Microphiopholis gracillima*. *American Zoologist* 26:42A.

Dobson, W. E., and R. L. Turner. 1989. Morphology and histology of the disc autotomy plane in *Ophiophragmus filograneus* (Echinodermata, Ophiurida). *Zoomorphology* (Berlin) 108:323–332.

Dobson, W. E., S. E. Stancyk, L. A. J. Clements, and R. M. Showman. 1991. Nutrient translocation during early disc regeneration in the brittlestar *Microphiopholis gracillima* (Stimpson) (Echinodermata: Ophiuroidea). *Biological Bulletin* (Woods Hole) 180:167–184.

Döderlein, L. 1911. Über japanische und andere Euryalae. Beiträge zur Naturgeschichte Ostasiens Herausgeben von Dr. F. Doflein. *Abhandlungen der Bayerischen Akademie der Wissenschaften mathematische-physikalisch Klasse* Suppl.-Bd. 2:1–123, pls. 1–9.

Donachy, J. E., and N. Watanabe. 1986. Effects

of salinity and calcium concentration on arm regeneration by *Ophiothrix angulata* (Echinodermata: Ophiuroidea). *Marine Biology* (Berlin) 91:253–257.

Donovan, S. K. 1993a. Contractile tissues in the cirri of ancient crinoids: Criteria for recognition. *Lethaia* 26:163–169.

———. 1993b. Jamaican Cenozoic Echinoidea. In *Biostratigraphy of Jamaica,* eds. R. M. Wright and E. Robinson, 371–412. Geological Society of America Memoir 182.

Donovan, S. K., and C. M. Gordon. 1993. Echinoid taphonomy and the fossil record: Supporting evidence from the Plio-Pleistocene of the Caribbean. *Palaios* 8:304–306.

Donovan, S. K., and B. Jones. 1994. Pleistocene echinoids (Echinodermata) from Bermuda and Barbados. *Proceedings of the Biological Society of Washington* 107:109–113.

Donovan, S. K., H. L. Dixon, R. K. Pickerill, and E. N. Doyle. 1994. Pleistocene echinoid (Echinodermata) fauna from southeast Jamaica. *Journal of Paleontology* 68:351–358.

Downey, M. E. 1973. Starfishes from the Caribbean and the Gulf of Mexico. *Smithsonian Contributions to Zoology* No. 126:vi + 1–158.

Drifmeyer, J. E. 1981. Urchin *Lytechinus variegatus* grazing on eelgrass, *Zostera marina*. *Estuaries* 4:374–375.

Dubois, R. 1975. A comparison of the distribution of the Echinodermata of a coral community with that of a nearby rock outcrop on the Texas continental shelf. M.S. thesis, Texas A&M University, College Station, Texas. ix + 153 pp.

Dugan, P. J., and R. J. Livingston. 1982. Long-term variation of macroinvertebrate assemblages in Apalachee Bay, Florida. *Estuarine Coastal and Shelf Science* 14:391–403.

Durham, J. W., H. B. Fell, A. G. Fisher, P. M. Kier, R. V. Melville, D. L. Pawson, and C. D. Wagner. 1966. Echinoids. In *Treatise on Invertebrate Paleontology, Part U, Echinodermata 3,* Volumes 1 and 2, ed. R. C. Moore, U211–U640. The Geological Society of America and The University of Kansas Press, Lawrence, Kansas.

Eckelbarger, K. J., and C. M. Young. 1992. Ovarian ultrastructure and vitellogenesis in ten species of shallow-water and bathyal sea cucumbers (Echinodermata: Holothuroidea). *Journal of the Marine Biological Association of the United Kingdom* 72:759–781.

Edwards, A., and R. Lubbock. 1983. The ecology of Saint Paul's Rock (Equatorial Atlantic). *Journal of Zoology* (London) 200:51–69.

Edwards, C. L. 1889. Notes on the embryology of *Mülleria agassizi,* Sel., a holothurian common at Green Turtle Cay, Bahamas. *Johns Hopkins University Circulars* 8:37.

———. 1909. The development of *Holothuria floridana* Pourtalès with especial reference to the ambulacral appendages. *Journal of Morphology* 20:211–230, pls. 1–3.

Ellers, O., and M. Telford. 1991. Forces generated by the jaws of clypeasteroids (Echinodermata: Echinoidea). *Journal of Experimental Biology* 155:585–604.

Ellington, W. R. 1976. Lactate dehydrogenase in the longitudinal muscle of the sea cucumber *Sclerodactyla briareus* (Echinodermata: Holothuroidea). *Marine Biology* (Berlin) 36:31–36.

Ellington, W. R., and J. M. Lawrence. 1974. Coelomic fluid volume regulation and isosmotic intracellular regulation by *Luidia clathrata* (Echinodermata: Asteroidea) in response to hyposmotic stress. *Biological Bulletin* (Woods Hole) 146:20–31.

Ely, C. A. 1942. Shallow-water Asteroidea and Ophiuroidea of Hawaii. *Bulletin of the Bernice P. Bishop Museum* 176:i + 1–63, pls. 1–13.

Emlet, R. B. 1986. Facultative planktotrophy in the tropical echinoid *Clypeaster rosaceus* (Linnaeus) and a comparison with obligate planktotrophy in *Clypeaster subdepressus* (Gray) (Clypeasteroida: Echinoidea). *Journal of Experimental Marine Biology and Ecology* 95:183–202.

———. 1988. Larval form and metamorphosis of a "primitive" sea urchin, *Eucidaris thouarsi* (Echinodermata: Echinoidea: Cidaroida), with implications for developmental and phylogenetic studies. *Biological Bulletin* (Woods Hole) 174:4–19.

Emson, R. H., and J. Foote. 1980. Environmental tolerances and other adaptive features of two intertidal rock pool echinoderms. In *Echinoderms: Present and Past*. Proceedings of the European Colloquium on Echinoderms, Brussels, 3–8 September 1979, ed. M. Jangoux, 163–169. Balkema, Rotterdam.

Emson, R., and J. Herring. 1985. Bioluminescence in deep and shallow water brittlestars. In *Echinodermata*. Proceedings of the Fifth International Echinoderm Conference, Galway, 24–29 September 1984, eds. B. F. Keegan and B. D. S. O'Connor, 656. Balkema, Rotterdam.

Emson, R. H., and P. V. Mladenov. 1987a. Studies of the fissiparous holothurian *Holothuria parvula* (Selenka) (Echinodermata: Holothuroidea). *Journal of Experimental Marine Biology and Ecology* 111:195–211.

———. 1987b. Brittlestar host specificity and apparent host discrimination by the parasitic copepod *Ophiopsyllus reductus*. *Parasitology* 94:7–15.

———. 1992. Field and laboratory observations on the feeding of *Macrophiothrix variabilis* (Lamarck) with notes on the feeding of *Ophiactis savignyi* (Müller & Troschel). In *The Marine Flora and Flora of Hong Kong and Southern China III*. Proceedings of the Fourth Marine Biological Workshop, 11–29 April 1989, ed. B. Morton, 769–778. Hong Kong University Press, Hong Kong.

Emson, R. H., and P. J. Whitfield. 1989. Aspects of the life history of a tide pool population of *Amphipholis squamata* (Ophiuroidea) from South Devon. *Journal of the Marine Biological Association of the United Kingdom* 69:27–41.

Emson, R. H., and I. C. Wilkie. 1980. Fission and autotomy in echinoderms. *Oceanography and Marine Biology Annual Review* 18:155–250.

———. 1982. The arm-coiling response of *Amphipholis squamata* (Delle Chiaje). In *Echinoderms*: Proceedings of the International Conference, Tampa Bay, 14–17 September 1981, ed. J. M. Lawrence, 11–18. Balkema, Rotterdam.

———. 1984. An apparent instance of recruitment following sexual reproduction in the fissiparous brittlestar *Ophiactis savignyi* Muller & Troschel. *Journal of Experimental Marine Biology and Ecology* 77:23–28.

Emson, R. H., M. B. Jones, and P. J. Whitfield. 1989. Habitat and latitude differences in reproductive pattern and life-history in the cosmopolitan brittle-star *Amphipholis squamata* (Echinodermata). In *Reproduction, Genetics and Distribution of Marine Organisms*. 23rd European Marine Biology Symposium, eds. J. S. Ryland and P. A. Tyler, 75–82. Olsen & Olsen, Fredensborg, Denmark.

Emson, R. H., P. V. Mladenov, and I. C. Wilkie. 1985a. Patterns of reproduction in small Jamaican brittle stars: Fission and brooding predominate. In *The Ecology of Coral Reefs*. Symposium Series for Undersea Research, NOAA's Undersea Research Program, Volume 3, ed. M. L. Reaka, 87–100. NOAA Undersea Research Program, Rockville, Maryland.

———. 1985b. Studies of the biology of the West Indian copepod *Ophiopsyllus reductus* (Siphonostomatoida: Cancerillidae) parasitic upon the brittlestar *Ophiocomella ophiactoides*. *Journal of Natural History* 19:151–171.

Emson, R., P. Whitfield, and P. Blake. 1988. The influence of parasitization on the population dynamics of *Amphipholis squamata*. In *Echinoderm Biology*. Proceedings of the Sixth International Echinoderm Conference, Victoria, 23–28 August 1987, eds. R. D. Burke, P. V. Mladenov, P. Lambert, and R. L. Parsley, 737–744. Balkema, Rotterdam.

Engstrom, N. A. 1980a. Development, natural history and interstitial habits of the apodous holothurian *Chiridota rotifera* (Pourtales, 1851) (Echinodermata: Holothuroidea). *Brenesia* 17:85–96.

———. 1980b. Hybridization and the systematic status of the aspidochirote holothurians *Holothuria* (*Halodeima*) *floridana* Pourtales, 1851 and *H. (H.) mexicana* Ludwig, 1875. *Brenesia* 17:69–84.

———. 1980c. Reproductive cycles of *Holothuria* (*Halodeima*) *floridana*, *H. (H.) mexicana* and their hybrids (Echinodermata: Holothuroidea) in southern Florida, U.S.A. *International Journal of Invertebrate Reproduction* 2:237–244.

———. 1982. Immigration as a factor in maintaining populations of the sea urchin *Lytechinus variegatus* (Echinodermata: Echinoidea) in seagrass beds on the south west coast of Puerto Rico. *Studies on the Neotropical Fauna and Environment* 17:51–60.

Ernest, R. G., and N. J. Blake. 1981. Reproductive patterns within sub-populations of *Lytechi-*

nus variegatus (Lamarck) (Echinodermata: Echinoidea). *Journal of Experimental Marine Biology and Ecology* 55:25–37.

Espinosa, J. 1982. *Astropecten articulatus* y *A. duplicatus* (Echinodermata: Asteroidea), dos importantes depredadores de bivalvos. *Poeyana* 249:1–12.

Farmanfarmaian, A. 1969a. Intestinal absorption and transport in *Thyone*. I. Biological aspects. *Biological Bulletin* (Woods Hole) 137:118–131.

———. 1969b. Intestinal absorption and transport in *Thyone*. II. Observations on sugar transport. *Biological Bulletin* (Woods Hole) 137:132–145.

Fell, H. B. 1940. Culture in vitro of the excised embryo of an ophiuroid. *Nature* (London) 146:173–175.

———. 1946. The embryology of the viviparous ophiuroid *Amphipholis squamata* Delle Chiaje. *Transactions of the Royal Society of New Zealand* 75:419–464.

———. 1960. Synoptic keys to the genera of Ophiuroidea. *Zoology Publications from Victoria University of Wellington* 26:1–44.

———. 1962. A revision of the major genera of amphiurid Ophiuroidea. *Transactions of the Royal Society of New Zealand, Zoology* 2:1–26, 1 pl.

Fell, P. E., E. H. Parry, and A. M. Balsamo. 1984. The life histories of sponges in the Mystic and Thames Estuaries (Connecticut), with emphasis on larval settlement and postlarval reproduction. *Journal of Experimental Marine Biology and Ecology* 78:127–141.

Ferguson, A. H. 1948. Experiments on the tolerance of several marine invertebrates to reduced salinity. *Proceedings of the Louisiana Academy of Sciences* 11:16–17.

Ferguson, J. C. 1982a. A comparative study of the net metabolic benefits derived from the uptake and release of free amino acids by marine invertebrates. *Biological Bulletin* (Woods Hole) 162:1–17.

———. 1982b. Support of metabolism of superficial structures through direct net uptake of dissolved primary amines in echinoderms. In *Echinoderms*: Proceedings of the International Conference, Tampa Bay, 14–17 September 1981, ed. J. M. Lawrence, 345–351. Balkema, Rotterdam.

———. 1985. Hemal transport of ingested nutrients by the ophiuroid, *Ophioderma brevispinum*. In *Echinodermata*. Proceedings of the Fifth International Echinoderm Conference, Galway, 24–29 September 1984, eds. B. F. Keegan and B. D. S. O'Connor, 623–626. Balkema, Rotterdam.

Fielman, K. T., S. E. Stancyk, W. E. Dobson, and L. A. J. Clements. 1991. Effects of disc and arm loss on regeneration by *Microphiopholis gracillima* (Echinodermata: Ophiuroidea) in nutrient-free seawater. *Marine Biology* (Berlin) 111:121–127.

Fontaine, A. 1953. The shallow-water echinoderms of Jamaica. Part II. The brittle-stars (Class Ophiuroidea). *Natural History Notes. Natural History Society of Jamaica* 5:197–206.

Fontaine, A. R., and F.-S. Chia. 1968. Echinoderms: An autoradiographic study of assimilation of dissolved organic molecules. *Science* (Washington, D.C.) 161:1153–1155.

Forcucci, D. 1994. Population density, recruitment and 1991 mortality event of *Diadema antillarum* in the Florida Keys. *Bulletin of Marine Science* 54:917–928.

Forskål, P. 1775. *Descriptiones Animalium Avium, Amphibiorum, Piscium, Insectorum, Vermium; quae in itinere Orientali observavit Petrus Forskål*. Post mortem auctoris edidit Carsten Niebuhr. Adjuncta est materia medica Kahirina atque tabula Maris Rubri geographica. Hauniae. 19 + xxiv + 164 pp., 1 map.

Fox, D. J., and M. Gilbert. 1991. Role of the neural ring in integrating brittlestar behavior. *Biology of Echinodermata*. Proceedings of the Seventh International Echinoderm Conference, Atami, 9–14 September 1990, eds. T. Yanagisawa, I. Yasumasu, C. Oguro, N. Suzuki, and T. Motokawa, 565. Balkema, Rotterdam.

Frazer, T. K., W. J. Lindberg, and G. R. Stanton. 1991. Predation on sand dollars by gray triggerfish, *Balistes capriscus*, in the northeastern Gulf of Mexico. *Bulletin of Marine Science* 48:159–164.

Fricke, H.-W. 1968. Beiträge zur Biologie der Gorgonenhäupter *Astrophyton muricatum* (Lamark) und *Astroboa nuda* (Lyman) (Ophiuroidea, Gorgonocephalidae). Inaugural-Dissertation zur Erlangung der Doctorwürde der Mathematisch-Naturwissenschaftlichen Fakultät der Freien Universität Berlin. Ernst-Reuter-Gesellschaft Berlin. ii + 106 pp.

Frizzell, D. L., and H. Exline. 1966. Holothuroidea—fossil record. In *Treatise on Invertebrate Paleontology, Part U, Echinodermata 3*, Volume 2, ed. R. C. Moore, U646–U672. The Geological Society of America and The University of Kansas Press, Lawrence, Kansas.

Gage, J. D., M. Pearson, A. M. Clark, G. L. G. Paterson, and P. A. Tyler. 1983. Echinoderms of the Rockall Trough and adjacent areas. I. Crinoidea, Asteroidea and Ophiuroidea. *Bulletin of the British Museum (Natural History) Zoology* 45:263–308.

Gallo, J. 1988. Contribucion al conocimiento de los equinodermos del Parque Nacional Natural Tayrona I. Echinoidea. *Trianea: Organo de Publicación Científica y Tecnológica del INDERENA* 1:99–110.

García-Arrarás, J. E., I. Torres-Avillán, and S. Ortíz-Miranda. 1991. Cells in the intestinal system of holothurians (Echinodermata) express cholecystokinin-like immunoreactivity. *General and Comparative Endocrinology* 83:233–242.

Gardiner, S. L., and R. M. Rieger. 1980. Rudimentary cilia in muscle cells of annelids and echinoderms. *Cell and Tissue Research* 213:247–252.

Ghiold, J. 1979. Spine morphology and its significance in feeding and burrowing in the sand dollar, *Mellita quinquiesperforata* (Echinodermata: Echinoidea). *Bulletin of Marine Science* 29:481–490.

Gilliland, P. M. 1992. Holothurians in the Blue Lias of southern Britain. *Paleontology* 35:159–216.

Gladfelter, W. B. 1978. General ecology of the cassiduloid urchin *Cassidulus caribbearum*. *Marine Biology* (Berlin) 47:149–160.

Glaser, O. C. 1907. Movement and problem solving in *Ophiura brevispina*. *Journal of Experimental Zoölogy* 4:203–220.

Glynn, P. W. 1965. Active movements and other aspects of the biology of *Astichopus* and *Leptosynapta* (Holothuroidea). *Biological Bulletin* (Woods Hole) 129:106–127.

———. 1968. Mass mortalities of echinoids and other reef flat organisms coincident with midday, low water exposures in Puerto Rico. *Marine Biology* (Berlin) 1:226–243.

Gmelin, J. F. 1791. *Caroli Linnaei . . . Systema Naturae per regna tria Naturae, secundum Classes, Ordines, Genera, Species, cum characteribus, differentiis, synonymis, locis*. Editio decima tertia, aucta, reformata, cura J. F. Gmelin. Tom. 1, Regnum Animale: pt. 6, Vermes: pp. 3021–3910. G. E. Beer, Lipsiae.

Goodbody, I. 1960. The feeding mechanism in the sand dollar *Mellita sexiesperforata* (Leske). *Biological Bulletin* (Woods Hole) 119:80–86.

Gooding, R. U. 1974. Animals associated with the sea urchin *Diadema antillarum*. In *Biota of the West Flower Garden Bank*, eds. T. J. Bright and L. H. Pequegnat, 333–336. Gulf Publishing Company, Houston, Texas.

Gordon, C. M. 1991. The poor fossil record of *Echinometra* (Echinodermata: Echinoidea) in the Caribbean region. *Journal of the Geological Society of Jamaica* 28:37–41.

Gosner, K. L. 1979. *A Field Guide to the Atlantic Seashore: Invertebrates and Seaweeds of the Atlantic Coast from the Bay of Fundy to Cape Hatteras*. Peterson Field Guide Series, no. 24. Houghton Mifflin Company, Boston. xvi + 329 pp., 64 pls.

Goudey-Perrière, F. 1979. *Amphiurophilus amphiurae* (Herouard), crustacé copépode parasite des bourses génitales de l'ophiure *Amphipholis squamata* Della Chiaje, Échinoderme: Morphologie des adultes et étude des stades juveniles. *Cahiers de Biologie Marine* 20:201–230.

Grave, C. 1898a. Embryology of *Ophiocoma echinata*, Agassiz. Preliminary Note. *Johns Hopkins University Circulars* 18:6–7.

———. 1898b. Notes on the ophiurids collected in Jamaica during June and July, 1897. *Johns Hopkins University Circulars* 18:7–8.

———. 1899. Notes on the development of *Ophiura olivacea*, Lyman. *Zoologischer Anzeiger* 22:92–96.

———. 1900. *Ophiura brevispina*. *Memoirs from the Biological Laboratory of the Johns Hopkins University* 4:79–100, pls. 1–3.

———. 1902a. Feeding habits of a spatangoid, *Moera atropos*, a brittle-star fish, *Ophiophragma wurdmanni*, and a holothurian, *Thyone briareus*. *Science* (New York) 15:579.

———. 1902b. A method of rearing marine larvae. *Science* (New York) 15:579–580.

———. 1914. A solution of the problem of yolk manipulation by *Ophiura*. *Science* (New York) 39:438.

———. 1916. *Ophiura brevispina*. II. An embryological contribution and a study of the effect of yolk substance upon development and developmental processes. *Journal of Morphology* 27:413–445, pls. 1–3.

Gray, I. E., M. E. Downey, and M. J. Cerame-Vivas. 1968a. Sea-stars of North Carolina. *U.S. Fish and Wildlife Service Fishery Bulletin* 67:127–163.

Gray, I. E., L. R. McCloskey, and S. C. Weihe. 1968b. The commensal crab *Dissodactylus mellitae* and its reaction to sand dollar host-factor. *Journal of the Elisha Mitchell Scientific Society* 84:472–481.

Gray, J. E. 1825. An attempt to divide the Echinida, or sea eggs, into natural families. *Annals of Philosophy*, n.s. 26:423–431.

———. 1840. A synopsis of the genera and species of the class Hypostomata (Asterias, Linnaeus). *Annals and Magazine of Natural History*, ser. 1, 6:175–184, 275–290.

Green, J. D. 1978. The annual reproductive cycle of an apodous holothurian, *Leptosynapta tenuis*: A bimodal breeding season. *Biological Bulletin* (Woods Hole) 154:68–78.

Greenstein, B. J. 1989. Mass mortality of the West-Indian echinoid *Diadema antillarum* (Echinodermata: Echinoidea): A natural experiment in taphonomy. *Palaios* 4:487–492.

———. 1991. An integrated study of echinoid taphonomy: Predictions for the fossil record of four echinoid families. *Palaios* 6:519–540.

———. 1993. Is the fossil record of regular echinoids really so poor? A comparison of living and subfossil assemblages. *Palaios* 8:587–601.

Griffith, H. 1987. Taxonomy of the genus *Dissodactylus* (Crustacea: Brachyura: Pinnotheridae) with descriptions of three new species. *Bulletin of Marine Science* 40:397–422.

Grober, M. S. 1988a. Responses of tropical reef fauna to brittle-star luminescence (Echinodermata: Ophiuroidea). *Journal of Experimental Marine Biology and Ecology* 115:157–168.

———. 1988b. Brittle-star bioluminescence functions as an aposematic signal to deter crustacean predators. *Animal Behavior* 36:493–501.

———. 1989. Starlight on the reef. *Natural History* 10:72–80.

Grunbaum, H., G. Bergman, D. P. Abbott, and J. C. Ogden. 1978. Intraspecific agonistic behavior in the rock-boring sea urchin *Echinometra lucunter* (L.) (Echinodermata: Echinoidea). *Bulletin of Marine Science* 28:181–188.

Guille, A., and S. Ribes. 1981. Échinodermes associés aux Scleractiniaires d'un récif frangeant de l'île de La Réunion (océan Indien). *Bulletin du Museum National d'Histoire Naturelle*, 4e sér., Section A Zoologie Biologie et Ecologie Animales 3A:73–92.

Haburay, K., R. W. Hastings, D. DeVries, and J. Massey. 1974. Tropical marine fishes from Pensacola, Florida. *Quarterly Journal of the Florida Academy of Sciences* 37:105–109.

Hajduk, S. L. 1992. Ultrastructure of the tube-foot of an ophiuroid echinoderm, *Hemipholis elongata*. Tissue and Cell 24:111–120.

Hajduk, S. L., and W. B. Cosgrove. 1975. Hemoglobin in an ophiuroid, *Hemipholis elongata*. *American Zoologist* 15:808.

Halpern, J. A. 1970. Growth rate of the tropical sea star *Luidia senegalensis* (Lamarck). *Bulletin of Marine Science* 20:626–633.

Hammond, L. S. 1981. An analysis of grain size modification in biogenic carbonate sediments by deposit-feeding holothurians and echinoids (Echinodermata). *Limnology and Oceanography* 26:896–906.

———. 1982a. Patterns of feeding and activity in deposit-feeding holothurians and echinoids (Echinodermata) from a shallow back-reef lagoon, Discovery Bay, Jamaica. *Bulletin of Marine Science* 32:549–571.

———. 1982b. Analysis of grain-size selection by deposit-feeding holothurians and echinoids (Echinodermata) from a shallow reef lagoon, Discovery Bay, Jamaica. *Marine Ecology Progress Series* 8:25–36.

———. 1983. Nutrition of deposit-feeding holothuroids and echinoids (Echinodermata) from a shallow reef lagoon, Discovery Bay, Jamaica. *Marine Ecology Progress Series* 10:297–305.

Harold, A. S., and M. Telford. 1990. Systematics, phylogeny and biogeography of the genus *Mellita* (Echinoidea: Clypeastroida). *Journal of Natural History* 24:987–1026.

Harriott, V. J. 1985. Reproductive biology of three congeneric sea cucumber species, *Holothuria atra*, *H. impatiens* and *H. edulis*, at Heron Reef, Great Barrier Reef. *Australian*

Journal of Marine and Freshwater Research 36:51–57.

Harvey, E. B. 1956. *The American Arbacia and Other Sea Urchins*. Princeton University Press, Princeton, New Jersey. xiv + 298 pp., 16 pls.

Hawkins, C. M. 1981. Efficiency of organic matter absorption by the tropical echinoid *Diadema antillarum* Philippi fed non-macrophytic algae. *Journal of Experimental Marine Biology and Ecology* 49:245–253.

Hay, M. E. 1984. Patterns of fish and urchin grazing on Caribbean coral reefs: Are previous results typical? *Ecology* 65:446–454.

Hay, M. E., and P. R. Taylor. 1985. Competition between herbivorous fishes and urchins on Caribbean reefs. *Oecologia* (Berlin) 65:591–598.

Hay, M. E., R. R. Lee, Jr., R. A. Guieb, and M. M. Bennett. 1986. Food preference and chemotaxis in the sea urchin *Arbacia punctulata* (Lamarck) Philippi. *Journal of Experimental Marine Biology and Ecology* 96:147–153.

Heatwole, D. W. 1981. Spawning and respiratory adaptations of the ophiuroid, *Hemipholis elongata* (Say) (Echinodermata). M.S. thesis, University of South Carolina, Columbia, South Carolina. vi + 84 pp.

Heatwole, D. W., and S. E. Stancyk. 1982. Spawning and functional morphology of the reproductive system in the ophiuroid, *Hemipholis elongata* (Say). In *Echinoderms:* Proceedings of the International Conference, Tampa Bay, 14–17 September 1981, ed. J. M. Lawrence, 469–474. Balkema, Rotterdam.

Heding, S. G. 1928. Papers from Dr. Th. Mortensen's Pacific Expedition 1914–16. XLVI. Synaptidae. *Videnskabelige Meddelelser fra den naturhistorisk Forening i Kjøbenhavn* 85:105–323.

Heding, S. G., and A. Panning. 1954. Phyllophoridae: Eine Bearbeitung der polytentaculaten dendrochiroten Holothurien des zoologischen Museums in Kopenhagen. *Spolia Zoologica Musei Hauniensis* 13:1–209.

Heffernan, J. M., and S. A. Wainwright. 1974. Locomotion of the holothurian *Euapta lappa* and redefinition of peristalsis. *Biological Bulletin* (Woods Hole) 147:95–104.

Hendler, G. 1973. Northwest Atlantic amphiurid brittlestars, *Amphioplus abditus* (Verrill), *Amphioplus macilentus* (Verrill), *Amphioplus se-pultus* n. sp. (Ophiuroidea: Echinodermata): Systematics, zoogeography annual periodicities, and larval adaptations. Ph.D. dissertation, University of Connecticut, Storrs, Connecticut. xii + 255 pp.

———. 1975. Adaptational significance of the patterns of ophiuroid development. *American Zoologist* 15:691–715.

———. 1977a. Development of *Amphioplus abditus* (Verrill) (Echinodermata: Ophiuroidea). I. Larval biology. *Biological Bulletin* (Woods Hole) 152:51–63.

———. 1977b. The differential effects of seasonal stress and predation on the stability of reef-flat echinoid populations. In *Proceedings: Third International Coral Reef Symposium*. Volume 1 (Biology), ed. D. L. Taylor, 217–223. Rosenstiel School of Marine and Atmospheric Science, University of Miami, Miami, Florida.

———. 1979a. Reproductive periodicity of ophiuroids (Echinodermata: Ophiuroidea) on the Atlantic and Pacific coasts of Panamá. In *Reproductive Ecology of Marine Invertebrates*, ed. S. E. Stancyk, 145–156. University of South Carolina Press, Columbia, South Carolina.

———. 1979b. Sex-reversal and viviparity in *Ophiolepis kieri*, n. sp., with notes on viviparous brittlestars from the Caribbean (Echinodermata: Ophiuroidea). *Proceedings of the Biological Society of Washington* 92:783–795.

———. 1982a. Slow flicks show star tricks: Elapsed-time analysis of basketstar *(Astrophyton muricatum)* feeding behavior. *Bulletin of Marine Science* 32:909–918.

———. 1982b. The feeding biology of *Ophioderma brevispinum* (Ophiuroidea: Echinodermata). In *Echinoderms:* Proceedings of the International Conference, Tampa Bay, 14–17 September 1981, ed. J. M. Lawrence, 21–27. Balkema, Rotterdam.

———. 1984a. The association of *Ophiothrix lineata* and *Callyspongia vaginalis*: A brittlestar-sponge cleaning symbiosis? *Marine Ecology (Pubblicazioni della Stazione Zoologica di Napoli I)* 5:9–27.

———. 1984b. Brittlestar color-change and phototaxis (Echinodermata: Ophiuroidea: Ophiocomidae). *Marine Ecology (Pubblicazioni della Stazione Zoologica di Napoli I)* 5:379–401.

———. 1988a. Ophiuroid skeleton ontogeny reveals homologies among skeletal plates of adults: A study of *Amphiura filiformis, Amphiura stimpsonii* and *Ophiophragmus filograneus* (Echinodermata). *Biological Bulletin* (Woods Hole) 174:20–29.

———. 1988b. Western Atlantic *Ophiolepis* (Echinodermata: Ophiuroidea): A description of *O. pawsoni* new species, and a key to the species. *Bulletin of Marine Science* 42:265–272.

———. 1991. Echinodermata: Ophiuroidea. In *Reproduction of Marine Invertebrates*, Volume VI, *Echinoderms and Lophophorates,* eds. A. C. Giese, J. S. Pearse, and V. B. Pearse, 355–511. The Boxwood Press, Pacific Grove, California.

———. In press (1995). New species of brittle stars from the western Atlantic: *Ophionereis vittata, Amphioplus sepultus,* and *Ophiostigma siva* (Echinodermata: Ophiuroidea). *Contributions in Science* (Los Angeles).

Hendler, G., and M. Byrne. 1987. Fine structure of the dorsal arm plate of *Ophiocoma wendti*: Evidence for a photoreceptor system (Echinodermata, Ophiuroidea). *Zoomorphology* (Berlin) 107:261–272.

Hendler, G., and B. S. Littman. 1986. The ploys of sex: Relationships among the mode of reproduction, body size and habits of coral-reef brittlestars. *Coral Reefs* 5:31–42.

Hendler, G., and D. L. Meyer. 1982. An association of a polychaete, *Branchiosyllis exilis* with an ophiuroid, *Ophiocoma echinata*, in Panama. *Bulletin of Marine Science* 32:736–744.

Hendler, G., and J. E. Miller. 1984a. Feeding behavior of *Asteroporpa annulata*, a gorgonocephalid brittlestar with unbranched arms. *Bulletin of Marine Science* 34:449–460.

———. 1984b. *Ophioderma devaneyi* and *Ophioderma ensiferum*, new brittlestar species from the western Atlantic. *Proceedings of the Biological Society of Washington* 97:442–461.

Hendler, G., and R. W. Peck. 1988. Ophiuroids off the deep end: Fauna of the Belizean forereef slope. In *Echinoderm Biology*. Proceedings of the Sixth International Echinoderm Conference, Victoria, 23–28 August 1987, eds. R. D. Burke, P. V. Mladenov, P. Lambert, and R. L. Parsley, 411–419. Balkema, Rotterdam.

Hendler, G., and R. L. Turner. 1987. Two new species of *Ophiolepis* (Echinodermata: Ophiuroidea) from the Caribbean Sea and Gulf of Mexico: With notes on ecology, reproduction and morphology. *Contributions in Science* (Los Angeles) 395:1–14.

Hendler, G., and P. A. Tyler. 1986. The reproductive cycle of *Ophioderma brevispinum* (Echinodermata: Ophiuroidea). *Marine Ecology (Pubblicazioni della Stazione Zoologica di Napoli I)* 7:115–122.

Henkel, D. H. 1982. Echinoderms of Enmedio Reef, southwestern Gulf of Mexico. M.S. thesis, Corpus Christi State University, Corpus Christi, Texas. vi + 78 pp.

Hertel, L., D. W. Duszynski, and J. E. Ubelaker. 1990. Turbellarians (Umagillidae) from Caribbean urchins with a description of *Syndisyrinx collongistyla*, n. sp. *Transactions of the American Microscopical Society* 109:272–281.

Hess, S. C. 1978. *Guide to the Commoner Shallow-water Asteroids (Starfish) of Florida, the Gulf of Mexico, and the Caribbean Region,* ed. G. Voss. University of Miami Sea Grant Program, Sea Grant Field Guide Series No. 7:1–39.

Hines, G. A., J. B. McClintock, and S. A. Watts. 1992. Levels of estradiol and progesterone in male and female *Eucidaris tribuloides* (Echinodermata, Echinoidea) over an annual gametogenic cycle. *Florida Scientist* 55:85–91.

Hoffman, S. G., and D. R. Robertson. 1983. Foraging and reproduction of two Caribbean reef toadfishes (Batrachoididae). *Bulletin of Marine Science* 33:919–927.

Hoggett, A. K., and F. W. E. Rowe. 1986. A reappraisal of the family Comasteridae A. H. Clark, 1908 (Echinodermata: Crinoidea), with the description of a new subfamily and a new genus. *Zoological Journal of the Linnean Society* 88:103–142.

Holland, N. D. 1967. Some observations on the saccules of *Antedon mediterranea* (Echinodermata, Crinoidea). *Pubblicazioni della Stazione Zoologica di Napoli* 35:257–262.

———. 1971. The fine structure of the ovary of the feather star *Nemaster rubiginosa* (Echinodermata: Crinoidea). *Tissue and Cell* 3:161–175.

———. 1991. Echinodermata: Crinoidea. In *Reproduction of Marine Invertebrates,* Volume VI, *Echinoderms and Lophophorates,* eds. A. C. Giese, J. S. Pearse, and V. B. Pearse, 247–299. The Boxwood Press, Pacific Grove, California.

Holmquist, J. G. 1994. Benthic microalgae as a dispersal mechanism for fauna: Influence of a marine tumbleweed. *Journal of Experimental Marine Biology and Ecology* 180:235–251.

Hopkins, T. S. 1988. A review of the distribution and proposed morphological groupings of extant species of the genus *Clypeaster* in the Caribbean Sea and Gulf of Mexico. In *Echinoderm Biology.* Proceedings of the Sixth International Echinoderm Conference, Victoria, 23–28 August 1987, eds. R. D. Burke, P. V. Mladenov, P. Lambert, and R. L. Parsley, 337–345. Balkema, Rotterdam.

Hopkins, T. S., D. R. Blizzard, S. A. Brawley, S. A. Earle, D. E. Grimm, D. K. Gilbert, P. G. Johnson, E. H. Livingston, C. H. Lutz, J. K. Shaw, and B. B. Shaw. 1977. A preliminary characterization of the biotic components of composite strip tracts on the Florida Middlegrounds, northeastern Gulf of Mexico. In *Proceedings. Third International Coral Reef Symposium.* Volume 1 (Biology), ed. D. L. Taylor, 31–37. Rosenstiel School of Marine and Atmospheric Science, University of Miami, Miami, Florida.

Hoppe, W. F. 1988. Growth, regeneration and predation in three species of large coral reef sponges. *Marine Ecology Progress Series* 50:117–125.

Horta-Puga, G., and J. P. Carricart-Ganivet. 1990. *Stylaster roseus* (Pallas, 1766): First record of a stylasterid (Cnidaria: Hydrozoa) in the Gulf of Mexico. *Bulletin of Marine Science* 47:575–576.

Hoskin, C. M., and J. K. Reed. 1985. Carbonate sediment production by the rock-boring urchin *Echinometra lucunter* and associated endolithic infauna at Black Rock, Little Bahama Bank. *Symposia Series Underwater Research* 3:151–161.

Hotchkiss, F. H. C. 1982. Ophiuroidea (Echinodermata) from Carrie Bow Cay, Belize. In *The Atlantic Barrier Reef Ecosystem at Carrie Bow Cay, Belize, I. Structure and Communities,* eds. K. Rützler and I. G. Macintyre. *Smithsonian Contributions to the Marine Sciences* No. 12:387–412.

Hughes, T. P. 1994. Catastrophes, phase shifts, and large-scale degradation of a Caribbean coral reef. *Science* (Washington, D.C.) 265:1547–1551.

Hughes, T. P., D. C. Reed, and M.-J. Boyle. 1987. Herbivory on coral reefs: Community structure following mass mortalities of sea urchins. *Journal of Experimental Marine Biology and Ecology* 113:39–59.

Humann, P. 1992. *Reef Creature Identification. Florida, Caribbean, Bahamas.* New World Publications, Jacksonville, Florida. 344 pp.

Humes, A. G., and G. Hendler. 1972. New cyclopoid copepods associated with the ophiuroid genus *Amphioplus* on the eastern coast of the United States. *Transactions of the American Microscopical Society* 91:539–555.

Humes, A. G., and J. H. Stock. 1973. A revision of the family Lichomolgidae Kossman, 1877, cyclopoid copepods mainly associated with marine invertebrates. *Smithsonian Contributions to Zoology* No. 127:v + 368.

Hunte, W., and D. Younglao. 1988. Recruitment and population recovery of *Diadema antillarum* (Echinodermata: Echinoidea) in Barbados. *Marine Ecology Progress Series* 45:109–119.

Hunte, W., I. Cote, and T. Tomascik. 1986. On the dynamics of the mass mortality of *Diadema antillarum* in Barbados. *Coral Reefs* 4:135–139.

Hunter, R. D., and H. Y. Elder. 1967. Analysis of burrowing mechanism in *Leptosynapta tenuis* and *Golfingia gouldi. Biological Bulletin* (Woods Hole) 133:470.

Hylander, B. L., and R. G. Summers. 1975. An ultrastructural investigation of the spermatozoa of two ophiuroids, *Ophiocoma echinata* and *Ophiocoma wendti*: Acrosomal morphology and reaction. *Cell and Tissue Research* 158:151–168.

Hyman, L. H. 1955. *The Invertebrates: Echinodermata. The Coelomate Bilateria.* McGraw-Hill, New York. vii + 763 pp.

Iliffe, T. M., and J. S. Pearse. 1982. Annual and

lunar reproductive rhythms of the sea urchin *Diadema antillarum* Philippi in Bermuda. *International Journal of Invertebrate Reproduction* 5:139–148.

Irimura, S. 1982. *The Brittle-stars of Sagami Bay. Collected by His Majesty The Emperor of Japan.* Biological Laboratory, Imperial Household, Tokyo. xii + 148 pp., 15 pls., 1 map.

———. 1988. Ossicles of the stomach wall of Ophiuroidea and their taxonomic significance. In *Echinoderm Biology.* Proceedings of the Sixth International Echinoderm Conference, Victoria, 23–28 August 1987, eds. R. D. Burke, P. V. Mladenov, P. Lambert, and R. L. Parsley, 315–322. Balkema, Rotterdam.

———. 1991. Ossicles of the stomach wall of Ophiuroidea and their taxonomic significance II. In *Biology of Echinodermata.* Proceedings of the Seventh International Echinoderm Conference, Atami, 9–14 September 1990, eds. T. Yanagisawa, I. Yasumasu, C. Oguro, N. Suzuki, and T. Motokawa, 215–220. Balkema, Rotterdam.

Itzkowitz, M., and T. Koch. 1991. Relationships between damselfish egg loss and brittlestars. *Bulletin of Marine Science* 48:164–166.

Jackson, J. B. C. 1992. Pleistocene perspectives on coral reef community structure. *American Zoologist* 32:719–731.

Jackson, J. B., and K. W. Kaufmann. 1987. *Diadema antillarum* was not a keystone predator in cryptic reef environments. *Science* (Washington, D.C.) 235:687–689.

Jackson, R. T. 1912. Phylogeny of the Echini, with a revision of Palaeozoic species. *Memoirs of the Boston Society of Natural History* 7:1–491 + pls. 1–76.

Jangoux, M. 1984. Diseases of echinoderms. *Helgoländer Meeresuntersuchungen* 37:207–216.

Jangoux, M., and J. M. Lawrence. 1982. *Echinoderm Nutrition.* Balkema, Rotterdam. xiii + 654 pp.

Jans, D., and M. Jangoux. 1989. Functional morphology of vibratile urnae in the synaptid holothuroid *Leptosynapta inhaerens* (Echinodermata). *Zoomorphology* (Berlin) 109:165–171.

———. 1992. Rejection of intracoelomic invading material by *Leptosynapta inhaerens* (Echinodermata: Holothuroida): A process of ecological significance? *Marine Ecology (Pubblicazioni della Stazione Zoologica di Napoli I)* 13:225–231.

Jensen, M. 1981. An alleged stirodont lantern in an irregular echinoid. In *Echinoderms: Present and Past.* Proceedings of the European Colloquium on Echinoderms, Brussels, 3–8 September 1979, ed. M. Jangoux, 31–35. Balkema, Rotterdam.

John, D. D., and A. M. Clark. 1954. The "Rosaura" Expedition. 3. The Echinodermata. *Bulletin of the British Museum (Natural History) Zoology* 2:139–162, pl. 6.

Johnson, J. 1972. The biology of *Amphipholis squamata* Delle Chiaje (Echinodermata: Ophiuroidea). Ph.D. dissertation, University of Newcastle-upon-Tyne, England. v + 187 pp.

Jones, M. B., and G. Smaldon. 1989. Aspects of the biology of a population of the cosmopolitan brittlestar *Amphipholis squamata* (Echinodermata) from the Firth of Forth, Scotland. *Journal of Natural History* 23:613–625.

Jordan, H. E. 1908. The germinal spot in echinoderm eggs. Carnegie Institution of Washington Publication No. 102, *Papers from the Tortugas Laboratory of the Carnegie Institution of Washington* 1:3–12.

Kampfer, S., and W. Tertschnig. 1992. Feeding biology of *Clypeaster rosaceus* (Echinoidea, Clypeasteridae) and its impact on shallow lagoon sediments. In *Echinoderm Research 1991.* Proceedings of the Third European Conference on Echinoderms, Lecce, Italy, 9–12 September 1991, eds. L. Scalera-Liaci and C. Canicattì, 197–198. Balkema, Rotterdam.

Kaplan, E. H. 1982. *A Field Guide to Coral Reefs of the Caribbean and Florida.* Peterson Field Guide Series, no. 27. Houghton Mifflin Company, Boston. xv + 289 pp., 37 pls.

Karlson, R. H. 1983. Disturbance and monopolization of a spatial resource by *Zoanthus sociatus* (Coelenterata: Anthozoa). *Bulletin of Marine Science* 33:118–131.

Karlson, R. H., and D. R. Levitan. 1990. Recruitment-limitation in open populations of *Diadema antillarum*: An evaluation. *Oecologia* (Berlin) 82:40–44.

Keller, B. D. 1976. Sea urchin abundance patterns in seagrass meadows: The effects of predation and competitive interactions. Ph.D. dissertation, The Johns Hopkins University, Baltimore, Maryland. i + 73 pp.

———. 1983. Coexistence of sea urchins in seagrass meadows: An experimental analysis of competition and predation. *Ecology* 64:1581–1598.

Kessel, R. G. 1964. Electron microscope studies on oocytes of an echinoderm *Thyone briareus*, with special reference to the origin of the annulate lamellae. *Journal of Ultrastructure Research* 10:498–514.

Kier, P. M. 1963. Tertiary echinoids from the Caloosahatchee and Tamiami formations of Florida. *Smithsonian Miscellaneous Collections* 145:1–63, pls. 1–18.

———. 1975. The echinoids of Carrie Bow Cay, Belize. *Smithsonian Contributions to Zoology* No. 206:1–45.

Kier, P. M., and R. E. Grant. 1965. Echinoid distribution and habits, Key Largo Coral Reef Preserve, Florida. *Smithsonian Miscellaneous Collections* 149:1–68.

Kille, F. R. 1935. Regeneration in *Thyone briareus* Lesueur following induced autotomy. *Biological Bulletin* (Woods Hole) 69:82–108.

———. 1937. Regeneration of the genus *Holothuria*. *Year Book of the Carnegie Institution of Washington* No. 36 (1936–1937):93–94.

Kissling, D. L., and G. T. Taylor. 1977. Habitat factors for reef-dwelling ophiuroids in the Florida Keys. In *Proceedings. Third International Coral Reef Symposium. Volume 1 (Biology)*, ed. D. L. Taylor, 225–231. Rosenstiel School of Marine and Atmospheric Science, University of Miami, Miami, Florida.

Kleitman, N. 1941. The effect of temperature on the righting of echinoderms. *Biological Bulletin* (Woods Hole) 80:282–298.

Klinger, T. S., S. A. Watts, and D. A. Forcucci. 1988. Effect of short-term feeding and starvation on storage and synthetic capacities of gut tissues of *Lytechinus variegatus* (Lamarck) (Echinodermata: Echinoidea). *Journal of Experimental Marine Biology and Ecology* 117:187–195.

Koehler, R. 1907. Revision de la collection des ophiures du muséum d'histoire naturelle de Paris. *Bulletin Scientifique de la France et de la Belgique* 41:279–351, pls. 10–14.

———. 1913. Ophiures. *Zoologische Jahrbücher Supplement* 11:351–380, pls. 20–21.

———. 1914a. A contribution to the study of ophiurans of the United States National Museum. *United States National Museum Bulletin* 84:viii + 173, pls. 1–18.

———. 1914b. Echinodermata I. Asteroidea, Ophiuroidea et Echinoidea. *Beiträge zur Kenntnis der Meeresfauna Westafrikas* 1(2), ed. W. Michaelsen, 129–303, pls. 1–12. L. Friederichsen & Co., Hamburg, Germany.

———. 1922. Contributions to the biology of the Philippine Archipelago and adjacent regions. Ophiurans of the Philippine Seas and adjacent waters. *United States National Museum Bulletin* 100:1–486, pls. 1–103.

———. 1926. Révision de quelques ophiures de Ljungman. Appartenant au Musée d'Histoire Naturelle de Stockholm. *Arkiv för Zoologi* 19A:1–29, pls. 1–5.

Komatsu, M., F.-S. Chia, and R. Koss. 1991a. Sensory neurons of the bipinnaria larva of the sea star, *Luidia senegalensis*. *Invertebrate Reproduction and Development* 19:203–211.

Komatsu, M., C. Oguro, and J. M. Lawrence. 1991b. A comparison of the development in three species of the genus, *Luidia* (Echinodermata: Asteroidea) from Florida. In *Biology of Echinodermata*. Proceedings of the Seventh International Echinoderm Conference, Atami, 9–14 September 1990, eds. T. Yanagisawa, I. Yasumasu, C. Oguro, N. Suzuki, and T. Motokawa, 489–498. Balkema, Rotterdam.

Koster, F., and I. E. Caycedo. 1979. Primer hallazgo de *Astichopus multifidus* (Echinodermata: Holothuroidea, Stichopodiidae) y *Carapus bermudensis* (Pisces: Gadiformes, Carapidae) en el Caribe colombiano, con notas sobre esta nueva asociación. *Boletin del Museo del Mar, Universidad de Bogota* 9:30–36.

Lacalli, T. C. 1988. Ciliary patterns and pattern rearrangements in the development of the doliolaria larva. In *Echinoderm Biology*. Proceedings of the Sixth International Echinoderm Conference, Victoria, 23–28 August 1987, eds. R. D. Burke, P. V. Mladenov, P. Lambert, and R. L. Parsley, 273–275. Balkema, Rotterdam.

Lamarck, J. B. A. de. 1816. *Histoire Naturelle des Animaux sans Vertèbres, présentant les caractères généraux et particuliers de ces animaux, leur distribution, leurs classes, leurs familles, leurs genres, et la citation des principales espèces qui s'y rapportent; Précédés d'une Introduction offrant la Détermination des caractères essentiels de l'Animal, sa distinction du végétal et des autres corps naturels, enfin, l'Exposition des Principes fondamentaux de la Zoologie.* Tome Second. Libraire Verdière, Paris. iii + 568 pp.

Lampert, K. 1885. Die Seewalzen—Holothurioidea: Eine systematische Monographie mit Bestimmungs- und Verbreitungs-Tabellen. *Reisen im Archipel der Philippinen von Dr. C. Semper.* II. *Wissenschaftliche Resultate* 4(3):1–310 pp., 1 pl. C. W. Kreidel's Verlag, Weisbaden.

Lares, M. T., and J. B. McClintock. 1991. The effects of food quality and temperature on the nutrition of the carnivorous sea urchin *Eucidaris tribuloides* (Lamarck). *Journal of Experimental Marine Biology and Ecology* 149:279–286.

Larrauri, L. R. A. 1981. Listado preliminar de los equinodermos de la costa Atlantica Colombiano. *Boletin Museo del Mar* (Bogotá) 10:24–39.

Lawrence, J. M. 1973. Level, content, and caloric equivalents of the lipid, carbohydrate, and protein in the body components of *Luidia clathrata* (Echinodermata: Asteroidea: Platyasterida) in Tampa Bay. *Journal of Experimental Marine Biology and Ecology* 11:263–274.

———. 1975. On the relationships between marine plants and sea urchins. *Oceanography and Marine Biology Annual Review* 13:213–286.

Lawrence, J. M., and F. Dehn. 1979. Biological characteristics of *Luidia clathrata* (Echinodermata: Asteroidea) from Tampa Bay and the shallow waters of the Gulf of Mexico. *Florida Scientist* 42:9–13.

Leske, N. G. 1778. *Additamenta ad Jacobi Theodori Klein naturalem dispositionem Echinodermatum. Accessit lucubratiuncula de Aculeis Echinorum marinorum, cum spicilegio de Belemnitis.* Edita et aucta a N. G. Leske. Lipsiae. 278 pp., 54 pls.

Lesser, M. P. 1986. Bacterial endosymbionts of *Amphipolis squamata*: Potential contribution to developmental and embryonic feeding biology. *American Zoologist* 26:22A.

Lesser, M. P., and C. W. Walker. 1992. Comparative study of the uptake of dissolved amino acids in sympatric brittlestars with and without endosymbiotic bacteria. *Comparative Biochemistry and Physiology, Part B, Comparative Biochemistry* 101B:217–223.

Lessios, H. A. 1981a. Reproductive periodicity of the echinoids *Diadema* and *Echinometra* on the two coasts of Panama. *Journal of Experimental Marine Biology and Ecology* 50:47–61.

———. 1981b. Divergence in allopatry: Molecular and morphological differentiation between sea urchins separated by the Isthmus of Panama. *Evolution* 35:618–638.

———. 1984. Possible prezygotic reproductive isolation in sea urchins separated by the Isthmus of Panama. *Evolution* 38:1144–1148.

———. 1985a. Genetic consequences of mass mortality in the Caribbean sea urchin *Diadema antillarum*. In *Proceedings of the Fifth International Coral Reef Congress,* Tahiti, 27 May–1 June 1985, Volume 4: Symposia and Seminars (B), eds. C. Gabrie and B. Salvat, 119–126. Antenne Museum—E.P.H.E., Moorea, French Polynesia.

———. 1985b. Annual reproductive periodicity in eight echinoid species on the Caribbean coast of Panama. In *Echinodermata.* Proceedings of the Fifth International Echinoderm Conference, Galway, 24–29 September 1984, eds. B. F. Keegan and B. D. S. O'Connor, 303–311. Balkema, Rotterdam.

———. 1987. Temporal and spatial variation in egg size of 13 Panamanian echinoids. *Journal of Experimental Marine Biology and Ecology* 114:217–239.

———. 1988a. Population dynamics of *Diadema antillarum* (Echinodermata: Echinoidea) following mass mortality in Panama. *Marine Biology* (Berlin) 99:515–526.

———. 1988b. Mass mortality of *Diadema antillarum* in the Caribbean: What have we learned? *Annual Review of Ecology and Systematics* 19:371–393.

———. 1990. Adaptation and phylogeny as determinants of egg size in echinoderms from the two sides of the Isthmus of Panama. *American Naturalist* 135:1–13.

———. 1991. Presence and absence of monthly reproductive rhythms among eight Caribbean echinoids off the coast of Panama. *Journal of Experimental Marine Biology and Ecology* 153:27–47.

Lessios, H. A., and C. W. Cunningham. 1990. Gametic incompatibility between species of the sea urchin *Echinometra* on the two sides of the Isthmus of Panama. *Evolution* 44:933–941.

Lessios, H. A., J. D. Cubit, D. R. Robertson, M. J. Shulman, M. R. Parker, S. D. Garrity, and S. C. Levings. 1984a. Mass mortality of *Diadema antillarum* on the Caribbean coast of Panama. *Coral Reefs* 3:173–182.

Lessios, H. A., D. R. Robertson, and J. D. Cubit. 1984b. Spread of *Diadema* mass mortality through the Caribbean. *Science* (Washington, D.C.) 226:335–337.

Lesueur, C. A. 1824. Descriptions of several new species of *Holothuria*. *Journal of the Academy of Natural Sciences of Philadelphia* 4:155–163.

Levitan, D. R. 1988a. Asynchronous spawning and aggregative behavior in the sea urchin *Diadema antillarum* (Philippi). In *Echinoderm Biology*. Proceedings of the Sixth International Echinoderm Conference, Victoria, 23–28 August 1987, eds. R. D. Burke, P. V. Mladenov, P. Lambert, and R. L. Parsley, 181–186. Balkema, Rotterdam.

———. 1988b. Density-dependent size regulation and negative growth in the sea urchin *Diadema antillarum* Philippi. *Oecologia* (Berlin) 76:627–629.

———. 1988c. Algal–urchin biomass responses following mass mortality of *Diadema antillarum* Philippi at St. John, U.S. Virgin Islands. *Journal of Experimental Marine Biology and Ecology* 119:167–178.

———. 1989. Density-dependent size regulation in *Diadema antillarum*: Effects on fecundity and survivorship. *Ecology* 70:1414–1424.

———. 1991. Skeletal changes in the test and jaws of the sea urchin *Diadema antillarum* in response to food limitation. *Marine Biology* (Berlin) 111:431–435.

Levitan, D. R., and S. J. Genovese. 1989. Substratum-dependent predator–prey dynamics: Patch reefs as refuges from gastropod predation. *Journal of Experimental Marine Biology and Ecology* 130:111–118.

Lewis, J. B. 1958. The biology of the tropical sea urchin *Tripneustes esculentus* Leske in Barbados, British West Indies. *Canadian Journal of Zoology* 36:607–621.

———. 1960. The fauna of rocky shores of Barbados, West Indies. *Canadian Journal of Zoology* 38:391–435.

———. 1964. Feeding and digestion in the tropical sea urchin *Diadema antillarum* Philippi. *Canadian Journal of Zoology* 42:549–557.

Lewis, J. B., and R. D. Bray. 1983. Community structure of ophiuroids (Echinodermata) from three different habitats on a coral reef in Barbados, West Indies. *Marine Biology* (Berlin) 73:171–176.

Lewis, J. B., and G. S. Storey. 1984. Differences in morphology and life history traits of the echinoid *Echinometra lucunter* from different habitats. *Marine Ecology Progress Series* 15:207–211.

Lewis, S. M., and P. C. Wainwright. 1985. Herbivore abundance and grazing intensity on a Caribbean coral reef. *Journal of Experimental Marine Biology and Ecology* 87:215–228.

Liddell, W. D. 1979. Shallow-water comatulid crinoids (Echinodermata) from Barbados, West Indies. *Canadian Journal of Zoology* 57:2413–2420.

———. 1982. Suspension feeding by Caribbean comatulid crinoids. In *Echinoderms*: Proceedings of the International Conference, Tampa Bay, 14–17 September 1981, ed. J. M. Lawrence, 33–39. Balkema, Rotterdam.

Liddell, W. D., and S. L. Ohlhorst. 1982. Morphological and electrophoretic analyses of Caribbean comatulid crinoid populations. In *Echinoderms*: Proceedings of the International Conference, Tampa Bay, 14–17 September 1981, ed. J. M. Lawrence, 173–182. Balkema, Rotterdam.

———. 1986. Changes in benthic community composition following the mass mortality of *Diadema* at Jamaica. *Journal of Experimental Marine Biology and Ecology* 95:271–278.

Linnaeus, C. 1758. *Systema Naturae per Regna Tria Naturae, secundum Classes, Ordines, Genera, Species, cum Characteribus, Differentiis,*

Synonymis, Locis. Tomus I. Editio Decima, Reformata. Impensis Direct. Laurentii Salvii, Holmiae. iv + 824 pp.

Ljungman, A. 1867. Om några nya arter af Ophiurider. *Öfversigt af Kongl. Vetenskaps-Akademiens Förhandlingar* 1866:163–166.

———. 1871. Förteckning öfver uti Vestindien af Dr. A. Goës samt under korvetten Josefinas expedition i Atlantiska Oceanen samlade Ophiurider. *Öfversigt af Kongl. Vetenskaps-Akademiens Förhandlingar* 1871:615–658.

Llewellyn, G., and D. L. Meyer. 1991. Biotic associates and ecology of reef dwelling unstalked crinoids of Bonaire, Netherlands Antilles. In *Biology of Echinodermata*. Proceedings of the Seventh International Echinoderm Conference, Atami, 9–14 September 1990, eds. T. Yanagisawa, I. Yasumasu, C. Oguro, N. Suzuki, and T. Motokawa, 584. Balkema, Rotterdam.

Ludwig, H. 1875. Beitrage zur Kenntniss der Holothurien. *Arbeiten aus dem Zoologisch-zootomisches Institut in Würzburg* 2:77–120, pls. 6–7.

———. 1881. Über eine lebendiggebärende Synaptide und zwei andere neue Holothurienarten der Brasilianischen Küste. *Archives de Biologie* 2:41–58, pl. 3.

———. 1886. Die von G. Chierchia auf der Fahrt der Kgl. Ital. Corvette Vittor Pisani gesammelten Holothurian. *Zoologische Jahrbücher. Abtheilung für Systematik, Geographie und Biologie der Thiere* 2:1–36, pls. 1–2.

———. 1904. Brutpflege bei Echinodermen. *Zoologische Jahrbücher Supplement* 7:683–699.

Lütken, C. 1856. Bidrag til Kundskaben Slangestjerner. II. Oversigt over de vestindiske Ophiurer. III. Bidrag til Kundskab om Ophiurerne ved Central-Amerikas Vestkyst. *Videnskabelige Meddelelser fra den naturhistorisk Forening i Kjøbenhavn* 1856:1–19, 20–26.

———. 1859a. Bidrag til kundskab om de ved kysterne af Mellenog Syd-Amerika levende arten af sostjerner. 1. *Videnskabelige Meddelelser fra den naturhistorisk Forening i Kjøbenhavn* 1859:25–97.

———. 1859b. Additamenta ad historiam Ophiuridarum. Beskrivelser af nye eller hidtil kun ufuldstaendigt kjendte Arter af Slangestjerner. Anden Afdeling. *Kongelige Danske Videnskabernes Selskabs Skrifter* 5(1861):177–271, pls. 1–5.

———. 1869. Additamenta ad historiam Ophiuridarum. Beskrivende og kritiske Bidrag til kundskab om Slangestjernerne. Tredie Afdeling. *Kongelige Danske Videnskabernes Selskabs Skrifter* 8(1870):23–109.

———. 1872. Ophiuridarum novarum vel minus cognitarum descriptiones nonnullae. Beskrivelser af nogle nye eller mindre bekjendte Slangestjerner. Med nogle Bemaerkninger om Selvdelingen hos Straaledyrene. *Oversigt over det Kongelige Danske Videnskabernes Selskabs Skrifter Forhandlinger* 2(1872):75–158, pls. 1–2.

Lyman, T. 1860. Descriptions of new Ophiuridae, belonging to the Smithsonian Institution and to the Museum of Comparative Zoölogy at Cambridge. *Proceedings of the Boston Society of Natural History* 7:193–205, 252–262, 424–425.

———. 1865. Ophiuridae and Astrophytidae. *Illustrated Catalogue of the Museum of Comparative Zoölogy at Harvard College* 1:1–200, pls. 1–2.

———. 1869. Preliminary report on the Ophiuridae and Astrophytidae dredged in deep water between Cuba and the Florida Reef, by L. E. Pourtales. *Bulletin of the Museum of Comparative Zoölogy at Harvard College* 1:309–354.

———. 1871. Supplement to the Ophiuridae and Astrophytidae. *Illustrated Catalogue of the Museum of Comparative Zoölogy at Harvard College* 6:1–17.

———. 1875. Zoölogical results of the Hassler Expedition. II. Ophiuridae and Astrophytidae, including those dredged by the late Dr. William Stimpson. *Illustrated Catalogue of the Museum of Comparative Zoölogy at Harvard College* 8:1–34, pls. 1–5.

———. 1877. Mode of forking among *Astrophytons*. *Proceedings of the Boston Society of Natural History* 19:102–108, pls. 4–7.

———. 1880. *A Preliminary List of the Known Genera and Species of Living Ophiuridae and Astrophytidae. With their Localities, and the Depths at which They Have Been Found; And References to the Principal Synonymies and Authorities.* Cambridge, Massachusetts. v + 45 pp.

———. 1882. Report on the Ophiuroidea dredged by H.M.S. Challenger, during the years 1873–1876. *Report on the Scientific Results of the Voyage of H.M.S. Challenger during the Years 1873–76. *Zoology,* Volume 5 (Part 14):1–386, pls. 1–48.

Macurda, D. B., Jr. 1973. Ecology of comatulid crinoids at Grand Bahama Island. *Hydro-Lab Journal* 2:9–24.

———. 1975. The bathymetry and zoogeography of shallow-water crinoids in the Bahama Islands. *Hydro-Lab Journal* 3:5–24.

———. 1976. Skeletal modifications related to food capture and feeding behavior of the basketstar *Astrophyton. Paleobiology* 2:1–7.

Macurda, D. B., Jr., and D. L. Meyer. 1977. Crinoids of West Indian coral reefs. *Studies in Geology. American Association of Petroleum Geologists* No. 4:231–237.

———. 1983. Sea lilies and feather stars. *American Scientist* 71:354–365.

Maddocks, R. R. 1987. An ostracode commensal of an ophiuroid and other new species of *Pontocypria* (Podocopida: Cypridacea). *Journal of Crustacean Biology* 7:727–737.

Madsen, F. J. 1941. On *Thyone warbergi* n. sp., a new holothurian from the Skagerrak, with remarks on *T. fusus* (O.F.M.) and other related species. *Göteborgs Kungl. Vetenskaps- och Vitterhets-Samhälles Handlingar* Ser. B, 1:1–31.

———. 1970. West African ophiuroids. Scientific results of the Danish Expedition to the Coast of Tropical West Africa 1945–1946. *Atlantide Report* No. 11:151–243.

Mallefet, J., F. Baguet, and M. Jangoux. 1989. Origin of the light emission to potassium chloride in *Amphipholis squamata* (Echinodermata). *Archives Internationales de Physiologie et de Biochemie* 97:P37.

Mallefet, J., P. Vanhoutte, and F. Baguet. 1992. Study of *Amphipholis squamata* luminescence. In *Echinoderm Research 1991.* Proceedings of the Third European Conference on Echinoderms, Lecce, Italy, 9–12 September 1991, eds. L. Scalera-Liaci and C. Canicattì, 125–130. Balkema, Rotterdam.

Mangum, C., and W. Van Winkle. 1973. Responses of aquatic invertebrates to declining oxygen conditions. *American Zoologist* 13:529–541.

Manwell, C. 1966. Sea cucumber sibling species: Polypeptide chain types and oxygen equilibrium of hemoglobin. *Science* (Washington, D.C.) 152: 1393–1396.

Manwell, C., and C. M. A. Baker. 1963. A sibling species of sea cucumber discovered by starch gel electrophoresis. *Comparative Biochemistry and Physiology* 10:39–53, 11 figs.

Marcus, N. H. 1980. Genetics of morphological variation of geographically distant populations of the sea urchin *Arbacia punctulata* (Lamarck). *Journal of Experimental Marine Biology and Ecology* 43:121–130.

Märkel, K. 1979. Structure and growth of the cidaroid socket-joint Lantern of Aristotle compared to the hinge-joint lantern of non-cidaroid regular urchins. *Zoomorphologie* 94:1–32.

———. 1981. Experimental morphology of coronar growth in regular echinoids. *Zoomorphology* (Berlin) 97:31–52.

Märkel, K., and U. Röser. 1983a. The spine tissues in the echinoid *Eucidaris tribuloides. Zoomorphology* (Berlin) 103:25–41.

———. 1983b. Calcite-resorption in the spine of the echinoid *Eucidaris tribuloides. Zoomorphology* (Berlin) 103:43–58.

Märkel, K., F. Kubanek, and A. Willgallis. 1971. Polykristalliner Calcit bei Seeigeln (Echinodermata, Echinoidea). *Zeitschrift für Zellforschung und Mikroskopische Anatomie* 119:355–377.

Märkel, K., U. Röser, U. Mackenstedt, and M. Klostermann. 1986. Ultrastructural investigation of matrix-mediated biomineralization in echinoids (Echinodermata, Echinoida). *Zoomorphology* (Berlin) 106:232–243.

Markle, D. F., and J. E. Olney. 1990. Systematics of the pearlfishes (Pisces: Carapidae). *Bulletin of Marine Science* 47:269–410.

Martens, E. von. 1867a. Ueber ostasiatische Echinodermen. *Archiv für Naturgeschichte,* Berlin 33:107–119, pl. 3.

———. 1867b. Über vier neue Schlangensterne, Ophiuren, des Kgl. zoologischen Museums vor. *Monatsberichte der Königlich-Preussischen Akademie der Wissenschaften zu Berlin* 1867:345–348.

Martin, D., and V. Alvà. 1988. Un polychète nouveau *Sphaerodorum ophiurophoretos* nov. sp. (Polychaeta: Sphaerodoridae), symbiotique de l'ophiure *Amphipholis squamata* (Delle Chiaje, 1828). *Bulletin van het Koninklijke Belgische Instituut voor Naruurwetenschappen, Biologie* 58:45–49.

Martin, R. B. 1968. Aspects of the ecology and behaviour of *Axiognathus squamata*. *Tane* 14: 65–81.

Martínez de Rodríguez, A. 1973. Contribucion al estudio de los holoturoideos de Venezuela. *Boletin Instituto Oceanografico Universidad de Oriente Cumana* 12:41–50.

Martínez de Rodriguez, A., and A. Mago Herminson. 1975. Contribucion al conocimiento de los holoturoideos (Holothuroidea: Echinodermata) de la region oriental de Venezuela. *Boletin Instituto Oceanografico Universidad de Oriente Cumana* 14:187–197.

Martínez M., A. 1989. Holoturoideos (Echinodermata, Holothuroidea) de la region nororiental de Venezuela y algunas dependencias federales. *Boletin Instituto Oceanografico Universidad de Oriente Cumana* 28:105–112.

———. 1991a. Holoturoideos Dendrochirotida (Holothuroidea: Echinodermata) I. Familia Cucumaridae. *Boletin Instituto Oceanografico Universidad de Oriente Cumana* 30:31–40.

———. 1991b. Holoturoideos Dendrochirotida (Holothuroidea: Echinodermata) I. Familia Sclerodactylidae. *Boletin Instituto Oceanografico Universidad de Oriente Cumana* 30:41–46.

Massin, C. 1993. On the taxonomic status of the genus *Parathyone* (Echinodermata, Holothuroidea, Dendrochirotida). *Bulletin de l'Institut Royal des Sciences Naturelles de Belgique Biologie* 63:257–258.

Matsumoto, H. 1917. A monograph of Japanese Ophiuroidea, arranged according to a new classification. *Journal of the College of Science, Imperial University of Tokyo* 38:1–408, pls. 1–7.

May, R. M. 1925. Les réactions sensorielles d'une ophiure (*Ophionereis reticulata*, Say). *Bulletin Biologique de la France et de la Belgique* 59:372–402.

Mayer, A. G. 1914. The effects of temperature upon tropical marine animals. Carnegie Institution of Washington Publication No. 183, *Papers from the Tortugas Laboratory of the Carnegie Institution of Washington* 6:1–24.

Mazur, J. E., and J. W. Miller. 1971. A description of the complete metamorphosis of the sea urchin *Lytechinus variegatus* cultured in synthetic sea water. *The Ohio Journal of Science* 71:30–36.

McClanahan, T. R. 1992. Epibenthic gastropods of the Middle Florida Keys: The role of habitat and environmental stress on assemblage composition. *Journal of Experimental Marine Biology and Ecology* 160:169–190.

McClintock, J. B. 1983. Escape response of *Argopecten irradians* (Mollusca: Bivalvia) to *Luidia clathrata* and *Echinaster* sp. (Echinodermata: Asteroidea). *Florida Scientist* 46:95–100.

McClintock, J. B., and J. M. Lawrence. 1981. An optimization study on the feeding behavior of *Luidia clathrata* Say (Echinodermata: Asteroidea). *Marine Behavior and Physiology* 7:263–275.

———. 1982. Photoresponse and associative learning in *Luidia clathrata* Say (Echinodermata: Asteroidea). *Marine Behavior and Physiology* 9:13–21.

———. 1984. Ingestive conditioning in *Luidia clathrata* (Say) (Echinodermata: Asteroidea): Effect of nutritional condition on selectivity, teloreception, and rates of ingestion. *Marine Behavior and Physiology* 10:167–181.

McClintock, J. B., and K. R. Marion. 1993. Predation by the king helmet *(Cassis tuberosa)* on six-holed sand dollars *(Leodia sexiesperforata)* at San Salvador, Bahamas. *Bulletin of Marine Science* 52:1013–1017.

McClintock, J. B., and S. A. Watts. 1990. The effects of photoperiod on gametogenesis in the tropical sea urchin *Eucidaris tribuloides* (Lamarck) (Echinodermata: Echinoidea). *Journal of Experimental Marine Biology and Ecology* 139:175–184.

McClintock, J. B., T. Hopkins, K. Marion, S. Watts, and G. Schinner. 1993. Population structure, growth, and reproductive biology of the gorgonocephalid brittlestar *Asteroporpa annulata*. *Bulletin of Marine Science* 52:925–936.

McClintock, J. B., T. S. Klinger, and J. M. Lawrence. 1982. Feeding preferences of echinoids for plant and animal food models. *Bulletin of Marine Science* 32:365–369.

———. 1983. Extraoral feeding in *Luidia clathrata* (Say) (Echinodermata: Asteroidea). *Bulletin of Marine Science* 33:171–172.

———. 1984. Chemoreception in *Luidia clathrata* (Echinodermata: Asteroidea): Qualitative and quantitative aspects of chemotactic responses to low molecular weight compounds. *Marine Biology* (Berlin) 84:47–52.

McGehee, M. A. 1992. Distribution and abundance of two species of *Echinometra* (Echinoidea) on coral reefs near Puerto Rico. *Caribbean Journal of Science* 28:173–183.

McLean, R. F. 1967. Erosion of burrows in beachrock by the tropical sea urchin *Echinometra lucunter*. *Canadian Journal of Zoology* 45:586–588.

McNulty, J. K. 1961. Ecological effects of sewage pollution in Biscayne Bay, Florida: Sediments and the distribution of benthic and fouling macro-organisms. *Bulletin of Marine Science of the Gulf and Caribbean* 11:394–447.

McNulty, J. K., R. C. Work, and H. B. Moore. 1962. Level sea bottom communities in Biscayne Bay and neighboring area. *Bulletin of Marine Science of the Gulf and Caribbean* 12:204–233.

McPherson, B. F. 1965. Contributions to the biology of the sea urchin *Tripneustes ventricosus*. *Bulletin of Marine Science* 15:228–244.

———. 1968a. Contributions to the biology of the sea urchin *Eucidaris tribuloides* (Lamarck). *Bulletin of Marine Science* 18:400–443.

———. 1968b. Feeding and oxygen uptake of the tropical sea urchin *Eucidaris tribuloides* (Lamarck). *Biological Bulletin* (Woods Hole) 135:308–321.

———. 1969. Studies on the biology of the tropical sea urchins *Echinometra lucunter* and *Echinometra viridis*. *Bulletin of Marine Science* 19:194–213.

Menton, D. N., and A. Z. Eisen. 1970. The structure of the integument of the sea cucumber, *Thyone briareus*. *Journal of Morphology* 141:17–36.

———. 1973. Cutaneous wound healing in the sea cucumber, *Thyone briareus*. *Journal of Morphology* 141:185–204.

Messing, C. G. 1978. A revision of the comatulid genus *Comactinia* A. H. Clark (Crinoidea: Echinodermata). *Bulletin of Marine Science* 28:49–80.

Messing, C. G., and J. H. Dearborn. 1990. Marine flora and fauna of the northeastern United States. Echinodermata: Crinoidea. *NOAA Technical Report, National Marine Fisheries Service* 91:ii, 1–30.

Mettrick, D. F., and J. B. Jennings. 1969. Nutrition and chemical composition of the rhabdocoel turbellarian *Syndesmis franciscana*, with notes on the taxonomy of *S. antillarum*. *Journal of the Fisheries Research Board of Canada* 26:2669–2679.

Meyer, D. L. 1972. *Ctenantedon*, a new antedonid crinoid convergent with comasterids. *Bulletin of Marine Science* 22:53–66.

———. 1973a. Coral Reef Project—Papers in memory of Dr. Thomas F. Goreau. 10. Distribution and living habits of comatulid crinoids near Discovery Bay, Jamaica. *Bulletin of Marine Science* 23:244–259.

———. 1973b. Feeding behavior and ecology of shallow-water unstalked crinoids (Echinodermata) in the Caribbean Sea. *Marine Biology* (Berlin) 22: 105–129.

Meyer, D. L., and W. I. Ausich. 1983. Biotic interactions among Recent and among fossil crinoids. In *Biotic Interactions in Recent and Fossil Benthic Communities*, eds. M. J. S. Tevesz and P. L. McCall, 377–427. Plenum Publishing Company, New York.

Meyer, D. L., and D. B. Macurda, Jr. 1977. Adaptive radiation of the comatulid crinoids. *Paleobiology* 3:74–82.

Meyer, D. L., C. G. Messing, and D. B. Macurda, Jr. 1978. Biological results of the University of Miami Deep-Sea Expeditions. 129. Zoogeography of tropical Western Atlantic Crinoidea (Echinodermata). *Bulletin of Marine Science* 28:412–441.

Michel, H. B. 1984. Culture of *Lytechinus variegatus* (Lamarck) (Echinodermata: Echinoidea) from egg to young adult. *Bulletin of Marine Science* 34:312–314.

Miller, J. E. 1984. Systematics of the ophidiasterid sea stars *Copidaster lymani* A. H. Clark, and *Hacelia superba* H. L. Clark (Echinodermata: Asteroidea) with a key to species of Ophidiasteridae from the western Atlantic. *Proceedings of the Biological Society of Washington* 97:194–208.

Miller, J. E., and D. L. Pawson. 1984. Holothurians (Echinodermata: Holothuroidea). *Memoirs of the Hourglass Cruises* 7 (Pt. 1):1–79.

Milliman, J. D. 1969. Four southwestern Caribbean atolls: Courtown Cays, Albuquerque Cays, Roncador Bank and Serrana Bank. *Atoll Research Bulletin* 129:iv + 26, 17 figs., 13 pls.

Millott, N. 1953. A remarkable association between *Ophionereis reticulata* (Say) and *Harmothoe lunulata* (Delle Chiaje). *Bulletin of Ma-*

rine Science of the Gulf and Caribbean 3:96–99.

———. 1956. The covering reaction of sea urchins. 1. A preliminary account of covering in the tropical echinoid *Lytechinus variegatus* (Lamarck), and its relation to light. *Journal of Experimental Biology* 33:508–523.

———. 1965. The enigmatic echinoids. In *Light as an Ecological Factor:* A Symposium held in Cambridge on March 30 to April 1st, 1965, eds. R. Bainbridge, G. C. Evans, and R. Rackham, 265–291. Blackwell, Oxford.

———. 1966. Light production. In *Physiology of Echinodermata*, ed. R. A. Boolootian, 487–501. Interscience Publishers, New York.

———. 1968. The dermal light sense. *Symposia of the Zoological Society of London* No. 23:1–36.

Millott, N. B., and R. Coleman. 1969. The podial pit—a new structure in the echinoid *Diadema antillarum* Philippi. *Zeitschrift für Zellforschung und Mikroskopische Anatomie* 95:187–197.

Mladenov, P. V. 1979. Unusual lecithotrophic development of the Caribbean brittle star *Ophiothrix oerstedi*. *Marine Biology* (Berlin) 55:55–62.

———. 1983. Breeding patterns of three species of Caribbean brittlestars (Echinodermata: Ophiuroidea). *Bulletin of Marine Science* 33:363–372.

———. 1985a. Development and metamorphosis of the brittlestar *Ophiocoma pumila*: Evolutionary and ecological implications. *Biological Bulletin* (Woods Hole) 168:285–295.

———. 1985b. Observations on reproduction and development of the Caribbean brittle star *Ophiothrix suensoni* (Echinodermata: Ophiuroidea). *Bulletin of Marine Science* 36:384–388.

Mladenov, P. V., and K. Brady. 1987. Reproductive cycle of the Caribbean feather star *Nemaster rubiginosa* (Echinodermata: Crinoidea). *Marine Ecology (Pubblicazioni della Stazione Zoologica di Napoli I)* 8:313–325.

Mladenov, P. V., and R. H. Emson. 1984. Divide and broadcast: Sexual reproduction in the West Indian brittle star *Ophiocomella ophiactoides* and its relationship to fissiparity. *Marine Biology* (Berlin) 81:273–282.

———. 1988. Density, size structure and reproductive characteristics of fissiparous brittle stars in algae and sponges: Evidence for interpopulational variation in levels of sexual and asexual reproduction. *Marine Ecology Progress Series* 42:181–194.

———. 1990. Genetic structure of populations of two closely related brittle stars with contrasting sexual and asexual life histories, with observations on the genetic structure of a second species. *Marine Biology* (Berlin) 104:265–274.

Mladenov, P. V., R. H. Emson, L. V. Colpitts, and I. C. Wilkie. 1983. Asexual reproduction in the West Indian brittle star *Ophiocomella ophiactoides* (H. L. Clark) (Echinodermata: Ophiuroidea). *Journal of Experimental Marine Biology and Ecology* 72:1–23.

Montague, J. R., J. A. Aguinaga, K. I. Ambrisco, D. L. Vassil, and W. Collazo. 1991. Laboratory measurement of ingestion rate for the sea urchin *Lytechinus variegatus* (Lamarck) (Echinodermata: Echinoidea). *Florida Scientist* 54:129–134.

Monteiro, A. M. G. 1992. Morfologia comparativa e distribuição batimétrica de duas espécies de Ophiuroidea, na região costeira de Ubatuba. *Boletim do Instituto Oceanográfico, São Paulo* 40:39–53.

Monteiro, A. M. G., and E. V. Pardo. 1994. Dieta alimentar de *Astropecten marginatus* e *Luidia senegalensis* (Echinodermata—Asteroidea). *Revista Brasileira de Biologia* 54:49–54.

Mooi, R. 1986. Structure and function of clypeasteroid miliary spines (Echinodermata, Echinoides). *Zoomorphology* (Berlin) 106:212–223.

Moore, D. R. 1956. Observations of predation on echinoderms by three species of Cassididae. *The Nautilus* 69:73–76.

Moore, H. B. 1965. The correlation of symmetry, color and spination in an urchin. *Bulletin of Marine Science* 15:245–254.

———. 1966. Ecology of echinoids. In *Physiology of Echinodermata*, ed. R. A. Boolootian, 73–85. John Wiley Interscience, New York.

Moore, H. B., and N. N. Lopez. 1966. The ecology and productivity of *Moira atropos* (Lamarck). *Bulletin of Marine Science* 16:648–667.

Moore, H. B., and B. F. McPherson. 1965. A contribution to the study of the productivity of the urchins *Tripneustes esculentus* and *Lytechinus variegatus*. Bulletin of Marine Science 15:855–871.

Moore, H. B., T. Jutare, J. C. Bauer, and J. A. Jones. 1963a. The biology of *Lytechinus variegatus*. Bulletin of Marine Science of the Gulf and Caribbean 13:23–53.

Moore, H. B., T. Jutare, J. A. Jones, B. F. McPherson, and C. F. E. Roper. 1963b. A contribution to the biology of *Tripneustes esculentus*. Bulletin of Marine Science 13:267–281.

Moore, R. C., and C. Teichert (eds.). 1978. *Treatise on Invertebrate Paleontology Part T, Echinodermata 2*, Volumes 1–3. The Geological Society of America and The University of Kansas Press, Lawrence, Kansas. xxxviii + 1027 pp.

Morales, M., C. Sierra, A. Vidal, J. del Castillo, and D. S. Smith. 1993. Pharmacological sensitivity of the articular capsule of the primary spines of *Eucidaris tribuloides*. Comparative Biochemistry and Physiology, Part C, Comparative Pharmacology and Toxicology 105C:25–30.

Morgulis, S. 1909. Regeneration in the brittlestar *Ophiocoma pumila*, with reference to the influence of the nervous system. Proceedings of the American Academy of Arts and Sciences, Boston 44:655–659, 1 pl.

Morrison, D. 1988. Comparing fish and urchin grazing in shallow and deeper coral reef algal communities. Ecology 69:1367–1382.

Mortensen, T. 1907. *The Danish Ingolf-Expedition*. Volume IV, Part 2. Echinoidea. (II.). H. Hagerup, Copenhagen. 200 pp., 1–19 pls., 1 tab.

———. 1917. On the development of some West Indian echinoderms. Year Book of the Carnegie Institution of Washington No. 15 (1916–1917):193–194.

———. 1920. On hermaphroditism in viviparous ophiurids. Acta Zoologica (Stockholm) 1:1–18, pl. 1.

Mortensen, T. 1921. *Studies of the Development and Larval Forms of Echinoderms*. G. E. C. Gad, Copenhagen, Denmark. xxxiii + 266 pp.

———. 1924. Echinoderms of New Zealand and the Auckland-Campbell Islands. II. Ophiuroidea. Videnskabelige Meddelelser fra den naturhistorisk Forening i Kjøbenhavn 77:91–177, pls. 3–4.

———. 1928. *A Monograph of the Echinoidea*. Volume I. *Cidaroidea*. C. A. Reitzel, Copenhagen. v + 1–551 pp., pls. 1–88.

———. 1931. Contribution to the study of the development and larval forms of some echinoderms. I. The development and larval forms of some tropical echinoderms. II. Observation on some Scandinavian echinoderm larvae. Kongelige Danske Videnskabernes Selskab Skrifter, Naturvidenskabelig og Mathematisk 41:1–39, pls. 1–7.

———. 1933a. Papers from Dr. Th. Mortensen's Pacific Expedition, 1914–16. LXV. Echinoderms of South Africa (Asteroidea and Ophiuroidea). Videnskabelige Meddelelser fra den naturhistorisk Forening i Kjøbenhavn 93:215–400, pls. 8–19.

———. 1933b. Papers from Dr. Th. Mortensen's Pacific Expedition, 1914–16. LXVI. The echinoderms of St. Helena (other than crinoids). Videnskabelige Meddelelser fra den naturhistorisk Forening i Kjøbenhavn 93:401–472, pls. 20–22.

———. 1940. *A Monograph of the Echinoidea*. Volume III. *(1). Aulodonta. With additions to Vol. II (Lepidocentra and Stirodonta)*. C. A. Reitzel, Copenhagen. iv + 370 pp., 77 pls.

———. 1941. Echinoderms of Tristan da Cunha. Results of the Norwegian Scientific Expedition to Tristan da Cunha 1937–1938 7:1–10, pl. 1.

———. 1943a. *A Monograph of the Echinoidea*. Volume III. *(2). Camarodonta. I. Orthopsidae, Glyphocyphidae, Temnopleuridae and Toxopneustidae*. C. A. Reitzel, Copenhagen. vii + 553 pp., 56 pls.

———. 1943b. *A Monograph of the Echinoidea*. Volume III. *(3). Camarodonta. II. Echinidae, Strongylocentrotidae, Parasaleniidae, Echinometridae*. C. A. Reitzel, Copenhagen. vi + 446 pp., 66 pls.

———. 1948a. *A Monograph of the Echinoidea*. Volume IV. *(1). Holectypoida, Cassiduloida*. C. A. Reitzel, Copenhagen. viii + 371 pp., 15 pls.

———. 1948b. *A Monograph of the Echinoidea*. Volume IV. *(2). Clypeastroida. Clypeastridae, Arachnoididae, Fibulariidae, Laganidae and Scutellidae*. C. A. Reitzel, Copenhagen. viii + 471 pp., 72 pls.

———. 1951. *A Monograph of the Echinoidea*. Volume V. *(2). Spatangoida. II. Amphisternata. II. Spatangidae, Loveniidae, Pericosmi-*

dae, Schizasteridae, Brissidae. C. A. Reitzel, Copenhagen. viii + 593 pp., 64 pls.

Mosher, C. 1956. Observations on evisceration and visceral regeneration in the sea-cucumber *Actinopyga agassizi* Selenka. *Zoologica, Contributions of the New York Zoological Society, New York* 41:17–26.

———. 1980. Distribution of *Holothuria arenicola* Semper in the Bahamas with observations on habitat, behavior, and feeding activity (Echinodermata: Holothuroidea). *Bulletin of Marine Science* 30: 1–12.

———. 1982. Spawning behavior of the aspidochirote holothurian *Holothuria mexicana* Ludwig. In *Echinoderms*: Proceedings of the International Conference, Tampa Bay, 14–17 September 1981, ed. J. M. Lawrence, 467–468. Balkema, Rotterdam.

Müller, J. 1840. Über den Bau des *Pentacrinus Caput Medusae. Bericht über die zur Bekanntmachung geeigneten Verhandlungen der Königlich-Preussischen Akademie der Wissenschaften zu Berlin* 1840:88–106.

———. 1850. Anatomische Studien über die Echinodermen. *Archiv für Anatomie, Physiologie und Wissenschaftliche Medecin* 1850:117–155.

Müller, J., and F. H. Troschel. 1842. *System der Asteriden*. Friedrich Vieweg und Sohn, Braunschweig, Germany. xx + 134 pp., 12 pls.

Muñoz, M. G. de V., and J. Ellies. 1982. The effect of ferrous sulfate and sodium hypochlorite on fertilization and development of *Echinometra lucunter*. In *Echinoderms*: Proceedings of the International Conference, Tampa Bay, 14–17 September 1981, ed. J. M. Lawrence, 525–527. Balkema, Rotterdam.

Murakami, S. 1963. The dental and oral plates of Ophiuroidea. *Transactions of the Royal Society of New Zealand* 4:1–48.

Myers, A. C. 1977a. Sediment processing in a marine subtidal sandy bottom community: I. Physical aspects. *Journal of Marine Research* 35: 609–632.

———. 1977b. Sediment processing in a marine subtidal sandy bottom community: II. Biological consequences. *Journal of Marine Research* 35:633–647.

Nappi, A. J., and J. A. Crawford. 1984. The occurrence and distribution of a syndesmid (Turbellaria: Umagillidae) in Jamaican sea urchins. *Journal of Parasitology* 70:595–597.

Nielsen, E. 1932. Papers from Dr. Th. Mortensen's Pacific Expedition, 1914–16. LIX. Ophiurans from the Gulf of Panama, California and the Strait of Georgia. *Videnskabelige Meddelelser fra den naturhistorisk Forening i Kjøbenhavn* 91:241–346.

Nigrelli, R. F., and S. Jakowska. 1960. Effects of holothurin, a steroid saponin from the Bahaman sea cucumber *(Actinopyga agassizi)* on various biological systems. *Annals of the New York Academy of Sciences* 90:884–892.

Nutting, C. C. 1919. Barbados-Antigua Expedition. Narrative and preliminary report of a zoological expedition from the University of Iowa to the Lesser Antilles under the auspices of the graduate college. *University of Iowa Studies in Natural History* 8:1–274, pls. 1–39.

Ogden, J. C. 1976. Some aspects of herbivore-plant relationships on Caribbean reefs and sea-grass beds. *Aquatic Botany* 2:103–116.

———. 1977. Carbonate-sediment production by parrot fish and sea urchins on Caribbean reefs. *Studies in Geology* 4:281–288.

Ogden, J. C., D. P. Abbott, and I. A. Abbott. 1973a. *Studies on the Activity and Food of the Echinoid Diadema antillarum Philippi on a West Indian Patch Reef*. West Indies Laboratory Special Publication No. 2. Fairleigh Dickinson University, Christiansted, St. Croix, U.S. Virgin Islands. 96 pp.

Ogden, J. C., R. A. Brown, and N. Salesky. 1973b. Grazing by the echinoid *Diadema antillarum* Philippi: Formation of halos around West Indian patch reefs. *Science* (Washington, D.C.) 182:713–716.

O'Gower, A. K., and J. W. Wacasey. 1967. Animal communities associated with *Thalassia, Diplanthera*, and sand beds in Biscayne Bay. I. Analysis of communities in relation to water movements. *Bulletin of Marine Science* 17: 175–210.

Oguro, C., T. Shōsaku, and M. Komatsu. 1982. Development of the brittle-star, *Amphipholis japonica* Matsumoto. In *Echinoderms*: Proceedings of the International Conference, Tampa Bay, 14–17 September 1981, ed. J. M. Lawrence, 491–496. Balkema, Rotterdam.

Ohshima, H. 1925. Notes on the development of

the sea-cucumber, *Thyone briareus*. *Science* (New York) 61:420–422.

Olmsted, J. M. D. 1917. The comparative physiology of *Synaptula hydriformis* (Lesueur). *Journal of Experimental Zoology* 24:333–379.

Pagett, R. M. 1985. Some observations upon the distribution of strontium in four species of ophiuroids—*Ophiocomina nigra* (Abildgaard), *Ophiura albida* (Forbes), *Ophiothrix suensoni* (Lütken) and *Ophioderma cinereum* (Müller and Troschel). *Journal of the Marine Biological Association of the United Kingdom* 65: 293–303.

Panning, A. 1949. Versuch einer Neuordnung der Familie Cucumariidae (Holothurioidea, Dendrochirota). *Zoologische Jahrbücher Abteilung für Systematik, Ökologie und Geographie der Tiere* 78:404–470.

Parker, D. A., and M. J. Shulman. 1986. Avoiding predation: Alarm responses of Caribbean sea urchins to simulated predation on conspecific and heterospecific sea urchins. *Marine Biology* (Berlin) 93:201–208.

Parslow, R. E., and A. M. Clark. 1963. Ophiuroidea of the Lesser Antilles. *Studies on the Fauna of Curaçao and Other Caribbean Islands* 15:24–50.

Patton, W. K., R. J. Patton, and A. Barnes. 1985. On the biology of *Gnathophylloides mineri*, a shrimp inhabiting the sea urchin *Tripneustes ventricosus*. *Journal of Crustacean Biology* 5: 616–626.

Pawson, D. L. 1976. Shallow-water sea cucumbers (Echinodermata: Holothuroidea) from Carrie Bow Cay, Belize. *Proceedings of the Biological Society of Washington* 89:369–382.

———. 1977. Marine flora and fauna of the northeastern United States. Echinodermata: Holothuroidea. *NOAA Technical Report, National Marine Fisheries Service Circular* No. 405:iii + 15 pp.

———. 1978. The echinoderm fauna of Ascension Island, South Atlantic Ocean. *Smithsonian Contributions to the Marine Sciences* No. 2:iv + 31.

———. 1986. Phylum Echinodermata. In *Marine Fauna and Flora of Bermuda: A Systematic Guide to the Identification of Marine Organisms,* ed. W. Sterrer, 522–541, text pl. 173–178, color pl. 14. John Wiley & Sons, New York.

Pawson, D. L., and I. E. Caycedo. 1980. *Holothuria (Thymiosycia) thomasi* new species, a large Caribbean coral reef inhabiting sea cucumber (Echinodermata: Holothuroidea). *Bulletin of Marine Science* 30:454–459.

Pawson, D. L., and C. A. Gust. 1981. *Holothuria (Platyperona) rowei,* a new sea cucumber from Florida (Echinodermata: Holothuroidea). *Proceedings of the Biological Society of Washington* 94: 873–877.

Pawson, D. L., and J. E. Miller. 1981. Western Atlantic sea cucumbers of the genus *Thyone*, with descriptions of two new species (Echinodermata: Holothuroidea). *Proceedings of the Biological Society of Washington* 94:391–403.

———. 1982. Studies of genetically controlled phenotypic characters in laboratory-reared *Lytechinus variegatus* (Lamarck) (Echinodermata: Echinoidea). In *Echinoderms:* Proceedings of the International Conference, Tampa Bay, 14–17 September 1981, ed. J. M. Lawrence, 165–171. Balkema, Rotterdam.

———. 1992. *Phyllophorus (Urodemella) arenicola,* a new sublittoral sea cucumber from the southeastern United States (Echinodermata: Holothuroidea). *Proceedings of the Biological Society of Washington* 105:483–489.

Pearse, A. S. 1908. Observations on the behavior of the holothurian, *Thyone briareus* (Lesueur). *Biological Bulletin* (Woods Hole) 15:259–288.

Pearse, J. S., and R. A. Cameron. 1991. Echinodermata: Echinoidea. In *Reproduction of Marine Invertebrates,* Volume VI, *Echinoderms and Lophophorates,* eds. A. C. Giese, J. S. Pearse, and V. B. Pearse, 513–662. The Boxwood Press, Pacific Grove, California.

Pearson, J. F. W. 1937. Studies on the life zones of marine waters adjacent to Miami: I. The distribution of the Ophiuroidea. *Proceedings of the Florida Academy of Sciences* 1:66–72.

Penchaszadeh, P. E., and M. E. Lera. 1983. Alimentación de tres especies tropicales de *Luidia* (Echinodermata, Asteroidea) en Golfo Triste, Venezuela. *Caribbean Journal of Science* 19:1–6.

Pentreath, R. J. 1970. Feeding mechanisms and the functional morphology of podia and spines in some New Zealand ophiuroids (Echinodermata). *Journal of Zoology* (London) 161: 395–429.

Pérez Ruzafa, A. 1989. Estudio ecológio y bionómico de los poblamientos bentónicos del Mar Menor (Murcia, SE de España). Ph.D. dissertation, Universidad de Murcia, Murcia, Spain. 751 pp.

Perrier, E. 1881. Reports on the results of dredging under the supervision of Alexander Agassiz, in the Gulf of Mexico, 1877–1878, by the United States Coast Survey Steamer "Blake," Lieut.-Commander C. D. Sigsbee, U.S.N., commanding, and in the Caribbean Sea, 1878–79, by the U.S.C.S.S. "Blake," Commander J. R. Bartlett, U.S.N. commanding. XIV. Description sommaire des espèces nouvelles d'astéries. *Bulletin of the Museum of Comparative Zoölogy at Harvard College* 9:1–31.

———. 1884. Mémoire sur les étoiles de mer recueillis dans la Mer des Antilles et la Golfe de Mexique. *Nouvelles archives du Muséum d'histoire naturelle, Paris* 6:127–276.

Petersen, J. A. 1976. Aspects of gas exchange in ophiuroids from the coast of Brazil. *Thalassia Jugoslavica* 12:295–296.

Petersen, J. A., and A. M. Almeida. 1976. Effects of salinity and temperature on the development and survival of the echinoids *Arbacia*, *Echinometra* and *Lytechinus*. *Thalassia Jugoslavica* 12:297–298.

Pettibone, M. H. 1993. Scaled polychaetes (Polynoidae) associated with ophiuroids and other invertebrates and review of species referred to *Malmgrenia* McIntosh and replaced by *Malmgreniella* Hartman, with descriptions of new taxa. *Smithsonian Contributions to Zoology* 538:vi + 92.

Pfister, C. A., and M. E. Hay. 1988. Associational plant refuges: Convergent patterns in marine and terrestrial communities result from differing mechanisms. *Oecologia* (Berlin) 77: 118–129.

Phelan, T. F. 1972. Comments on the echinoid genus *Encope*, and a new subgenus. *Proceedings of the Biological Society of Washington* 85: 109–130.

Philippi, R. A. 1845. Beschreibung einiger neuen Echinodermen nebst kritischen Bemerkungen über einige weniger bekannte Arten. *Archiv für Naturgeschichte* 11:344–359.

Pina Albuquerque, J. A. 1985. Presencia de *Amphipholis squamata* (Delle Chiaje, 1828) (Echinodermata: Ophiuroidea) en el Mar Menor (Murcia). *Anales de Biologia 3 Seccion Biologia Animal* 1:121–122.

Pina Albuquerque, J. A., and A. Pérez Ruzafa. 1984. Aportación catálogo de Equinodermos del Litoral Murciano. *Actas do IV° Simpósio Ibérico de Estudos do Benthos Marinho 1984* 3:269–276.

Poddobiuk, R. H. 1985. Evolution and adaptation in some Caribbean Oligo-Miocene clypeasters. In *Echinodermata*. Proceedings of the Fifth International Echinoderm Conference, Galway, 24–29 September 1984, eds. B. F. Keegan and B. D. S. O'Connor, 75–80. Balkema, Rotterdam.

Polson, E. S., L. L. Robbins, and J. Lawrence. 1993. Intraskeletal matrix proteins in Echinodermata. *Journal of the Marine Biological Association of the United Kingdom* 73:727–730.

Pomory, C. 1989. Range extensions for *Isostichopus badionotus* Selenka, 1867 and *Holothuria* (*Halodeima*) *grisea* Selenka, 1867 (Echinodermata: Holothuroidea). *Texas Journal of Science* 41:330–331.

Pompa, L., A. S. Prieto, and R. Manrique. 1990. Abundance and spatial distribution pattern in a population of the urchin *Echinometra lucunter* (L.) in the Gulf of Cariaco, Venezuela. *Acta Cientifica Venezolana* 40:289–294.

Porter, J. W., and O. W. Meier. 1992. Quantification of loss and change in Floridian reef coral populations. *American Zoologist* 32:625–640.

Pourtalès, L. F. 1851. On the Holothuriae of the Atlantic coast of the United States. *Proceedings of the American Association for the Advancement of Science,* Fifth meeting, held at Cincinnati, Ohio, May 1851:8–16.

———. 1869. List of the crinoids obtained on the coasts of Florida and Cuba, by the United States Coast Survey Gulf Stream Expeditions, in 1867, 1868, 1869. *Bulletin of the Museum of Comparative Zoölogy at Harvard College* 1:355–358.

Powell, E. N. 1977. Particle size selection and sediment reworking in a funnel feeder, *Leptosynapta tenuis* (Holothuroidea, Synaptidae). *Internationale Revue der Gesamten Hydrobiologie* 62:385–408.

Prosser, C. L., R. A. Nystrom, and T. Nagai. 1965. Electrical and mechanical activity in intestinal muscles of several invertebrate animals. *Comparative Biochemistry and Physiology* 14:53–70.

Rader, D. N. 1982. Orthonectid parasitism: Effects on the ophiuroid, *Amphipholis squamata*. In *Echinoderms:* Proceedings of the International Conference, Tampa Bay, 14–17 September 1981, ed. J. M. Lawrence, 395–401. Balkema, Rotterdam.

Randall, J. E. 1967. Food habits of reef fishes of the West Indies. *Institute of Marine Sciences University of Miami. Studies in Tropical Oceanography* No. 5:665–847.

Randall, J. E., R. E. Schroeder, and W. A. Starck II. 1964. Notes on the biology of the echinoid *Diadema antillarum. Caribbean Journal of Science* 4:421–433.

Rathbun, R. 1879. A list of the Brazilian echinoderms, with notes on their distribution, etc. *Transactions Connecticut Academy of Arts and Sciences* 5:139–158.

Reimer, R. D., and A. A. Reimer. 1975. Chemical control of feeding in four species of tropical ophiuroids of the genus *Ophioderma. Comparative Biochemistry and Physiology, Part A, Comparative Physiology* 51A:915–927.

Reinthal, P. N., B. Kensley, and S. M. Lewis. 1984. Dietary shifts in the queen triggerfish, *Balistes vetula*, in the absence of its primary food item, *Diadema antillarum. Marine Ecology (Pubblicazioni della Stazione Zoologica di Napoli I)* 5:191–195.

Renaud, P. E., M. E. Hay, and T. M. Schmitt. 1990. Interactions of plant stress and herbivory: Intraspecific variation in the susceptibility of a palatable versus an unpalatable seaweed to sea urchin grazing. *Oecologia* (Berlin) 82:217–226.

Richmond, E. A. 1962. The fauna and flora of Horn Island, Mississippi. *Gulf Research Reports* 1:59–106.

Roberts, M. S., R. C. Terwilliger, and N. B. Terwilliger. 1984. Comparison of sea cucumber hemoglobin structures. *Comparative Biochemistry and Physiology, Part B, Comparative Biochemistry* 77B:237–243.

Robertson, D. R. 1987. Responses of two coral reef toadfishes (Batrachoididae) to the demise of their primary prey, the sea urchin *Diadema antillarum. Copeia* 1987:637–642.

Rockstein, M. 1971. The distribution of phosphoarginine and phosphocreatine in marine invertebrates. *Biological Bulletin* (Woods Hole) 141:167–175.

Roller, R. A., and W. B. Stickle. 1993. Effects of temperature and salinity acclimation of adults on larval survival, physiology, and early development of *Lytechinus variegatus* (Echinodermata: Echinoidea). *Marine Biology* (Berlin) 116:583–591.

Rosén, N. 1910. Zur Kenntniss der parasitischen Schnecken. *Lunds Universitets Arsskrift*, Ny Földj, Afdeling 2, 6:1–67.

Rosenberg, V. A., and R. P. Wain. 1982. Isozyme variation and genetic differentiation in the decorator sea urchin *Lytechinus variegatus* (Lamarck, 1816). In *Echinoderms:* Proceedings of the International Conference, Tampa Bay, 14–17 September 1981, ed. J. M. Lawrence, 193–197. Balkema, Rotterdam.

Rothans, T. C., and A. C. Miller. 1991. A link between biologically imported particulate organic nutrients and the detritus food web in reef communities. *Marine Biology* (Berlin) 110:145–150.

Rowe, F. W. E. 1969. A review of the family Holothuriidae (Holothuroidea: Aspidochirotida). *Bulletin of the British Museum (Natural History) Zoology* 18:119–170.

Rumrill, S. S. 1982. Constrasting reproductive patterns among ophiuroids (Echinodermata) from southern Monterey Bay, U.S.A. M.S. thesis, University of California, Santa Cruz, Santa Cruz, California. xi + 260 pp.

Ruppert, E., and R. Fox. 1988. *Seashore Animals of the Southeast. A Guide to Common Shallow-water Invertebrates of the Southeastern Atlantic Coast*. University of South Carolina Press, Columbia, South Carolina. 429 pp.

Rylaarsdam, K. W. 1983. Life histories and abundance patterns of colonial corals on Jamaican reefs. *Marine Ecology Progress Series* 13:249–260.

Sambrano, A., H. Díaz, and J. E. Conde. 1990. Caracterización de la ingesta en *Isostichopus badionotus* (Salenka) y *Holothuria mexicana*

Ludwig (Echinodermata: Holothuroidea). *Caribbean Journal of Science* 26:45–51.

Sammarco, P. W. 1980. *Diadema* and its relationship to coral spat mortality: Grazing, competition, and biological disturbance. *Journal of Experimental Marine Biology and Ecology* 45:245–272.

———. 1982a. Echinoid grazing as a structuring force in coral communities: Whole reef manipulations. *Journal of Experimental Marine Biology and Ecology* 61:31–55.

———. 1982b. Effects of grazing by *Diadema antillarum* Philippi (Echinodermata: Echinoidea) on algal diversity and community structure. *Journal of Experimental Marine Biology and Ecology* 65:83–105.

Sammarco, P. W., J. S. Levinton, and J. C. Ogden. 1974. Grazing and control of coral reef community structure by *Diadema antillarum* Philippi (Echinodermata: Echinoidea): A preliminary study. *Journal of Marine Research* 32:47–53.

Say, T. 1825. On the species of the Linnean genus *Asterias*, inhabiting the coast of the United States. *Journal of the Academy of Natural Sciences of Philadelphia* 5:141–154.

Scheibling, R. E. 1980a. Homing movements of *Oreaster reticulatus* (L.) (Echinodermata: Asteroidea) when experimentally translocated from a sand patch habitat. *Marine Behavior and Physiology* 7:213–223.

———. 1980b. The microphagous feeding behavior of *Oreaster reticulatus* (Echinodermata: Asteroidea). *Marine Behavior and Physiology* 7:225–232.

———. 1980c. Abundance, spatial distribution, and size structure of populations of *Oreaster reticulatus* (L.) (Echinodermata: Asteroidea) in seagrass beds. *Marine Biology* (Berlin) 57:95–105.

———. 1980d. Abundance, spatial distribution, and size structure of populations of *Oreaster reticulatus* (L.) (Echinodermata: Asteroidea) on sand bottoms. *Marine Biology* (Berlin) 57:107–119.

———. 1981. Optimal foraging movements of *Oreaster reticulatus* (L.) (Echinodermata: Asteroidea). *Journal of Experimental Marine Biology and Ecology* 51:173–185.

———. 1982a. Feeding habits of *Oreaster reticulatus* (Echinodermata: Asteroidea). *Bulletin of Marine Science* 32:504–510.

———. 1982b. Habitat utilization and bioturbation by *Oreaster reticulatus* (Asteroidea) and *Meoma ventricosa* (Echinoidea) in a subtidal sand patch. *Bulletin of Marine Science* 32:624–629.

Scheibling, R. E., and P. V. Mladenov. 1987. The decline of the sea urchin, *Tripneustes ventricosus*, fishery of Barbados: A survey of fishermen and consumers. *U.S. National Marine Fisheries Service Marine Fisheries Review* 49:62–69.

Schneider, D. C. 1985. Predation on the urchin *Echinometra lucunter* (Linnaeus). *Journal of Experimental Marine Biology and Ecology* 92:19–27.

Schoener, A. 1972. Fecundity and possible mode of development of some deep-sea ophiuroids. *Limnology and Oceanography* 17:193–199.

Schoppe, S. 1991. *Echinometra lucunter* (Linnaeus) (Echinoidea, Echinometridae) als Wirt einer komplexen Lebensgemeinschaft im Karibischen Meer. *Helgoländer Meeresuntersuchungen* 45:373–379.

Schroeder, R. E. 1962. Urchin killer. *Sea Frontiers* 8:156–160.

Schroeder, T. E. 1981. Development of a "primitive" sea urchin *(Eucidaris tribuloides)*: Irregularities in the hyaline layer, micromeres, and primary mesenchyme. *Biological Bulletin* (Woods Hole) 161:141–151.

Schwartz, F. J., and H. J. Porter. 1977. Fishes, macroinvertebrates, and their ecological interrelationships with a calico scallop bed off North Carolina. *U.S. Fish and Wildlife Service Fishery Bulletin* 75:427–446.

Scoffin, T. P., C. W. Stearn, D. Boucher, P. Frydl, C. M. Hawkins, I. G. Hunter, and J. K. McGeachy. 1980. Calcium carbonate budget of a fringing reef on the west coast of Barbados. Part 2. Erosion, sediments and internal structures. *Bulletin of Marine Science* 30:475–508.

Sefton, N. 1987. Rings on coral fingers. *Sea Frontiers* 33:134–135.

Sefton, N., and S. K. Webster. 1986. *A Field Guide to Caribbean Reef Invertebrates*. Sea Challengers, Monterey, California; E. J. Brill, Leiden. 112 pp.

Selenka, E. 1867. Beitrage zur Anatomie und Systematik der Holothurien. *Zeitschrift für wis-*

senschaftliche Zoologie 17:291–374, pls. 17–20.

Semper, C. 1868. Holothurien. *Reisen im Archipel der Philippien von Dr. C. Semper in Würzburg.* II. *Wissenschaftliche Resultate* 1:iv + 288 pp., pls. 1–40. Verlag von Wilhelm Engelmann, Leipzig.

Serafy, D. K. 1970. A new species of *Clypeaster* from the Gulf and Caribbean and a key to the species in the tropical northwestern Atlantic (Echinodermata: Echinoidea). *Bulletin of Marine Science* 20:662–677.

———. 1973. Variation in the polytypic sea urchin *Lytechinus variegatus* (Lamarck, 1816) in the western Atlantic (Echinodermata: Echinoidea). *Bulletin of Marine Science* 23:525–534.

———. 1979. Echinoids (Echinodermata: Echinoidea). *Memoirs of the Hourglass Cruises* 5:1–120.

Serafy, D. K., and F. J. Fell. 1985. Marine flora and fauna of the northeastern United States. Echinodermata: Echinoidea. *NOAA Technical Report, National Marine Fisheries Service* 33:1–27.

Sharp, D. T., and I. E. Gray. 1962. Studies on factors affecting the local distribution of two sea urchins, *Arbacia punctulata* and *Lytechinus variegatus*. *Ecology* 43:309–313.

Sheridan, P. F., and A. C. Badger. 1981. Responses of experimental estuarine communities to continuous chlorination. *Estuarine Coastal and Shelf Science* 13:337–347.

Shick, J. M. 1983. Respiratory gas exchange in echinoderms. In *Echinoderm Studies*, Volume 1, eds. M. Jangoux and J. M. Lawrence, 67–110. Balkema, Rotterdam.

Shirley, T. C. 1982. The importance of echinoderms in the diet of fishes of a sublittoral rock reef. In *South Texas Fauna*, eds. B. R. Chapman and J. W. Tunnell, 49–55. Caesar Kleberg Wildlife Research Institute, College of Agriculture, Texas A&I University, Knoxville, Texas.

Shulman, M. J. 1990. Aggression among sea urchins on Caribbean coral reefs. *Journal of Experimental Marine Biology and Ecology* 140:197–208.

Sides, E. M. 1982. Estimates of partial mortality for eight species of brittle-stars. In *Echinoderms:* Proceedings of the International Conference, Tampa Bay, 14–17 September 1981, ed. J. M. Lawrence, 327. Balkema, Rotterdam.

———. 1985. Interference competition between brittle-stars? In *Echinodermata*. Proceedings of the Fifth International Echinoderm Conference, Galway, 24–29 September 1984, eds. B. F. Keegan and B. D. S. O'Connor, 639–644. Balkema, Rotterdam.

———. 1987. An experimental study of the use of arm regeneration in estimating rates of sublethal injury on brittle-stars. *Journal of Experimental Marine Biology and Ecology* 106:1–16.

Sides, E. M., and J. D. Woodley. 1985. Niche separation in three species of *Ophiocoma* (Echinodermata: Ophiuroidea) in Jamaica, West Indies. *Bulletin of Marine Science* 36:701–715.

Simms, M. J. 1988. The phylogeny of post-Palaeozoic crinoids. In *Echinoderm Phylogeny and Evolutionary Biology*, eds. C. R. C. Paul and A. B. Smith, 269–284. Clarendon Press, Oxford.

Singletary, R. L. 1970. The biology and ecology of *Amphioplus coniortodes*, *Ophionepthys limicola*, and *Micropholis gracillima* (Ophiuroidea: Amphiuridae). Ph.D. dissertation, University of Miami, Coral Gables, Florida. ix + 127 pp.

———. 1971. Thermal tolerance of ten shallow-water ophiuroids in Biscayne Bay, Florida. *Bulletin of Marine Science* 21:938–943.

———. 1973. A new species of brittlestar from Florida. *Florida Scientist* 36:175–178.

———. 1980. The biology and ecology of *Amphioplus coniortodes*, *Ophionepthys limicola*, and *Micropholis gracillima* (Ophiuroidea: Amphiuridae). *Caribbean Journal of Science* 16:39–55.

Singletary, R. L., and H. B. Moore. 1974. A redescription of the *Amphioplus coniortodes–Ophionepthys limicola* community of Biscayne Bay, Florida. *Bulletin of Marine Science* 24:690–699.

Sisak, M. M., and F. Sander. 1985. Respiratory behavior of the western Atlantic holothuroidian (Echinodermata) *Holothuria glaberrima* (Selenka) at various salinities, temperatures and oxygen tensions. *Comparative Biochemistry and Physiology, Part A, Comparative Physiology* 80A:25–29.

Sladen, W. P. 1889. Report on the Asteroidea collected during the Voyage of H.M.S. Challenger, during the years 1873–1876. *Report on the Scientific Results of the Voyage of H.M.S. Challenger during the Years 1873–76. Zoology,* Volume 30 (Part 51): xvi + 1–893, 117 pls., map.

Sloan, N. A. 1980. Aspects of the feeding biology of asteroids. *Oceanography and Marine Biology Annual Review* 18:57–124.

———. 1982. Size and structure of echinoderm populations associated with different coexisting coral species at Aldabra Atoll, Seychelles. *Marine Biology* (Berlin) 66:67–75.

———. 1985. Echinoderm fisheries of the world: A review. In *Echinodermata.* Proceedings of the Fifth International Echinoderm Conference, Galway, 24–29 September 1984, eds. B. F. Keegan and B. D. S. O'Connor, 109–124. Balkema, Rotterdam.

Sloan, N. A., and B. von Bodungen. 1980. Distribution and feeding of the sea cucumber *Isostichopus badionotus* in relation to shelter and sediment criteria of the Bermuda platform. *Marine Ecology Progress Series* 2:257–264.

Sluiter, C. P. 1910. Westindische Holothurien. *Zoologische Jahrbücher Supplement* 11:331–342.

Smiley, S., F. S. McEuen, C. Chaffee, and S. Krishnan. 1991. Echinodermata: Holothuroidea. In *Reproduction of Marine Invertebrates,* Volume VI, *Echinoderms and Lophophorates,* eds. A. C. Giese, J. S. Pearse, and V. B. Pearse, 663–760. The Boxwood Press, Pacific Grove, California.

Smith, A. B. 1984a. *Echinoid Palaeobiology.* George Allen & Unwin, London. xii + 190 pp. + 1 fig.

———. 1984b. Classification of the Echinodermata. *Palaeontology* 27:431–459.

Smith, A. B., and J. Gallemí. 1991. Middle Triassic holothurians from northern Spain. *Palaeontology* 34:49–76.

Smith, C. L., J. C. Tyler, and M. N. Feinberg. 1981. Population ecology and biology of the pearlfish *(Carapus bermudensis)* in the lagoon at Bimini, Bahamas. *Bulletin of Marine Science* 31:876–902.

Smith, D. S., J. del Castillo, M. Morales, and B. Luke. 1990. The attachment of collagenous ligament to stereom in primary spines of the sea-urchin, *Eucidaris tribuloides. Tissue and Cell* 22:157–176.

Smith, G. N., Jr. 1971a. Regeneration in the sea cucumber *Leptosynapta.* I. The process of regeneration. *Journal of Experimental Zoology* 177:319–330.

———. 1971b. Regeneration in the sea cucumber *Leptosynapta.* II. The regenerative capacity. *Journal of Experimental Zoology* 177:331–342.

Smith, G. N., Jr., and M. J. Greenberg. 1973. Chemical control of the evisceration process in *Thyone briareus. Biological Bulletin* (Woods Hole) 144:421–436.

Snyder, N., and H. Snyder. 1970. Alarm response of *Diadema antillarum. Science* (Washington, D.C.) 168:276–278.

Snyder, R. D. 1980. Commensal turbellarians from Bermuda holothurians. *Canadian Journal of Zoology* 58:1741–1744.

Spencer, W. K., and C. W. Wright. 1966. Asterozoans. In *Treatise on Invertebrate Paleontology, Part U, Echinodermata 3,* Volume 1, ed. R. C. Moore, U4–U107. The Geological Society of America and The University of Kansas Press, Lawrence, Kansas.

Stancyk, S. E. 1973. Development of *Ophiolepis elegans* (Echinodermata: Ophiuroidea) and its implications in the estuarine environment. *Marine Biology* (Berlin) 21:7–12.

———. 1974. Life history patterns of three estuarine brittlestars (Ophiuroidea) at Cedar Key, Florida. Ph.D. dissertation, University of Florida, Gainesville, Florida. vi + 77 pp.

———. 1975. The life history pattern of *Ophiothrix angulata* (Ophiuroidea). *American Zoologist* 15:793.

Stancyk, S. E., and P. L. Shaffer. 1977. The salinity tolerance of *Ophiothrix angulata* (Say) (Echinodermata: Ophiuroidea) in latitudinally separate populations. *Journal of Experimental Marine Biology and Ecology* 29:35–43.

Stancyk, S. E., H. M. Golde, P. A. Pape-Lindstrom, and W. E. Dobson. 1994. Born to lose. I. Measures of tissue loss and regeneration by the brittlestar *Microphiopholis gracillima* (Echinodermata: Ophiuroidea). *Marine Biology* (Berlin) 118:451–462.

Stephenson, T. A., and A. Stephenson. 1972. *Life between Tidemarks on Rocky Shores.* W. H. Freeman and Company, San Francisco, California. xiii + 425 pp.

Stickle, W. B. 1988. Patterns of nitrogen excretion in the phylum Echinodermata. *Comparative Biochemistry and Physiology, Part A, Comparative Physiology* 91A:317–321.

Stickle, W. B., and W. J. Diehl. 1987. Effects of salinity on echinoderms. In *Echinoderm Studies,* Volume 1, eds. M. Jangoux and J. M. Lawrence, 235–285. Balkema, Rotterdam.

Stickle, W. B., T. C. Shirley, and T. D. Sabourin. 1982. Patterns of nitrogen excretion in for [sic] species of echinoderms as a function of salinity. In *Echinoderms:* Proceedings of the International Conference, Tampa Bay, 14–17 September 1981, ed. J. M. Lawrence, 371–377. Balkema, Rotterdam.

Stier, T. T. B. 1933. Diurnal changes in activities and geotropism in *Thyone briareus. Biological Bulletin* (Woods Hole) 64:326–332.

Stimpson, W. 1852. Descriptions of two new species of *Ophiolepis* from the southern coast of the United States. *Proceedings of the Boston Society of Natural History* 4:224–226.

Stock, J. H., A. G. Humes, and R. H. Gooding. 1963. Copepoda associated with West Indian invertebrates—II. Cancerillidae, Micropontiidae (Siphonostoma). *Studies of the Fauna of Curaçao and Other Caribbean Islands* 15:1–23.

Stockard, C. R. 1909. The rate of regeneration and the effect of new tissue on the old body. *Science* (New York) 29:430.

Strathmann, M. F., and S. S. Rumrill. 1987. Phylum Echinodermata, Class Ophiuroidea. In *Reproduction and Development of Marine Invertebrates of the Northern Pacific Coast,* ed. M. Strathmann, 556–573. University of Washington Press, Seattle, Washington.

Stricker, S. A. 1985. The ultrastructure and formation of the calcareous ossicles in the body wall of the sea cucumber *Leptosynapta clarki* (Echinodermata: Holothuroida). *Zoomorphology* (Berlin) 105:209–222.

———. 1986. The fine structure and development of calcified skeletal elements in the body wall of holothurian echinoderms. *Journal of Morphology* 188:273–288.

Sullivan, K. M. 1988. Physiological ecology and energetics of regeneration in reef rubble brittlestars. In *Echinoderm Biology.* Proceedings of the Sixth International Echinoderm Conference, Victoria, 23–28 August 1987, eds. R. D. Burke, P. V. Mladenov, P. Lambert, and R. L. Parsley, 523–529. Balkema, Rotterdam.

Tabb, D. C., and R. B. Manning. 1961. A checklist of the flora and fauna of northern Florida Bay and adjacent brackish waters of the Florida mainland collected during the period July, 1957 through September 1960. *Bulletin of Marine Science of the Gulf and Caribbean* 11:552–649.

Telford, M. 1978. Distribution of two species of *Dissodactylus* (Brachyura: Pinnotheridae) among their echinoid host populations in Barbados. *Bulletin of Marine Science* 28:651–658.

———. 1981. A hydrodynamic interpretation of sand dollar morphology. *Bulletin of Marine Science* 31:605–622.

———. 1990. Computer simulation of deposit-feeding by sand dollars and sea biscuits (Echinoidea: Clypeasteroida). *Journal of Experimental Marine Biology and Ecology* 142:75–90.

Telford, M., and R. Mooi. 1986. Resource partitioning by sand dollars in carbonate and siliceous sediments: Evidence from podial and particle dimensions. *Biological Bulletin* (Woods Hole) 171:197–207.

Telford, M., R. Mooi, and O. Ellers. 1985. A new model of podial deposit feeding in the sand dollar, *Mellita quinquiesperforata* (Leske): The sieve hypothesis challenged. *Biological Bulletin* (Woods Hole) 169:431–448.

Telford, M., R. Mooi, and A. S. Harold. 1987. Feeding activities of two species of *Clypeaster* (Echinoides, Clypeasteroidea): Further evidence of clypeasteroid resource partitioning. *Biological Bulletin* (Woods Hole) 172:324–336.

Tennent, D. H., M. S. Gardiner, and D. E. Smith. 1931. A cytological and biochemical study of the ovaries of the sea-urchin *Echinometra lucunter. Carnegie Institution of Washington Publication* No. 27:1–46, pls. 1–7.

Tertschnig, W. P. 1984. Sea urchins in seagrass communities: Resource management as a functional perspective of adaptive strategies. In *Echinodermata.* Proceedings of the Fifth International Echinoderm Conference, Galway, 24–29 September 1984, eds. B. F. Keegan and B. D. S. O'Connor, 361–367. Balkema, Rotterdam.

———. 1989. Diel activity patterns and foraging

dynamics of the sea urchin *Tripneustes ventricosus* in a tropical seagrass community and a reef environment (Virgin Islands). *Marine Ecology (Pubblicazioni della Stazione Zoologica di Napoli I)* 10:3–21.

Terwilliger, R. C., and N. B. Terwilliger. 1988. Structure and function of holothurian hemoglobins. In *Echinoderm Biology*. Proceedings of the Sixth International Echinoderm Conference, Victoria, 23–28 August 1987, eds. R. D. Burke, P. V. Mladenov, P. Lambert, and R. L. Parsley, 589–595. Balkema, Rotterdam.

Tewes, R. 1984. The ecology and feeding biology of *Ophioderma cinereum* in a mangrove environment. M.S. thesis, University of Missouri-Kansas City, Kansas City, Missouri. viii + 67 pp.

Théel, H. 1886. Report on the Holothurioidea dredged by H.M.S. Challenger during the years 1873–76. Part I. *Report on the Scientific Results of the Voyage of H.M.S. Challenger during the Years 1873–76. Zoology*, Volume 14 (Part 39):1–290, pls. 1–16.

Thomas, L. P. 1960. A note on the feeding habits of the West Indian sea star *Oreaster reticulatus* (Linnaeus). *Quarterly Journal of the Florida Academy of Sciences* 23:167–168.

———. 1961. Distribution and salinity tolerance in the amphiurid brittlestar, *Ophiophragmus filograneus* (Lyman, 1875). *Bulletin of Marine Science of the Gulf and Caribbean* 11:158–160.

———. 1962a. Two large ophiodermatid brittlestars new to Florida. *Quarterly Journal of the Florida Academy of Sciences* 25:65–69.

———. 1962b. The shallow water amphiurid brittle stars (Echinodermata, Ophiuroidea) of Florida. *Bulletin of Marine Science of the Gulf and Caribbean* 12:623–694.

———. 1963. A redescription of the amphiurid brittlestar *Ophiocnida cubana*, A. H. Clark, 1917. *Proceedings of the Biological Society of Washington* 76:217–221.

———. 1964. *Amphiodia atra* (Stimpson) and *Ophionema intricata* Lütken, additions to the shallow water amphiurid brittlestar fauna of Florida (Echinodermata: Ophiuroidea). *Bulletin of Marine Science of the Gulf and Caribbean* 14:158–167.

———. 1965a. The rediscovery of *Amphiura kinbergi* Ljungman, 1871 (Ophiuroidea: Echinodermata). *Bulletin of Marine Science* 15:638–641.

———. 1965b. A new species of *Ophiophragmus* (Ophiuroidea: Echinodermata) from the Gulf of Mexico. *Bulletin of Marine Science* 15:850–854.

———. 1965c. A monograph of the amphiurid brittlestars of the western Atlantic. Ph.D. dissertation, University of Miami, Coral Gables, Florida. xiii + 504 pp.

———. 1966. A revision of the tropical American species of *Amphipholis* (Echinodermata: Ophiuroidea). *Bulletin of Marine Science* 16:827–833.

———. 1967. The systematic position of *Amphilimna* (Echinodermata: Ophiuroidea). *Proceedings of the Biological Society of Washington* 80:123–130.

———. 1973. Western Atlantic brittlestars of the genus *Ophionereis*. *Bulletin of Marine Science* 23:585–599.

Tiffany, W. J., III. 1978. Mass mortality of *Luidia senegalensis* (Lamarck, 1816) on Captiva Island, Florida, with a note on its occurrence in Florida Gulf coastal waters. *Florida Scientist* 41:63–64.

Tikasingh, E. S. 1963. The shallow water holothurians of Curaçao, Aruba and Bonaire. *Studies on the Fauna of Curaçao and Other Caribbean Islands* 14:77–99.

Tommasi, L. R. 1967. Sôbre dois Amphiuridae de fauna marinha do Sul do Brasil. *Contribuições avulsas do Instituto Oceanográfico Sao Paulo, Séries Oceanográfia Biológica* 12:1–5.

———. 1969. Lista dos Holothurioidea recentes do Brasil. *Contribuições avulsas do Instituto Oceanográfico Sao Paulo, Séries Oceanografia Biológica* 15:1–29.

———. 1970. Os ofiuróides recentes do Brasil e de regiões Vizinhas. *Contribuições avulsas do Instituto Oceanográfico Sao Paulo, Séries Oceanográfia Biológica* 20:1–146.

Tortonese, E. 1934. Gli Echinodermi del Museo di Torino Parte II—Ofiuroidi. *Bollettino del Museo di Zoologia dell'Universita di Torino* 44:1–53, pls. 1–8.

———. 1939. Sopra un'ofiura poco nota del mar delle Antille. *Bollettino del Museo di Zoologia dell'Universita di Torino* 46:1–3.

———. 1983. Remarks on the morphology and taxonomy of *Ophioderma longicaudum* (Retz.) from the Mediterranean. *Atti della Societa Italiana de Scienze Naturali e del Museo Civico di Storia Naturale de Milano* 124:21–28.

Tortonese, E., and M. Demir. 1960. The echinoderm fauna of the Sea of Marmara and the Bosporus. *İstanbul üniversitesi fen fakültesi hidrobiologi enstitüsü* 5B:5–16.

Trott, L. B. 1981. A general review of the pearlfishes (Pisces, Carapidae). *Bulletin of Marine Science* 31:623–628.

Trotter, J. A., and T. J. Koob. 1994. Biochemical characterization of fibrillar collagen from the mutable spine ligament of the sea-urchin *Eucidaris tribuloides*. *Comparative Biochemistry and Physiology, Part B, Comparative Biochemistry* 107B:125–134.

Tsuchiya, M., Y. Nakasone, and M. Nishihara. 1986. Community structure of coral associated invertebrates of the hermatypic coral, *Pavona frondifera*, in the Gulf of Thailand. *Galaxea* 5:129–140.

Turner, R. L. 1985. Annual recruitment in the brackish-water ophiuroid *Ophiophragmus filograneus*. In *Echinodermata*. Proceedings of the Fifth International Echinoderm Conference, Galway, 24–29 September 1984, eds. B. F. Keegan and B. D. S. O'Connor, 659. Balkema, Rotterdam.

Turner, R. L., and J. M. Lawrence. 1979. Volume and composition of echinoderm eggs: Implications for the use of egg size in life-history models. In *Reproductive Ecology of Marine Invertebrates*, ed. S. E. Stancyk, 25–40. University of South Carolina Press, Columbia, South Carolina.

Turner, R. L., and C. E. Meyer. 1980. Salinity tolerance of the brackish-water echinoderm *Ophiophragmus filograneus* (Ophiuroidea). *Marine Ecology Progress Series* 2:249–256.

Turner, R. L., and J. E. Miller. 1988. Post-metamorphic recruitment and morphology of two sympatric brittlestars. In *Echinoderm Biology*. Proceedings of the Sixth International Echinoderm Conference, Victoria, 23–28 August 1987, eds. R. D. Burke, P. V. Mladenov, P. Lambert, and R. L. Parsley, 493–502. Balkema, Rotterdam.

Turner, R. L., and J. D. Murdoch. 1976. Potential of arms as a nutrient source for disc regeneration in a brittlestar. *American Zoologist* 16:228.

Turner, R. L., and C. M. Norlund. 1988. Labral morphology in heart urchins of the genus *Brissopsis* (Echinodermata: Spatangoida), with an illustrated revised key to the western Atlantic species. *Proceedings of the Biological Society of Washington* 101:890–897.

Turner, R. L., D. W. Heatwole, and S. E. Stancyk. 1982. Ophiuroid discs in stingray stomachs: Evasive autotomy or partial consumption of prey? In *Echinoderms:* Proceedings of the International Conference, Tampa Bay, 14–17 September 1981, ed. J. M. Lawrence, 331–335. Balkema, Rotterdam.

Vadas, R. L., T. Fenchel, and J. C. Ogden. 1982. Ecological studies on the sea urchin *Lytechinus variegatus* and the algal-seagrass communities of the Miskito Cays, Nicaragua. *Aquatic Botany* 14:109–115.

Valentine, J. F. 1991a. Temporal variation in populations of the brittlestars *Hemipholis elongata* (Say, 1825) and *Microphiopholis atra* (Stimpson, 1852) (Echinodermata: Ophiuroidea) in eastern Mississippi Sound. *Bulletin of Marine Science* 48:597–605.

———. 1991b. The reproductive periodicity of *Microphiopholis atra* (Stimpson, 1852) and *Hemipholis elongata* (Say, 1825) (Echinodermata: Ophiuroidea) in eastern Mississippi Sound. *Ophelia* 33:121–129.

Valentine, J. F., and K. L. Heck, Jr. 1991. The role of sea urchin grazing in regulating subtropical seagrass meadows: Evidence from field manipulations in the northern Gulf of Mexico. *Journal of Experimental Marine Biology and Ecology* 154:215–230.

VandenSpiegel, D., and M. Jangoux. 1987. Cuvierian tubules of the holothuroid *Holothuria forskali* (Echinodermata): A morphofunctional study. *Marine Biology* (Berlin) 96:263–275.

———. 1992. Cuvierian organs in the holothuroid genus *Actinopyga*. In *Echinoderm Research 1991*. Proceedings of the Third European Conference on Echinoderms, Lecce, Italy, 9–12 September 1991, eds. L. Scalera-Liaci and C. Canicattì, 131. Balkema, Rotterdam.

———. 1993. Fine structure and behaviour of

the so-called Cuvierian organs in the holothuroid genus *Actinopyga* (Echinodermata). *Acta Zoologica* (Stockholm) 74:43–50.

van Veghel, M. L. J. 1993. Multiple species spawning on Curacao reefs. *Bulletin of Marine Science* 52:1017–1021.

Verrill, A. E. 1873. Report upon the invertebrate animals of Vineyard Sound and the adjacent waters, with an account of the physical characters of the region. *United States Commission of Fish and Fisheries. Report of the Commissioner* 1(1871–2):295–778, pls. 1–38.

———. 1899a. Report on the Ophiuroidea collected by the Bahama Expedition from the University of Iowa in 1893. *Bulletin from the Laboratories of Natural History of the State University of Iowa* 5:1–86, pls. 1–8.

———. 1899b. North American Ophiuroidea. Part I. Revision of certain families and genera of West Indian ophiurans. Part II. A faunal catalogue of the known species of West Indian ophiurans. *Transactions Connecticut Academy of Arts and Sciences* 10:301–386, pls. 17–18.

———. 1915. Report on the starfishes of the West Indies, Florida, and Brazil, including those obtained by Bahama Expedition from the University of Iowa in 1893. *Bulletin from the Laboratories of Natural History of the State University of Iowa* 7:1–232, pls. 1–29.

Vodicka, M., G. R. Green, and D. L. Poccia. 1990. Sperm histones and chromatin structure of the "primitive" sea urchin *Eucidaris tribuloides*. *Journal of Experimental Zoology* 256:179–188.

Voss, G. L. 1976. *Seashore Life of Florida and the Caribbean: A Guide to the Common Marine Invertebrates of the Atlantic from Bermuda to the West Indies and of the Gulf of Mexico*. E. A. Seemann Publishing, Miami, Florida. 168 pp.

Voss, G. L., and N. A. Voss. 1955. An ecological survey of Soldier Key, Biscayne Bay, Florida. *Bulletin of Marine Science of the Gulf and Caribbean* 5:203–229.

Walenkamp, J. H. C. 1976. The asteroids of the coastal waters of Surinam. *Zoologische Verhandelingen* (Leiden) No. 147:1–91, pls. 1–18.

Walker, C. W., and J. Fineblit. 1982. Interactions between adult and embryonic tissues during brooding in *Axiognathus squamata* (Echinodermata, Ophiuroidea). In *Echinoderms:* Proceedings of the International Conference, Tampa Bay, 14–17 September 1981, ed. J. M. Lawrence, 523. Balkema, Rotterdam.

Walker, C. W., and M. P. Lesser. 1989. Nutrition and development of brooded embryos in the brittlestar *Amphipholis squamata*: Do endosymbiotic bacteria play a role? *Marine Biology* (Berlin) 103:519–530.

Walker, C. W., and F. Smith. 1985. Use of dissolved organic matter during brooding in *Amphipholis squamata*: Implications for editing the developmental program of the ophiopluteus. In *Echinodermata*. Proceedings of the Fifth International Echinoderm Conference, Galway, 24–29 September 1984, eds. B. F. Keegan and B. D. S. O'Connor, 660. Balkema, Rotterdam.

Walsh, G. E., L. L. McLaughlin, M. K. Louie, C. H. Deans, and E. M. Lores. 1986. Inhibition of arm regeneration by *Ophioderma brevispina* (Echinodermata, Ophiuroidea) by tributyltin oxide and triphenyltin oxide. *Ecotoxicology and Environmental Safety* 12:95–100.

Warén, A. 1980. Descriptions of new taxa of Eulimidae (Mollusca, Prosobranchia), with notes on some previously described genera. *Zoologica Scripta* 9:283–306.

———. 1984. A generic revision of the family Eulimidae (Gastropoda, Prosobranchia). *Journal of Molluscan Studies Supplement* 13:1–96.

Warén, A., and C. Mifsud. 1990. *Nanobalcis*, a new eulimid genus parasitic on cidaroid sea urchins with two new species, and comments on *Sabinella bonifaciae* (Nordsieck). *Bolletino Malacologico* 26:37–46.

Warén, A., and R. Moolenbeek. 1989. A new eulimid gastropod, *Trochostilifer eucidaricola*, parasitic on the pencil urchin *Eucidaris tribuloides* from the southern Caribbean. *Proceedings of the Biological Society of Washington* 102:169–75.

Watts, S. A., and J. M. Lawrence. 1986. Seasonal effects of temperature and salinity on the organismal activity of the seastar *Luidia clathrata* (Say) (Echinodermata: Asteroidea). *Marine Behavior and Physiology* 12:161–169.

———. 1990. The effect of temperature and salinity interactions on righting, feeding and growth in the sea star *Luidia clathrata* (Say). *Marine Behavior and Physiology* 17:159–165.

Webb, C. M. 1989. Larval swimming and substrate selection in the brittle star *Ophioderma brevispinum*. In *Reproduction, Genetics and Distribution of Marine Organisms*. 23rd European Marine Biology Symposium, eds. J. S. Ryland and P. A. Tyler, 217–224. Olsen & Olsen, Fredensborg, Denmark.

Weihe, S. C., and I. E. Gray. 1968. Observations on the biology of the sand dollar *Mellita quinquiesperforata* (Leske). *Journal of the Elisha Mitchell Scientific Society* 84:315–327.

Wells, H. W., and M. J. Wells. 1961. Observations on *Pinnaxodes floridensis,* a new species of pinnotherid crustacean commensal in holothurians. *Bulletin of Marine Science* 11:267–279.

Wells, H. W., M. J. Wells, and I. E. Gray. 1961. Food of the sea-star *Astropecten articulatus*. *Biological Bulletin* (Woods Hole) 120:265–271.

Wendt, P. H., R. F. Van Dolah, and C. B. O'Rourke. 1985. A comparative study of the invertebrate macrofauna associated with seven sponge and coral species collected from the South Atlantic Bight. *Journal of the Elisha Mitchell Scientific Society* 101:187–203.

Werding, B., and H. Sanchez. 1989. Pinnotherid crabs of the genus *Dissodactylus* Smith, 1870, associated with irregular sea urchins at the Caribbean coast of Colombia (Crustacea: Decapoda: Pinnotheridae). *Zoologische Mededelingen uitgegevin door het Rijksmuseum van Natuurlijke Historie te Leiden* 63:35–42.

Westergren, A. M. 1911. Reports on the scientific results of the expedition to the Tropical Pacific, in charge of Alexander Agassiz, by the U.S. Fish Commission Steamer "Albatross," from August, 1899, to March, 1900, Commander Jefferson F. Moser, U.S.N., Commanding. XV. Echini. *Echinonëus* and *Micropetalon*. *Memoirs of the Museum of Comparative Zoölogy at Harvard College* 39:35–68, pls. 1–31.

Whitfield, P. J., and R. H. Emson. 1983. Presumptive ciliated receptors associated with the fibrillar glands of the spines of the echinoderm *Amphipholis squamata*. *Cell and Tissue Research* 232:609–624.

———. 1988. *Parachordeumium amphiurae:* A cuckoo copepod? *Hydrobiologia* 167/168:523–531.

Wilkie, I. C., R. H. Emson, and P. V. Mladenov. 1984. Morphological and mechanical aspects of fission in *Ophiocomella* (Echinodermata, Ophiuroida). *Zoomorphology* (Berlin) 104:310–322.

Williams, A. H. 1981. An analysis of competitive interactions in a patchy back-reef environment. *Ecology* 62:1107–1120.

Williams, E. H., Jr., and T. J. Wolfe-Waters. 1990. An abnormal incidence of the commensal copepod, *Doridicola astrophyticus* Humes, associated with injury of its host, the basketstar, *Astrophyton muricatum* (Lamarck). *Crustaceana* (Leiden) 59:302.

Williams, L. B., E. H. Williams, Jr., and A. G. Bunkley. 1986. Isolated mortalities of the sea urchins *Astropyga magnifica* and *Eucidaris tribuloides* in Puerto Rico. *Bulletin of Marine Science* 38:391–393.

Williams, S. L., and R. C. Carpenter. 1988. Nitrogen-limited primary productivity of coral reef algal turfs: Potential contribution of ammonium excreted by *Diadema antillarum*. *Marine Ecology Progress Series* 47:145–152.

Wilson, H. V. 1900. Marine biology at Beaufort. *American Naturalist* 34:339–360.

Windom, H. L., and D. R. Kendall. 1979. Accumulation and biotransformation of mercury in coastal and marine biota. In *The Biogeochemistry of Mercury in the Environment,* ed. J. O. Nriagu, 303–323. Elsevier/North-Holland Biomedical Press, Amsterdam, Netherlands.

Witman, J. D. 1982. Population structure of three sympatric ophiuroids *(Ophiopholis aculeata, Ophiura robusta, Axiognathus squamata)* from subtidal habitats at the Isles of Shoals, N. H. In *Echinoderms:* Proceedings of the International Conference, Tampa Bay, 14–17 September 1981, ed. J. M. Lawrence, 323. Balkema, Rotterdam.

Wolfe, T. J. 1978. Aspects of the biology of *Astrophyton muricatum* (Lamarck, 1816) (Ophiuroidea: Gorgonocephalidae). M.S. thesis, University of Puerto Rico, Mayaguez. xiv + 1–142 pp., pls. 1–6.

———. 1982. Habits, feeding, and growth of the basketstar *Astrophyton muricatum*. In *Echinoderms:* Proceedings of the International Conference, Tampa Bay, 14–17 September 1981, ed. J. M. Lawrence, 299–304. Balkema, Rotterdam.

Woodley, J. D. 1975. The behavior of some amphiurid brittle-stars. *Journal of Experimental Marine Biology and Ecology* 18:29–46.

———. 1982. Photosensitivity of *Diadema antillarum:* Does it show scototaxis? In *Echinoderm Biology.* Proceedings of the Sixth International Echinoderm Conference, Victoria, 23–28 August 1987, eds. R. D. Burke, P. V. Mladenov, P. Lambert, and R. L. Parsley, 61. Balkema, Rotterdam.

Woodley, J. D., and R. H. Emson. 1988. Submersible and laboratory observations on *Asteroporpa annulata* from the island slope of north Jamaica. In *Echinoderm Biology.* Proceedings of the Sixth International Echinoderm Conference, Victoria, 23–28 August 1987, eds. R. D. Burke, P. V. Mladenov, P. Lambert, and R. L. Parsley, 818. Balkema, Rotterdam.

Woodley, J. D., E. A. Chornesky, A. Clifford, J. B. C. Jackson, L. S. Kaufman, N. Knowlton, J. C. Lang, M. Pearson, J. W. Porter, M. C. Rooney, K. W. Rylaarsdam, V. J. Tunnicliffe, C. M. Wahle, J. L. Wulff, A. S. G. Curtis, M. D. Dallmeyer, B. Jupp, M. A. R. Koehl, J. Neigel, and E. M. Sides. 1981. Hurricane Allen's impact on Jamaican coral reefs. *Science* (Washington, D.C.) 214:749–755.

Wray, G. A., and D. R. McClay. 1988. The origin of spicule-forming cells in the "primitive" sea urchin *(Eucidaris tribuloides)* which appears to lack primary mesenchyme cells. *Development* 103:305–315.

Yamaguchi, M. 1975. Coral-reef asteroids of Guam. *Biotropica* 7:12–23.

Yoshida, M. 1966. Photosensitivity. In *Physiology of Echinodermata,* ed. R. A. Boolootian, 435–464. Interscience Publishers, New York.

Zavodnik, D. 1972. Amphiuridae (Echinodermata, Ophiuroidea) of the Adriatic Sea. *Zoologische Jahrbücher Abteilung für Systematik, Ökologie und Geographie der Tiere* 99:610–625.

Zeiller, W. 1974. *Tropical Marine Invertebrates of Southern Florida and the Bahama Islands.* Wiley-Interscience, New York. ix + 132 pp.

Ziesenhenne, F. C. 1955. A review of the genus *Ophioderma* M. & T. In *Essays in the Natural Science in Honor of Capt. A. Hancock,* 185–201. University of Southern California Press, Los Angeles, California.

Zimmerman, K. M., S. E. Stancyk, and L. A. J. Clements. 1988. Substrate selection by the burrowing brittlestar *Microphiopholis gracillima* (Stimpson) (Echinodermata: Ophiuroidea). *Marine Behaviour and Physiology* 13:239–255.

INDEX TO SCIENTIFIC NAMES

Names for which full or partial descriptions and discussions are provided are in **boldface** type; junior synonyms of those names are cross-referenced to main entries. Binominal names are cross-indexed to citations of subgenera. Italicized numbers refer to pages with species illustrations.

Abra aequalis, 71
Acanthaster planci, 65
Acropora, 16, 219, 330, 331
Acropora cervicornis, 16, 219, 225
Acropora palmata, 13, 14, 16
Actinometra discoidea, 52. See also *Nemaster discoideus*
Actinopyga, 256, 258, 284
Actinopyga agassizi, 40, 282–284, *283,* 317
Actinopyga parvula, 292. See also *Holothuria (Platyperona) parvula*
Agaricia, 16, 186, 219
Agelas, 147
Agelas clathrodes, 147
Amphiodia, 152, 153, 161
Amphiodia atra, 144, 153, 154, 156, *159,* 189
Amphiodia erecta, 177, 178. See also *Ophiophragmus septus*
Amphiodia gyraspis, 156. See also *Amphiodia atra*
Amphiodia limbata, 156. See also *Amphiodia atra*
Amphiodia planispina, 38, 144, 152–153, *155,* 156, *189,* 333
Amphiodia pulchella, 38, 112, 127, 153–154, 156, 180, *189,* 333

Amphiodia rhabdota, 175. See also *Ophiophragmus pulcher*
Amphiodia trychna, 38, 153, 154, *155*–156, *189*
Amphiodia tymbara, 156. See also *Amphiodia trychna*
Amphioplus, 151, 170
Amphioplus abditus, 158
Amphioplus albidus, 144, 153
Amphioplus coniortodes, 38, 156–157, 160, 170, *190*
Amphioplus macilentus, 158
Amphioplus sepultus, 38, 156, 157–158, *159,* 167, *190,* 333
Amphioplus thrombodes, 38, *93,* 156, 158–*159,* 167, 175, *190*
Amphipholis, 152, 153, 161
Amphipholis gracillima, 38, 144, 156, *159*–160, 166, *191,* 333
Amphipholis januarii, 38, 123, 161–162, 175, *191,* 333
Amphipholis pachybactra, 162. See also *Amphipholis januarii*
Amphipholis septa, 178. See also *Ophiophragmus septus*
Amphipholis squamata, 38, 110, 112, 127, 154, 162–164, 180, *191,* 333
Amphiura, 151, 167
Amphiura cordifera, 176. See also *Ophiophragmus riisei*

Amphiura fibulata, 38, 164–165, 167, *192,* 333
Amphiura intricata, see *Amphiura (Ophiocnema) intricata*
Amphiura kinbergi, 93, 165
Amphiura kuekenthali, 167. See also *Amphiura palmeri*
Amphiura (Ophiocnema) intricata, 38, 151, 162, *165*–166, *192*
Amphiura palmeri, 38, 164, 165, 166–167, *192,* 333
Amphiura planispina, 153. See also *Amphiodia planispina*
Amphiura pulchella, 154. See also *Amphiodia pulchella*
Amphiura repens, 154. See also *Amphiodia pulchella*
Amphiura riisei, 176, *195.* See also *Ophiophragmus riisei*
Amphiura scabriuscula, 169. See also *Ophiocnida scabriuscula*
Amphiura semiermis, 165
Amphiura septa, 178. See also *Ophiophragmus septus*
Amphiura stimpsoni, 168. See also *Amphiura stimpsonii*
Amphiura stimpsonii, 38, 122, 167, *167*–168, *192,* 333
Amphiura vivipara, 168. See also *Amphiura stimpsonii*

INDEX

Amphiura vivipara annulata, 168. See also *Amphiura stimpsonii*
Amphiura wurdemanni, 174. See also *Ophiophragmus wurdemanii*
Amphiuridae, 4, 38, 95, 96, 97, 151–180, *189–195*
Analcidometra armata, 36, 54–55
Analcidometra caribbea, 55. See also *Analcidometra armata*
Anisotremus surinamensis, 78, 240
Anoplodiera sp., 299
Antedon armata, 55. See also *Analcidometra armata*
Antedonidae, 36, 56–7, 340
Antedon rubiginosa, 54. See also *Nemaster rubiginosa*
Aphrodisiac, 258
Apodida, 41, 256, 257
Apogon sp., 209
Arbacia punctulata, 39, 205, *214–215*, 248, 333
Arbaciidae, 39, 214–215
Arbacioidea, 39
Argopecten gibbus, 69
Argopecten irradians, 69
Aspidochirotida, 40, 255, 256, 257
Asterias, 65
Asterias alternata, 67. See also *Luidia alternata*
Asterias articulatus, 72. See also *Astropecten articulatus*
Asterias clathrata, 69. See also *Luidia clathrata*
Asterias reticulata, 84. See also *Oreaster reticulatus*
Asterias senegalensis, 71. See also *Luidia senegalensis*
Asterias sentus, 87. See also *Echinaster (Othilia) sentus*
Asterias spinosa, 85. See also *Echinaster (Othilia) echinophorus*
Asterias squamata, 164. See also *Amphipholis squamata*
Asterina folium, 37, 74–75, 333
Asterinidae, 37, 74–75
Asterinides folium, 75. See also *Asterina folium*
Asteriscus folium, 75. See also *Asterina folium*
Asteroidea, 17, 35, 36, 59, 333

Asteroporpa, 101
Asteroporpa affinis, 101. See also *Asteroporpa annulata*
Asteroporpa annulata, 37, 100–101, 105
Asteropseidae, 37, 81–82
Astichopus multifidus, 40, 279–280, *324*, 333
Astropecten, 71
Astropecten articulatus, 36, 67, 71–72, 73, 74
Astropecten duplicatus, 36, 72–74, *73*, 333
Astropectinidae, 36, *61*, 71–74
Astrophyton muricatum, 2, 37, 91, 101–103, *102*, 104, 138, 334
Astrophyton muricatum caraibica, 103. See also *Astrophyton muricatum*
Astropyga magnifica, 39, 208–209, 333
Avicennia nitida, 14
Avrainvillea, 330
Axiognathus, 152
Axiognathus squamatus, 164. See also *Amphipholis squamata*

Balcis, 280, 282, 290
Balcis intermedia, 294
Balistes capriscus, 235
Balistes vetula, 84
Batophora, 311
Branchiosyllis, 103
Branchiosyllis exilis, 113
Brandothuria impatiens, 300. See also *Holothuria (Thymiosycia) impatiens*
Brissidae, 39, 241–247
Brissopsis, 241
Brissopsis elongata elongata, 39, 239, 241–242, 249, 333
Brissopsis elongata jarlii, 242
Brissus unicolor, 39, 242–243, 244, 249

Callianassa, 15
Callinectes sapidus, 236
Callyspongia vaginalis, 184
Cancerilla, 163
Carapus bermudensis, 280, 282, 283, 287, 290, 294, 297, 301
Carreta carreta, 271

Cassis, 213, 221
Cassis madagascariensis spinella, 246
Cassis tuberosa, 213, 218, 232, 235, 246
Caudina obesacauda, 303. See also *Paracaudina chilensis obesacauda*
Caudinidae, 41, 302–303
Caulerpa, 278
Charonia variegata, 84
Chiridotidae, 41, 313–314
Chiridota rotifera, 41, *313–314*, *325*, 333
Chordata, 17, 35
Cidaridae, 39, 206–208
Cidaris (Diadema) antillarum, 213. See also *Diadema antillarum*
Cidarite tribuloides, 208. See also *Eucidaris tribuloides*
Cidaroidea, 198
Clostridium perfringens, 212
Clostridium sordelli, 212
Clypeasteridae, 39, 228–232
Clypeaster luetkeni, 39, 228–229, 333
Clypeasteroida, 39
Clypeaster rosaceus, 39, 164, 229–230, 234, 244, 248, 333
Clypeaster (Stolonoclypus) Lütkeni, 228. See also *Clypeaster luetkeni*
Clypeaster (Stolonoclypus) subdepressus, 232. See also *Clypeaster subdepressus*
Clypeaster subdepressus, 2, 39, 228, 231–232, 246, 248, 333
Codium, 330
Colobometridae, 36, 54–55
Colochirus gemmatus, 264. See also *Thyonella gemmata*
Colochirus pygmaeus, 261. See also *Ocnus pygmaeus*
Comactinia echinoptera, 36, 48–49
Comactinia meridionalis, 49
Comactinia meridionalis meridionalis, 44
Comasteridae, 36, 45, 48–53, 56, 340
Comatula echinoptera, 49. See also *Comactinia echinoptera*
Comatulida, 36

Concentricycloidea, 18
Copidaster lymani, 37, 75–76, 80
Crinoidea, 17, 36, 43, 333
Crinometra brevipinna, 2
Critomolgus astrophyticus, 103
Ctenantodon kinziei, 36, 56–57
Cucumaria, 256
Cucumaria cognata, 276. See also *Stolus cognatus*
Cucumariidae, 40, 257, 259–265
Cypraecassis testiculus, 218, 224, 228, 243

Davidaster, 50
Dendrochirotida, 40, 255
Diadema antillarum, 39, 202, 205, 208, 209, 210–213, 220, 224, 225, 226, 250, 333
Diadema mexicanum, 213
Diadematidae, 39, 208–213
Diadematoida, 39
Diamphiodia, 152
Diploria, 16
Dissodactylus, 230, 232, 245
Dissodactylus crinitichelis, 235
Dissodactylus latus, 235
Dissodactylus mellitae, 238
Dissodactylus primitivus, 245, 247
Donax, 207
Donax variabilis, 217, 223
Duasmodactyla seguroensis, 40, 259–260, *315*

Echinanthus subdepressus, 232. See also *Clypeaster subdepressus*
Echinaster, 86
Echinaster echinophorus, see *Echinaster (Othilia) echinophorus*
Echinasteridae, 37, 84–87
Echinaster (Othilia) echinophorus, 37, 84–85, 86
Echinaster (Othilia) graminicola, 86
Echinaster (Othilia) guyanensis, 86
Echinaster (Othilia) paucispinus, 86
Echinaster (Othilia) sentus, 37, 85–87, 333

Echinaster (Othilia) serpentarius, 86
Echinaster (Othilia) spinulosus, 86
Echinaster sentus, see *Echinaster (Othilia) sentus*
Echinaster sp. C, 87. See also *Echinaster (Othilia) sentus*
Echinodermata, 17, 29, 35
Echinodiscus sexies perforatus, 235. See also *Leodia sexiesperforata*
Echinoida, 39
Echinoidea, 17, 39, 197, 198, 333
Echinometra, 203, 207, 211
Echinometra lucunter polypora, 34, 222
Echinometra lucunter lucunter, 34, 39, 113, 211, 222–225, *223,* 226, 248, 333
Echinometra viridis, 39, 113, 211, 222, 224, 225–226, *248,* 333
Echinometridae, 39, 222–226
Echinoneidae, 39, 227–228
Echinoneus cyclostomus, 39, 227 228, 244, 248, 333
Echinus grandis, 247. See also *Plagiobrissus grandis*
Echinus lucunter, 225. See also *Echinometra lucunter*
Echinus punctulatus, 215. See also *Arbacia punctulata*
Echinus rosaceus, 230. See also *Clypeaster rosaceus*
Echinus variegatus, 218. See also *Lytechinus variegatus*
Echinus ventricosus, 222. See also *Tripneustes ventricosus*
Ellisella barbadensis, 54
Encope aberrans, 39, 232–233, 250, 274
Encope michelini, 39, 233–234, 246, 249, 333
Encope michelini imperforata, 234
Epitomapta, 314
Epitomapta roseola, 41, 303–304, 313, *325, 333*
Ersilia stancyki, 106
Euapta, 314
Euapta lappa, 41, 304–306, *305,* 313, *325,* 333
Eucidaris thouarsii, 208

Eucidaris tribuloides, 2, 39, 203, 206–208, 248, 333
Eucidaris tribuloides tribuloides, 208. See also *Eucidaris tribuloides*
Eucidaris tribuloides var. *africana,* 206
Euechinoidea, 198
Eunice rubra, 186
Euryale muricatum, 103. See also *Astrophyton muricatum*
Euthyonidiella destichada, 40, 266–267, *318*
Euthyonidiella trita, 40, 267–268, *318*

Fistularia impatiens, 300. See also *Holothuria (Thymiosycia) impatiens*
Forcipulatacea, 64
Forcipulatida, 64
Fundulus diaphanus, 271

Gnathophylloides mineri, 222
Gnathophylum americanum, 280, 282
Goniaster tesselatus, 2
Gorgonocephalidae, 37, 96, 100–105
Gracilaria tikvahiae, 215

Hacelia superba, 2
Haemulon album, 239
Haemulon plumieri, 278
Halimeda, 16, 109, 154, 175, 219, 229, 311, 313, 330, 332
Halimeda opuntia, 286
Halodule, 154, 161, 263, 311, 332
Halodule wrightii, 83, 84, 158, 172
Halophila, 278, 279
Harmothoe lunulata, 126. See also *Malmgreniella variegata*
Heliaster kubiniji, 65
Hemieuryalidae, 38, 97, 142–143
Hemipholis cordifera, 145. See also *Hemipholis elongata*
Hemipholis elongata, 38, 143–145, *144,* 153
Hermenia verruculosa, 116
Holectypoida, 39

INDEX

Holothuria, 256, 258, 293, 294, 296, 299
Holothuria arenicola, see *Holothuria (Thymiosycia) arenicola*
Holothuria briareus, 271. See also *Sclerodactyla briareus*
Holothuria cubana, see *Holothuria (Cystipus) cubana*
Holothuria (Cystipus) cubana, 40, 284–285, 298, 300, 320, 333
Holothuria floridana, see *Holothuria (Halodeima) floridana*
Holothuria forskali, 292
Holothuria glaberrima, see *Holothuria (Selenkothuria) glaberrima*
Holothuria grisea, see *Holothuria (Halodeima) grisea*
Holothuria (Halodeima) floridana, 2, 40, 285–287, 288, 290, 320
Holothuria (Halodeima) grisea, 40, 287–288, 321
Holothuria (Halodeima) mexicana, 40, 286, 288–290, 289, 320, 333
Holothuria hydriformis, 313. See also *Synaptula hydriformis*
Holothuria impatiens, see *Holothuria (Thymiosycia) impatiens*
Holothuria mexicana, see *Holothuria (Halodeima) mexicana*
Holothuria parvula, see *Holothuria (Platyperona) parvula*
Holothuria (Platyperona) parvula, 40, 257, 291–292, 296, 322, 333
Holothuria (Platyperona) rowei, 35, 40, 292–293
Holothuria princeps, see *Holothuria (Theelothuria) princeps*
Holothuria rowei, see *Holothuria (Platyperona) rowei*
Holothuria (Selenkothuria) glaberrima, 40, 255, 293–294, 321
Holothuria (Semperothuria) surinamensis, 40, 257, 292, 294–296, 295, 322
Holothuria (Sporadipus) arenicola, 299. See also *Holothuria (Thymiosycia) arenicola*
Holothuria surinamensis, see *Holothuria (Semperothuria) surinamensis*
Holothuria (Theelothuria) princeps, 40, 296–297, 322, 333
Holothuria thomasae, 301. See also *Holothuria (Thymiosycia) thomasi*
Holothuria thomasi, see *Holothuria (Thymiosycia) thomasi*
Holothuria (Thymiosycia) arenicola, 40, 258, 285, 297–299, 298, 300, 323, 333
Holothuria (Thymiosycia) impatiens, 40, 285, 298, 299–300, 323
Holothuria (Thymiosycia) thomasi, 35, 40, 300–302, 324, 333
Holothuriidae, 40, 282–303
Holothuroidea, 18, 35, 40, 251, 255, 333

Isostichopus badionotus, 40, 164, 280–282, 281, 290, 324, 333

Lactophrys bicaudalis, 84
Lactophrys trigonus, 134, 299
Laurencia, 286
Leodia sexiesperforata, 39, 234–235, 249, 333
Lepidonopsis humilis, 185
Leptosynapta, 206, 256, 307, 309, 313, 314
Leptosynapta crassipatina, 41, 306–307, 308, 309
Leptosynapta inhaerens, 309
Leptosynapta multigranula, 41, 306, 307, 326, 333
Leptosynapta multigranulata, 307. See also *Leptosynapta multigranula*
Leptosynapta parvipatina, 41, 306, 307, 309
Leptosynapta roseola, 304. See also *Epitomapta roseola*
Leptosynapta tenuis, 2, 41, 306, 307, 308–309, 326
Linckia guildingii, 37, 76–78, 77, 79, 80
Lobophora, 16
Loxechinus albus, 204
Luidia, 68, 69

Luidia alternata alternata, 36, 66–67, 69, 72
Luidia alternata numidica, 67
Luidia clathrata, 36, 67, 68–69, 72, 173, 239
Luidia senegalensis, 36, 69–71, 70, 154, 158
Luidiidae, 36, 66–71
Lytechinus variegatus, 34, 35, 39, 202, 207, 216–218, 219, 221, 248, 333
Lytechinus variegatus atlanticus, 216. See also *Lytechinus variegatus*
Lytechinus variegatus carolinus, 216, 218. See also *Lytechinus variegatus*
Lytechinus variegatus variegatus, 216. See also *Lytechinus variegatus*
Lytechinus williamsi, 34, 35, 39, 219–220, 248

Macrogynium ovalis, 282
Malmgreniella galetaensis, 177
Malmgreniella hendleri, 171
Malmgreniella maccraryae, 158, 160
Malmgreniella puntotorensis, 156, 171, 172, 175, 177
Malmgreniella taylori, 160
Malmgreniella variegata, 125, 126
Marginaster echinulatus, 82. See also *Poraniella echinulata*
Megadenus holothuricola, 290
Melanella hypsela, 280
Mellita, 69, 72, 236
Mellita isometra, 39, 235, 236–238, 237, 250
Mellita quinquiesperforata, 235, 236
Mellita sexiesperforata, 235. See also *Leodia sexiesperforata*
Mellita tenuis, 234, 236
Mellitidae, 232–238
Meoma ventricosa grandis, 244, 245
Meoma ventricosa ventricosa, 40, 84, 243–245, 244, 249, 333
Microphiopholis, 152
Millepora, 16, 181, 187, 219

Moira atropos, 39, 69, *238–239,* 241, 249, 333
Moira atropus, 239. See also *Moira atropos*
Mollusca, 35
Molpadiida, 41
Monogamus minibulla, 224
Montacuta percompressa, 309
Montastrea, 330
Montastrea annularis, 13, 16
Mülleria Agassizii, 284. See also *Actinopyga agassizi*
Mülleria parvula, 292. See also *Holothuria parvula*
Mysella sp., 309
Mysella sp. C, 160

Nanobalcis worsfoldi, 208
Nassarius vibex, 71
Neaeromya sp., 245
Nemaster, 16, 49–54, 56
Nemaster discoideus, 36, 49, *50–52,* 53, 333
Nemaster grandis, 44, 49, 50, 51
Nemaster iowensis, 44, 49, 50
Nemaster mexicanensis, 49, 50
Nemaster rubiginosus, 36, 49, 50, 51, *52–54*
Neofibularia, 188
Neofibularia nolitangere, 147
Neothyonidium parvum, 40, 272–273, *316*

Oceanida sp., 175
Ocnus, 263
Ocnus braziliensis, 263
Ocnus pygmaeus, 261
Ocnus surinamensis, 40, 260–262, *261,* 263, *315*
Ocnus suspectus, 40, 261, 262–263, *315*
Oculina, 78, 280
Oculina varicosa, 295
Ophiacantha oligacantha, 119. See also *Ophiocomella ophiactoides*
Ophiacantha ophiactoides, 119. See also *Ophiocomella ophiactoides*
Ophiactidae, 38, 97, 143–151
Ophiactis, 96, 118, 145, 146, 181, 184, 188

Ophiactis algicola, 38, *145–146,* 334
Ophiactis cyanosticta, 150
Ophiactis loricata, 146
Ophiactis lymani, 146, 150
Ophiactis maculosa, 150
Ophiactis muelleri, 147, 150
Ophiactis muelleri var. *quinqueradia,* 147. See also *Ophiactis quinqueradia*
Ophiactis notabilis, 151
Ophiactis plana, 146, 150–151
Ophiactis quinqueradia, 38, 146–147, 150, 334
Ophiactis rubropoda, 38, *147–148*
Ophiactis savignyi, 38, *93,* 146, 147, 148–151, *149,* 334
Ophialcaea glabra, 99. See also *Ophioblenna antillensis*
Ophidiaster bayeri, 80
Ophidiaster guildingii, 37, 76, 78, *79–80,* 333
Ophidiasteridae, 37, 75–80
Ophidiaster trychnus, 80
Ophioblenna antillensis, 37, *98* 99
Ophiocnida, 151
Ophiocnida caribea, 123
Ophiocnida cubana, 172. See also *Ophiophragmus cubanus*
Ophiocnida filogranea, 175. See also *Ophiophragmus filograneus*
Ophiocnida scabriuscula, 38, 154, 162, 167, 168, 178, *193,* 334
Ophiocoma, 96, 97, 111, 114, 115, 117, 118, 119, 120, 140
Ophiocoma dentata, 164
Ophiocoma echinata, 37, *91, 93,* 110–113, *111,* 114, 115, 116, 117, 127, 130, 154, 164, 180, 334
Ophiocoma erinaceus, 164
Ophiocoma paucigranulata, 37, *113–114,* 116, 334
Ophiocoma pumila, 37, 112, 114–116, *115,* 118, 119, *126,* 180, 334
Ophiocoma riisei, 118
Ophiocoma wendti, 118. See also *Ophiocoma wendtii*
Ophiocoma wendtii, 37, *91,* 107,
112, 113, 114, 115, *116–118,* 130, 136, 334
Ophiocomella, 97, 111, 119, 139
Ophiocomella ophiactoides, 37, 113, 115, *118–119,* 334
Ophiocomidae, 37, 95, 97, 120, 111–123
Ophiocominae, 37, 113, 111–119
Ophiocryptus hexacanthus, 136. See also *Ophioderma cinereum*
Ophiocryptus stage, 132. See also *Ophioderma brevicauda*
Ophioderma, 92, 96, 130, 132, 134, 135, 137, 138, 139, 140, 141
Ophioderma anitae, 96, 138, 139
Ophioderma antillarum, 136. See also *Ophioderma cinereum*
Ophioderma appressum, 38, 112, *129–130,* 133, 136, 138, 140, 334
Ophioderma brevicauda, 132. See also *Ophioderma brevicaudum*
Ophioderma brevicaudum, 38, *131–132,* 133, 136, 139
Ophioderma brevispinum, 38, *132–134,* 136, 140, 186, 334
Ophioderma cinereum, 38, *91,* 130, *134–136, 135,* 140, 334
Ophioderma devaneyi, 96
Ophioderma ensiferum, 96
Ophioderma guttata, 138. See also *Ophioderma guttatum*
Ophioderma guttatum, 38, 103, 130, *137–138,* 139, 334
Ophioderma holmesii, 134. See also *Ophioderma brevispinum*
Ophioderma januarii, 133, 134
Ophioderma longicaudum, 130
Ophioderma olivacea, 134. See also *Ophioderma brevispinum*
Ophioderma phoenium, 38, 130, *138–139,* 334
Ophioderma rubicunda, 140. See also *Ophioderma rubicundum*
Ophioderma rubicundum, 2, 38, 99, 107, 129, 130, 133, 138, *139–140,* 334
Ophioderma saxatilis, 136. See also *Ophioderma cinereum*
Ophioderma serpens, 134. See also *Ophioderma brevispinum*
Ophioderma sp., juv., 140. See

INDEX

also *Ophioderma rubicundum*
Ophioderma sp., young, 136. See also *Ophioderma cinereum*
Ophioderma squamosissima, 142. See also *Ophioderma squamosissimum*
Ophioderma squamosissimum, 38, *91*, 137, *141*–142
Ophiodermatidae, 38, 97, 129–142
Ophioderma virescens, 130. See also *Ophioderma appressum*
Ophiodromus obscurus, 232, 245
Ophiolepis, 96, 109
Ophiolepis ailsae, 108
Ophiolepis atra, 156. See also *Amphiodia atra*
Ophiolepis elegans, 37, *91*, *93*, *105*–107
Ophiolepis gemma, 37, *106*–107, 109
Ophiolepis gracillima, 160. See also *Amphipholis gracillima*
Ophiolepis impressa, 37, 107–109, *108*, 110, 115, 126, 146, 334
Ophiolepis paucispina, 37, 109–110, 334
Ophiolepis savignyi, 151. See also *Ophiactis savignyi*
Ophiolepis uncinata, 144. See also *Hemipholis elongata*
Ophiomitrella glabra, 99. See also *Ophioblenna antillensis*
Ophiomyxa flaccida, 37, 98, 99–100, 139, 140, 334
Ophiomyxa tumida, 100
Ophiomyxidae, 37, 96, 98–100
Ophionema intricata, 167. See also *Amphiura (Ophionema) intricata*
Ophionephthys limicola, 38, 151, 156, 160, 165, 166, *169*–170, 190
Ophionereididae, 37, 97, 123–128
Ophionereis, 96, 123, *125*, 126, 127, 128
Ophionereis annulata, 126
Ophionereis olivacea, 37, 123–*124*, 126, 128
Ophionereis reticulata, 37, *91*,

109, 115, 123, *125*–127, 128, 167, 334
Ophionereis squamata, 128. See also *Ophionereis squamulosa*
Ophionereis squamulosa, 38, 112, 123, 126, *127*–128, 154, 180, 334
Ophionereis vittata, 38, *128*, 334
Ophionyx, 144. See also *Hemipholis elongata*
Ophiophragmus, 151, 152, 171, 173
Ophiophragmus brachyactis, 176. See also *Ophiophragmus riisei*
Ophiophragmus cubanus, 38, *171*–172, 173, 194
Ophiophragmus filograneus, 38, *172*–175, *173*, 194
Ophiophragmus luetkeni, 177. See also *Ophiophragmus septus*
Ophiophragmus moorei, 174–175, 194
Ophiophragmus pulcher, 38, *91*, *93*, 123, 153, 154, 159, 162, 174–175, *195*, 334
Ophiophragmus riisei, 38, 175–176, *195*
Ophiophragmus septus, 38, 176–178, *177*, *195*
Ophiophragmus wurdemani, 174. See also *Ophiophragmus wurdemanii*
Ophiophragmus wurdemanii, 144, 173–174, *194*
Ophiophragmus wurdemanni, 174. See also *Ophiophragmus wurdemanii*
Ophiopsila, 92, 95, 96, 97, 120, 121
Ophiopsila hartmeyeri, 37, *120*
Ophiopsila polysticta, 123
Ophiopsila riisei, 37, 120, *121*–122, 334
Ophiopsila vittata, 37, *122*–123, 162, 334
Ophiopsilinae, 37, 120–123
Ophiopsyllus reductus, 113, 119
Ophiostigma, 118, 151
Ophiostigma isacanthum, 179. See also *Ophiostigma isocanthum*
Ophiostigma isocanthum, 38, 112, 127, 154, 167, 178–179, 180, *193*, 334

Ophiostigma siva, 38, *179*–180, *193*, 334
Ophiostigma sp., 180. See also *Ophiostigma siva*
Ophiothrix, 96, 180, 181, 182, 183, 184, 185, 187, 188
Ophiothrix (Acanthophiothrix) suensoni, 188. See also *Ophiothrix suensonii*
Ophiothrix angulata, 38, *91*, 109, 133, 140, 147, 180, 180–182, *181*, 183, 187, 334
Ophiothrix angulata var. *atrolineata*, 182. See also *Ophiothrix angulata*
Ophiothrix angulata var. *megalaspis*, 182. See also *Ophiothrix angulata*
Ophiothrix angulata var. *phoinissa*, 182. See also *Ophiothrix angulata*
Ophiothrix angulata var. *phlogina*, 182. See also *Ophiothrix angulata*
Ophiothrix angulata var. *poecilia*, 182. See also *Ophiothrix angulata*
Ophiothrix angulata var. *violacea*, 182. See also *Ophiothrix angulata*
Ophiothrix brachyactis, 38, 182–183
Ophiothrix hartfordi, 183
Ophiothrix lineata, 39, 147, 183–185, *184*, 334
Ophiothrix oerstedii var. *lutea*, 185, 187. See also *Ophiothrix orstedii*
Ophiothrix orstedii, 39, 133, 139, 140, 147, 185–187, *186*, 334
Ophiothrix Örstedii, 187. See also *Ophiothrix orstedii*
Ophiothrix Ørstedii, 187. See also *Ophiothrix orstedii*
Ophiothrix pallida, 182. See also *Ophiothrix angulata*
Ophiothrix platyactis, 185
Ophiothrix sp., 188
Ophiothrix suensoni, 188. See also *Ophiothrix suensonii*
Ophiothrix suensonii, 39, *91*, *93*, 147, *187*–188, 334
Ophiotrichidae, 38, 97, 180–188

Ophiozona impressa, 109. See also *Ophiolepis impressa*
Ophiura angulata, 182. See also *Ophiothrix angulata*
Ophiura appressa, 130. See also *Ophioderma appressum*
Ophiura brevispina, 134. See also *Ophioderma brevispinum*
Ophiura echinata, 113. See also *Ophiocoma echinata*
Ophiura elongata, 145. See also *Hemipholis elongata*
Ophiura flaccida, 100. See also *Ophiomyxa flaccida*
Ophiura holmesii, 134. See also *Ophioderma brevispinum*
Ophiura isocantha, 179. See also *Ophiostigma isocanthum*
Ophiura olivacea, 134. See also *Ophioderma brevispinum*
Ophiura paucispina, 110. See also *Ophiolepis paucispina*
Ophiura reticulata, 127. See also *Ophionereis reticulata*
Ophiurida, 37, 94
Ophiuridae, 37, 97, 105–110
Ophiuroidea, 17, 37, 89, 333
Oreasteridae, 37, 82–84
Oreaster reticulatus, 37, 65, 82–84, *83*, 245, 333
Oxycomanthus japonicus, 47

Padina gymnospora, 215
Palythoa, 16
Panulirus argus, 213
Paracaudina chilensis chilensis, 302, 303
Paracaudina chilensis obesacauda, 41, 302–303, 324
Paracentrotus lividus, 204
Parachordeumium, 164
Paraster, 239
Paraster doederleini, 39, 239–240, 249, 333
Paraster floridiensis, 39, 240–241, 244, 249, 333
Paraster (Schizaster) floridiensis, 241. See *Paraster floridiensis*
Parastichopus, 257
Parathyone, 263
Parophiopsyllus ligatus, 158
Paxillosida, 36, 64
Pelseneeria sp., 213

Penicillus, 295, 311, 313, 329, 330
Periclimenes bowmani, 54
Periclimenes perryae, 103
Periclimenes sp., 55
Perknaster fuscus antarcticus, 65
Phragmatopoma lapidosa, 287, 313
Phrynophiurida, 37, 94
Phyllophoridae, 40, 272–278
Phyllophorus, 256
Phyllophorus arenicola, see *Phyllophorus (Urodemella) arenicola*
Phyllophorus destichadus, 267. See also *Euthyonidiella destichada*
Phyllophorus occidentalis, see *Phyllophorus (Urodemella) occidentalis*
Phyllophorus parvum, 273. See also *Neothyonidium parvum*
Phyllophorus seguroensis, 260. See also *Duasmodactyla seguroensis*
Phyllophorus tritus, 268. See also *Euthyonidiella trita*
Phyllophorus (Urodemella) arenicola, 40, 273–274, 319
Phyllophorus (Urodemella) occidentalis, 40, 260, 274–275, 318
Pinnaxodes floridensis, 297
Pisaster ochraceus, 65
Plagiobrissus grandis, 2, 40, 244, 245–247, *246*, 249, 333
Podarke obscura, 69, 71, 218
Pontocypria hendleri, 98, 100, 136
Poraniella echinulata, 37, 81–82
Poraniella regularis, 82. See also *Poraniella echinulata*
Porcellana sayana, 280
Porites, 295, 313, 330
Porites astreoides, 16
Porites porites, 16
Presynaptiphilus amphiopli, 158
Protankyra benedeni, 310
Protankyra ramiurna, 41, 310–311, 327
Pseudanthessius deficiens, 136
Pseudopterogorgia, 54, 55
Pseudothyone belli, 40, 268–269, 317, 333

Rhizophora, 311
Rhizophora mangle, 14
Rhopalura ophiocomae, 164

Sabinella troglodytes, 208
Sargassum, 221
Sargassum filipendula, 215
Schizasteridae, 39, 238–241
Schizostella bayeri, 105. See also *Schizostella bifurcata*
Schizostella bifurcata, 37, 103, 104–105
Sclerodactyla, 271
Sclerodactyla briareus, 40, 269–271, *270*, 317
Sclerodactylidae, 40, 266–271
Semperia cognata, 276. See also *Stolus cognatus*
Serranus subligarius, 154
Sigsbeia conifera, 38, 91, 142–143
Sigsbeia murrhina, 143
Spatangoida, 39
Spatangus atropos, 239. See also *Moira atropos*
Spatangus (Brissus) unicolor, 243. See also *Brissus unicolor*
Spatangus ventricosus, 245. See also *Meoma ventricosa ventricosa*
Spinulosacea, 64
Spinulosida, 37, 64
Stichopodidae, 40, 279–282
Stichopus, 256
Stichopus badionotus, 282. See also *Isostichopus badionotus*
Stichopus japonicus, 257
Stichopus multifidus, 280. See also *Astichopus multifidus*
Stolus cognatus, 40, 275–276, 319
Stolus cognitus, 276. See also *Stolus cognatus*
Strongylocentrotus droebachiensis, 205
Strongylocentrotus franciscanus, 204
Stylaster filogranus, 143
Stylaster roseus, 143
Synalpheus sp., 103
Synalpheus townsendi, 54
Synapta lappa, 306. See also *Euapta lappa*

INDEX

Synapta rotifera, 314. See also *Chiridota rotifera*
Synapta tenuis, 309. See also *Leptosynapta tenuis*
Synapta vivipara, 313. See also *Synaptula hydriformis*
Synaptidae, 41, 256, 309, 303–313
Synaptula, 314
Synaptula hydriformis, 41, 311–313, *312*, *314*, *327*, 333
Syndisyrinx antillarum, 213
Syndisyrinx collongistyla, 218, 220, 224, 226
Syndisyrinx evelinae, 224
Syringodium, 160, 276, 279, 280, 296, 307, 329, 330, 331
Syringodium filiforme, 15, 83

Tedania ignis, 86
Temnopleuroida, 39
Thalassia, 133, 155, 161, 167, 207, 217, 218, 221, 263, 276, 279, 280, 296, 307, 310, 311, 313, 329, 330, 331, 332
Thalassia testudinum, 15, 83, 217, 223
Theelothuria princeps, 297. See also *Holothuria (Theelothuria) princeps*
Thelenota, 258
Thyone, 256
Thyone belli, 269. See also *Pseudothyone belli*
Thyone briareus, 271. See also *Sclerodactyla briareus*
Thyone deichmannae, 40, 276–277, 278, *319*
Thyone inermis, 277
Thyonella gemmata, 40, 263–264, *316*
Thyonella pervicax, 40, 264–265, *316*
Thyone pervicax, 265. See also *Thyonella pervicax*
Thyone pseudofusus, 40, 277–278, *319*
Thyoneria cognata, 276. See also *Stolus cognatus*
Thyone surinamensis, 262. See also *Ocnus surinamensis*
Thyone suspecta, 263. See also *Ocnus suspectus*

Thyone trita, 268. See also *Euthyonidiella trita*
Thyonidium occidentale, 275. See also *Phyllophorus (Urodemella) occidentalis*
Thyonidium parvum, 273. See also *Neothyonidium parvum*
Toxopneustidae, 39, 216–222
Trachythyonidium occidentale, 275. See also *Phyllophorus (Urodemella) occidentalis*
Tripneustes ventricosus, 39, 84, 207, 218, 220–222, 248, 333
Trochostilifer eucidaricola, 208
Tropiometra carinata, 44
Tuleariocaris neglecta, 213

Udotea, 329, 330
Ulva, 311

Valvatacea, 64
Valvatida, 36, 64
Verongia lacunosa, 147, 184
Vitreolina sp., 134
Vitreolina arcuata, 182

Wahlia macrostylifera, 282

Zelinkiela synaptae, 159, 160, 175
Zostera, 133, 270
Zostera marina, 217

cm and mm